INTRODUCTION TO
THE THEORY OF
NEURAL COMPUTATION

INTRODUCTION TO THE THEORY OF NEURAL COMPUTATION

John Hertz
NORDITA

Anders Krogh
Niels Bohr Institute

Richard G. Palmer
Duke University and the Santa Fe Institute

Lecture Notes Volume I

SANTA FE INSTITUTE
STUDIES IN THE SCIENCES OF COMPLEXITY

Addison-Wesley Publishing Company
The Advanced Book Program
Redwood City, California • Menlo Park, California • Reading, Massachusetts
New York • Don Mills, Ontario • Wokingham, United Kingdom • Amsterdam
Bonn • Sydney • Singapore • Tokyo • Madrid • San Juan

Publisher: *Allan M. Wylde*
Production Manager: *Jan V. Benes*
Marketing Manager: *Laura Likely*

Director of Publications, Santa Fe Institute: *Ronda K. Butler-Villa*
Technical Assistant, Santa Fe Institute: *Della L. Ulibarri*

Hertz, John A.
   Introduction to the theory of neural computation / John A. Hertz,
Richard G. Palmer, Anders S. Krogh.
      p. cm.—(Santa Fe Institute studies in the sciences of complexity.
Lecture notes : v. 1) (Computation and neural systems series)
      Includes index.
      1. Neural computers. 2. Neural circuitry. I. Palmer, Richard G. II.
Krogh, Anders S. III. Title. IV. Series. V. Series: Computation and neural
systems series.
QA76.5.H475   1991   006.3—dc20   91-701
ISBN 0-201-50395-6.—ISBN 0-201-51560-1 (pbk.)

This volume was typeset using T$_E$Xtures on a Macintosh II computer. Camera-ready output
from an Apple LaserWriter IINT Printer.

2 3 4 5 6 7 8 9 10 - MA - 95 94 93 92 91

About the Santa Fe Institute

The *Santa Fe Institute* (SFI) is a private, independent organization dedicated to multidisciplinary scientific research and graduate education in the natural, computational, and social sciences. The driving force behind its creation in 1984 was the need to understand those complex systems that shape human life and much of our immediate world—evolution, the learning process, the immune system, the world economy. The intent is to make the new tools now being developed at the frontiers of the computational sciences and in the mathematics of nonlinear dynamics more readily available for research in the applied physical, biological, and social sciences.

All titles from the *Santa Fe Institute Studies in the Sciences of Complexity* series will carry this imprint, the SFI logo, which is based on a Mimbres pottery design (circa A.D. 950–1150), drawn by Betsy Jones.

Santa Fe Institute Studies in the Sciences of Complexity

PROCEEDINGS VOLUMES

Volume	Editor	Title
I	David Pines	Emerging Syntheses in Science, 1987
II	Alan S. Perelson	Theoretical Immunology, Part One, 1988
III	Alan S. Perelson	Theoretical Immunology, Part Two, 1988
IV	Gary D. Doolen et al.	Lattice Gas Methods of Partial Differential Equations, 1989
V	Philip W. Anderson et al.	The Economy as an Evolving Complex System, 1988
VI	Christopher G. Langton	Artificial Life: Proceedings of an Interdisciplinary Workshop on the Synthesis and Simulation of Living Systems, 1988
VII	George I. Bell & Thomas G. Marr	Computers and DNA, 1989
VIII	Wojciech H. Zurek	Complexity, Entropy, and the Physics of Information, 1990
IX	Alan S. Perelson & Stuart A. Kauffman	Molecular Evolution on Rugged Landscapes: Proteins, RNA and the Immune System, 1990

LECTURES VOLUMES

Volume	Editor	Title
I	Daniel L. Stein	Lectures in the Sciences of Complexity, 1989
II	Erica Jen	1989 Lectures in Complex Systems, 1990

LECTURE NOTES VOLUMES

Volume	Author	Title
I	John Hertz, Anders Krogh, & Richard Palmer	Introduction to the Theory of Neural Computation, 1991
II	Gérard Weisbuch	Complex Systems Dynamics, 1991

Contents

Series Foreword

We are witnessing the creation of new sciences of complexity, sciences that may well occupy the center of intellectual life in the twenty-first century. The Santa Fe Institute was founded to assist at the birth of these new sciences. Those involved in this activity are proceeding under the conviction that there is a common set of principles shared by the disparate complex systems under study, that the time is ripe to understand those principles, and that it is essential to develop them and the associated tools for dealing in a systematic way with complex systems.

Complex systems typically do not fit within the confines of one of the traditional disciplines, but require for their successful study knowledge and techniques from several disciplines. Thus one task of the Institute has been to find new ways to encourage cooperative research among scholars from different fields. The Studies in the Sciences of Complexity is one means that the Institute has adopted for accelerating the development of the sciences of complexity. These volumes make available to the scholarly community the results of conferences and workshops sponsored by the Institute, lectures presented in the Complex Systems Summer School, other lecture notes, and monographs by active researchers.

The sciences of complexity are emerging in part as a synthesis of some of the traditional sciences, including biology, computer science, physics, and mathematics. In part they are emerging as a result of new ideas, new questions, and new techniques only recently developed. Among these latter are the emergence of heretofore undreamed of computer power on the scientist's desktop and the not unrelated progress in nonlinear dynamics, computer graphics, and adaptive programs. These newly emerging tools and techniques also offer the prospect of new collaboration between the traditional sciences and the social sciences, a collaboration that will extend modeling techniques to incorporate realistic detailed models of human behavior. Thus, this Series is intended to range broadly across many fields of intellectual endeavor incorporating work in all the areas listed above. The apparently disparate topics, however, share common themes that relate them to the emergent sciences of complexity.

The Santa Fe Institute, and hence this Series, would not exist without the support of farsighted individuals in government funding agencies and private foundations who have recognized the promise of the new approaches to complex systems research being fostered here. It is a pleasure to acknowledge the broad research grants received by the Institute from the Department of Energy, the John D. and Catherine T. MacArthur Foundation, and the National Science Foundation that, together with numerous other grants, have made possible the work of the Institute.

L. M. Simmons, Jr.

Santa Fe, New Mexico
October 1, 1990

Foreword

The past decade has seen an explosive growth in studies of neural networks. In part this was the result of technological advances in personal and main-frame computing, enabling neural network investigators to simulate and test ideas in ways not readily available before 1980. Another major impulse was provided by Hopfield's work on neural networks with symmetric connections. Such networks had previously been dismissed as not brain-like and therefore not worth studying. I myself fell into this trap some twenty-five years ago when I formulated what are now termed the standard equations for studying neural networks, those using the so-called squashing or logistic function. It was to Hopfield's credit that he "stepped back from biological reality" as Toulouse has put it, and uncovered an interesting set of properties and uses for symmetric networks. What followed is an interesting episode in the sociology of science. Hopfield's papers triggered an explosion, particularly in the statistical physics community, leading to a whole series of dramatic advances in the understanding of symmetric networks and their properties, especially in respect of their utility as distributed memory stores, and as solvers of constrained optimization problems, e.g., small versions of the famous Traveling Salesman Problem.

At more-or-less the same time, other developments in neural networks, possibliely even more important, were taking place, culminating in the publication by Rumelhart, Hinton, and Williams of the now well-known "Back-Propagation Algorithm" for solving the fundamental problem of training neural networks to compute desired functions, a problem first formulated by Rosenblatt in the late 1950's in his now classical work on Perceptrons. Again this paper triggered a massive explosion of work on trainable neural networks which continues to this day.

The authors of this book, Palmer, Krogh, and Hertz, are statistical physicists who have experienced these developments. They have sought to provide an introduction to the theory behind all the hoopla, and to summarize the current state of the subject. They have, wisely, eschewed neurobiology from their coverage, and have concentrated on what they know best, statistical mechanics, and how it is applied to neural networks. In my opinion they have succeeded admirably in providing a clear and readable account of the statistical mechanical ideas underlying neural networks, including some account of the analogy between neural networks and spinglasses, and the famous Replica Method for analyzing such materials. They have also done justice to Back-Propagation in providing an up-to-date treatment of Recurrent Back-Propagation in its various manifestations. Readers who take the trouble to follow the mathematics outlined in this book will be rewarded with valuable insights into how neural networks really work. One cannot ask for much more in any scientific publication.

Jack Cowan

Mathematics Department
The University of Chicago
September 24, 1990

Foreword

It is quite clear, as convincingly illustrated in this textbook, that the theory underlying learning and computing in networks of linear threshold units has developed into a mature subfield existing somewhere between physics, computer science, and neurobiology. We have not only a growing number of examples where learning techniques are successfully applied to practical problems such as recognizing handwritten postal mail codes or protein structures or cases where theories of unsupervised Hebbian learning mimicking certain aspects of neuronal development but now possess a solid understanding of why these algorithms perform so well on certain types of processing or why they fail, why certain features—such as hidden units—are necessary and how these approaches to learning relate to more traditional methods used in statistics to estimate a poorly sampled or unknown function in the presence of noise. Thus, it appears that neural networks are here to stay after three consecutive cycles of enthusiasm and skepticism, first peaking in the 1940's with McCullough and Pitt's seminar work, then again in the 1960's with Rosenblatt's perceptron convergence theorem and its denouement by Minsky and Papert, and finally for a third time in the 1980's with Hopfield's energy approach and the modern era of multilayered networks ushered in by the backpropagation learning technique. The influence of the neural network learning paradigm on Artificial Intelligence will be profound, so much so that we will need to modify our basic notion of the Turing test as an operational definition of intelligence to encompass at least some rudimentary learning abilities. At this point in time, it is still too early to describe the long-term effect of neural networks on neurobiology and experimental neuroscience. While neural network analysis has been one of the key impulses behind "computational neuroscience," we have yet to develop specific instances where such an analysis has been used to successfully analyze and understand some real neurobiological circuits.

This monograph succinctly captures these trends and summarizes the current state of the art by way of highlighting the analogies to statistical mechanics and electric circuit theory as well as by discussing various practical applications. It is done without overdue emphasis on a formal mathematical treatment, appealing rather to the intuition of the reader. Throughout the book, the emphasis is on those features of neural networks relevant to information processing, storage and recall, that is to computation and function, linking physics to computing machines.

The Computation and Neural Systems Series—Over the past 600 million years, biology has solved the problem of processing massive amounts of noisy and highly redundant information in a constantly changing environment by evolving networks of billions of highly interconnected nerve cells. It is the task of scientists—be they mathematicians, physicists, biologists, psychologists, or computer scientists—to understand the principles underlying information processing in these complex structures. At the same time, researchers in machine vision, pattern recognition, speech understanding, robotics, and other areas of artificial intelligence can profit from understanding features of existing nervous systems. Thus, a new field is emerging: the study of how computations can be carried out in extensive networks of heavily interconnected processing elements, whether these networks are carbon- or silicon-based. Addison-Wesley's new "Computation and Neural Systems" series will reflect the diversity of this field with textbooks, course materials, and monographs on topics ranging from the biophysical modeling of dendrites and neurons, to computational theories of vision and motor control, to the implementation of neural networks using VLSI or optics technology, to the study of highly parallel computational architectures.

Christof Koch

Pasadena, California
September 21, 1990

Preface

We generally like our titles shorter than an *Introduction to the Theory of Neural Computation*, but all those words are important in understanding our purpose:

Neural Computation

Our subject matter is computation by artificial neural networks. The adjective "neural" is used because much of the inspiration for such networks comes from neuroscience, *not* because we are concerned with networks of real neurons. Brain modelling is a different field and, though we sometimes describe biological analogies, our prime concern is with what the *artificial* networks can do, and why. It is arguable that "neural" should be purged from the vocabulary of this field—perhaps *Network Computation* would have been more accurate in our title—but at present it is firmly ensconced. We do however avoid most other biological terms in non-biological contexts, including "neuron" (*unit*) and "synapse" (*connection*).

Theory

We emphasize the *theoretical* aspects of neural computation. Thus we provide little or no coverage of applications in engineering or computer science; implementations in hardware or software; or implications for cognitive science or artificial intelligence. There are recent books on all these topics and we prefer to complement rather than to compete. On the other hand, we feel that even those whose interest in the subject is completely practical may benefit from a broad theoretical perspective. We are no doubt biased by the fact that we are theorists by trade, but in our own experience we found this background to be essential in using neural networks for practical applications (not described in this book).

Introduction

Our book is intended as an introduction. This has implications at both extremes: where we start from, and how far we go. We try to start from the beginning, and assume little of the reader beyond some mathematical training. We do not assume any prior knowledge of neural networks, or of physics, engineering, or computer science. There are local exceptions to this ideal, but nothing that is central.

On the other hand we do *not* go to the end. The theory built up around neural networks is huge, and we cannot hope to cover it all. We do discuss most of the major architectures and theoretical concepts, but at varying depth. We stop short of very intricate or technical analysis, and of most directions that we consider dead ends. We aim not at mathematical rigor, but at conveying understanding through mathematics. Understanding should, we feel, consist not only of "knowing what," but also of "knowing how"; especially *knowing how to go on* [Wittgenstein, 1958]. With that in mind we are usually not satisfied with simply stating or deducing a given result, but instead try to show the reader how to think about it, how to handle and hold it.

Bibliography and Coverage

At the same time we try to provide access to the research literature for further reading. We selected the bibliography with the primary aim of assisting the reader, and only with the secondary aim of attributing credit. When there was a choice we picked the more readable source, especially in references to the associated areas of mathematics, computer science, and physics.

It may be tempting to consider this book as a comprehensive survey of what has been done in the theory of neural computation. It does have that character in some sections, particularly those near the forefront, where we try to describe just *who* has done *what* recently. But that is not our overall aim and we have *not* tried to be complete. Omissions may be for reasons of ignorance, complexity, irrelevance, obscurity, pedagogy, space, or many other reasons. We apologize only for our ignorance.

Approach

Our selection and treatment of material reflects our background as physicists. This background has helped us to understand how these complex systems function, often in terms of physical analogies. Others might find easier paths into the subject area from computer science, statistics, or psychology, and there could be written equally good or better books along these lines. We tell the story the way that we are best able to understand it, and hope that our readers find the perspective enlightening.

We often view the analysis of artificial neural networks as a statistical mechanics problem. Like many systems in condensed matter physics, these networks are large collections of interacting entities with emergent properties, and it should not be surprising that related mathematical tools are useful. Nevertheless, this is a new feature in the engineering world, where one normally expects every minute aspect of the operation of machinery (or software) to follow an explicit plan or program. But systems are becoming so complex that this kind of detailed engineering is neither possible nor desirable. Some of the networks described in this book illustrate how systems can design themselves with relatively little external guidance. The user will never know all the internal details, but will need methods like those we describe to analyze how the whole thing works.

In our experience some people, even among those working on neural networks, appear to be frightened by the prospect of having to learn statistical mechanics. They also question its necessity. Often, however, their equations turn out to be exactly the same ones that we would have written. Anyone trying to analyze the typical behavior of a many-component system is doing statistical mechanics whether it is called that or not. We hope that the doubters will not be put off by a few partition functions here and there, but will benefit by seeing many problems put in a more unified perspective.

This is not to say that we employ *only* a statistical mechanics or "physics" approach, or that one has to be a physicist to read this book. Far from it. Explicit statistical mechanics is used only rarely, and is explained where it arises (and in the Appendix). It often underlies and motivates our approach, but is not often visible at the surface. And one certainly does not need to be a physicist; we have tried hard not to assume anything (besides mathematics) of the reader, and to avoid or explain words and ideas specific to physics.

Prerequisites

There are no prerequisites besides mathematics. The mathematical level varies somewhat, with more required for the starred (⋆) sections and for Chapter 10 than for the rest. These sections may be omitted however without loss of continuity. On the whole we expect our readers to know something about multi-variate calculus, ordinary differential equations, basic probability and statistics, and linear algebra. But in most of these areas a general familiarity should be enough. The exception for some readers may be linear algebra, where we use vectors and matrices, inner products, matrix inversion, eigenvalues and eigenvectors, and diagonalization of quadratic forms. However, eigenvalues and eigenvectors are not used in any essential way until Chapter 8, and diagonalization appears only in starred sections. Kohonen [1989] provides an introduction to the linear algebra needed for neural network theory.

Acknowledgments

This book began as lecture notes for a half-semester course given at Duke University (and broadcast on the North Carolina teleclassroom network) in the spring of 1988. The audience for these lectures was very broad, including people from neurosciences and cognitive sciences as well as computer science, engineering, and physics. Later versions were used as the basis of summer school lectures at Santa Fe in June 1988 [Palmer, 1989], and for a one-semester course for physics and computer science students at the University of Copenhagen in the fall of 1989. We thank all of the students in all of these courses for constructive feedback that led to successive improvements.

We owe a debt of gratitude to many of our colleagues in Durham, Copenhagen, and Santa Fe who encouraged, supported and helped us in this work. These include Ajay, Alan, Benny, Corinna, Dave, Ingi, David, Frank, Gevene, Jack, John, Jun, Kurt, Lars, Marjorie, Mike, Per, Ronda, Søren, Stu, Tamas, and Xiang. Two of us (JH and AK) also thank the Physics Department at Duke for their hospitality in the spring of 1988, when this whole enterprise got started. AK thanks the Carlsberg Foundation for generous financial support.

Finally, we reserve our deepest appreciation for our wives and families. It is a hackneyed theme to thank loved ones for patience and understanding while a book was being written; but now we know why, and do give heartfelt thanks.

Richard Palmer
Anders Krogh
John Hertz

Durham and Copenhagen
August 1990

ONE

Introduction

Anyone can see that the human brain is superior to a digital computer at many tasks. A good example is the processing of visual information: a one-year-old baby is much better and faster at recognizing objects, faces, and so on than even the most advanced AI system running on the fastest supercomputer.

The brain has many other features that would be desirable in artificial systems.

- It is robust and fault tolerant. Nerve cells in the brain die every day without affecting its performance significantly.

- It is flexible. It can easily adjust to a new environment by "learning"—it does not have to be programmed in Pascal, Fortran or C.

- It can deal with information that is fuzzy, probabilistic, noisy, or inconsistent.

- It is highly parallel.

- It is small, compact, and dissipates very little power.

Only in tasks based primarily on simple arithmetic does the computer outperform the brain!

This is the real motivation for studying neural computation. It is an alternative computational paradigm to the usual one (based on a programmed instruction sequence), which was introduced by von Neumann and has been used as the basis of almost all machine computation to date. It is inspired by knowledge from neuroscience, though it does not try to be biologically realistic in detail. It draws its methods in large degree from statistical physics, and that is why the lectures on which this book is based originally formed part of a physics course. Its potential applications lie of course mainly in computer science and engineering. In addition it may be of value as a modelling paradigm in neuroscience and in sensory and cognitive psychology.

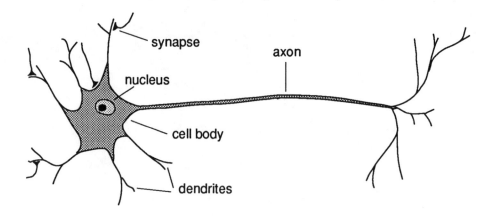

FIGURE 1.1 Schematic drawing of a typical neuron.

The field is also known as neural networks, neurocomputation, associative networks, collective computation, connectionism, and probably many other things. We will use all these terms freely.

1.1 Inspiration from Neuroscience

Today's research in neural computation is largely motivated by the possibility of making artificial computing networks. Yet, as the term "neural network" implies, it was originally aimed more towards modelling networks of real neurons in the brain. The models are extremely simplified when seen from a neurophysiological point of view, though we believe that they are still valuable for gaining insight into the principles of biological "computation." Just as most of the details of the separate parts of a large ship are unimportant in understanding the behavior of the ship (e.g., that it floats, or transports cargo), so many details of single nerve cells may be unimportant in understanding the *collective* behavior of a network of cells.

Neurons

The brain is composed of about 10^{11} **neurons** (nerve cells) of many different types. Figure 1.1 is a schematic drawing of a single neuron. Tree-like networks of nerve fiber called **dendrites** are connected to the **cell body** or **soma**, where the cell nucleus is located. Extending from the cell body is a single long fiber called the **axon**, which eventually branches or **arborizes** into strands and substrands. At the ends of these are the transmitting ends of the **synaptic junctions**, or **synapses**, to other neurons. The receiving ends of these junctions on other cells can be found

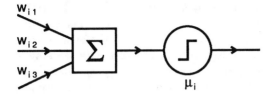

FIGURE 1.2 Schematic diagram of a McCulloch-Pitts neuron. The unit fires if the weighted sum $\sum_j w_{ij} n_j$ of the inputs reaches or exceeds the threshold μ_i.

both on the dendrites and on the cell bodies themselves. The axon of a typical neuron makes a few thousand synapses with other neurons.

The transmission of a signal from one cell to another at a synapse is a complex chemical process in which specific transmitter substances are released from the sending side of the junction. The effect is to raise or lower the electrical potential inside the body of the receiving cell. If this potential reaches a threshold, a pulse or **action potential** of fixed strength and duration is sent down the axon. We then say that the cell has "fired". The pulse branches out through the axonal arborization to synaptic junctions to other cells. After firing, the cell has to wait for a time called the **refractory period** before it can fire again.

McCulloch and Pitts [1943] proposed a simple model of a neuron as a binary threshold unit. Specifically, the model neuron computes a weighted sum of its inputs from other units, and outputs a one or a zero according to whether this sum is above or below a certain threshold:

$$n_i(t+1) = \Theta\Big(\sum_j w_{ij} n_j(t) - \mu_i\Big). \tag{1.1}$$

See Fig. 1.2. Here n_i is either 1 or 0, and represents the state of neuron i as *firing* or *not firing* respectively. Time t is taken as discrete, with one time unit elapsing per processing step. $\Theta(x)$ is the unit **step function**, or Heaviside function:

$$\Theta(x) = \begin{cases} 1 & \text{if } x \geq 0; \\ 0 & \text{otherwise.} \end{cases} \tag{1.2}$$

The weight w_{ij} represents the strength of the synapse connecting neuron j to neuron i. It can be positive or negative corresponding to an **excitatory** or **inhibitory** synapse respectively. It is zero if there is no synapse between i and j. The cell-specific parameter μ_i is the threshold value for unit i; the weighted sum of inputs must reach or exceed the threshold for the neuron to fire.

Though simple, a McCulloch-Pitts neuron is computationally a powerful device. McCulloch and Pitts proved that a synchronous assembly of such neurons is capable in principle of **universal computation** for suitably chosen weights w_{ij}. This means that it can perform any computation that an ordinary digital computer can, though not necessarily so rapidly or conveniently.

Real neurons involve many complications omitted from this simple description. The most significant ones include:

■ Real neurons are often not even approximately threshold devices as described above. Instead they respond to their input in a continuous way. This is sometimes referred to as a **graded response**. But the nonlinear relationship between the input and the output of a cell is a universal feature. Our working hypothesis is that it is the nonlinearity that is essential, not its specific form. In any case, continuous-valued units can be modelled too, and are sometimes more convenient to deal with than threshold units.

■ Many real cells also perform a *nonlinear* summation of their inputs, which takes us a bit further from the McCulloch-Pitts picture. There can even be significant logical processing (e.g., AND, OR, NOT) within the dendritic tree. This can in principle be taken care of by using several formal McCulloch-Pitts neurons to represent a single real one, though there has been little work along these lines so far. We will generally ignore this complication, since the simple McCulloch-Pitts picture is already very rich and interesting to study.

■ A real neuron produces a sequence of pulses, not a simple output level. Representing the firing rate by a single number like n_i, even if continuous, ignores much information—such as pulse phase—that might be carried by such a pulse sequence. The majority of experts do not think that phase plays a significant role in most neuronal circuits, but agreement is incomplete.

■ Neurons do not all have the same fixed delay ($t \rightarrow t + 1$). Nor are they updated synchronously by a central clock. We will in fact use asynchronous updating in much of this book.

■ The amount of transmitter substance released at a synapse may vary unpredictably. This sort of effect can be modelled, at least crudely, by a stochastic generalization of the McCulloch-Pitts dynamics.

A simple generalization of the McCulloch-Pitts equation (1.1) which includes some of these features is

$$n_i := g\Big(\sum_j w_{ij} n_j - \mu_i\Big). \tag{1.3}$$

The number n_i is now continuous-valued and is called the **state** or **activation** of unit i. The threshold function $\Theta(x)$ of (1.1) has been replaced by a more general nonlinear function $g(x)$ called the **activation function, gain function, transfer function,** or **squashing function**. Rather than writing the time t or $t+1$ explicitly as we did in (1.1), we now simply give a rule for updating n_i whenever that occurs.[1] Units are often updated *asynchronously*: in random order at random times.

Nowhere in this book do we attempt a detailed description of networks of real neurons, or of other neurobiological structures or phenomena. Kandel and Schwartz [1985] give an excellent introduction. We do sometimes appeal to biological realism, and do describe a few models of cortical organization in Chapters 8 and 9, but the emphasis is generally on the computational abilities of network models, not on their

---

[1]Note that we use the symbol ":=" to emphasize that the right-hand side is assigned to the left-hand side upon update; the equality is not continuously true.

direct applicability to brain modelling. Nevertheless, despite the intimidating detail and complexity of real brains, we do believe that the kind of theory discussed in this book is relevant to neuroscience. But the connection is not so much at the level of detailed modelling as at the level of *algorithms and representation* [Marr, 1982]. That is, this kind of approach can help in formulating and testing what sort of computational algorithms the brain is using in different tasks. While the biological and artificial implementations of the algorithms are very different, there can be many features in common at the algorithmic level.

When discussing artificial neural networks it remains commonplace to talk of "neurons" and "synapses", even though the network components are far simpler than their biological counterparts. We prefer to use the terms "units" and "connections" (or "weights") except when discussing networks that are intended as direct models of brain structures. Other terms in use for the units include "processing elements" and "neurodes".

Parallel Processing

In computer science terms, we can describe the brain as a parallel system of about 10^{11} processors. Using the simplified model (1.3) above, each processor has a very simple program: it computes a weighted sum of the input data from other processors and then outputs a single number, a nonlinear function of this weighted sum. This output is then sent to other processors, which are continually doing the same kind of calculation. They are using different weights and possibly different gain functions; the coefficients w_{ij} are in general different for different i, and we could also make $g(x)$ be site-dependent. These weights and gain functions can be thought of as local data stored by the processors.

The high connectivity of the network (i.e., the fact that there are many terms in the sum in (1.3)), means that errors in a few terms will probably be inconsequential. This tells us that such a system can be expected to be robust and its performance will degrade gracefully in the presence of noise or errors. In the brain itself cells die all the time without affecting the function, and this robustness of the biological neural networks has probably been essential to the evolution of intelligence.

The contrast between this kind of processing and the conventional von Neumann kind could not be stronger. Here we have very many processors, each executing a very simple program, instead of the conventional situation where one or at most a few processors execute very complicated programs. And in contrast to the robustness of a neural network, an ordinary sequential computation may easily be ruined by a single bit error.

It is worth remarking that the typical cycle time of neurons is a few milliseconds, which is about a million times slower than their silicon counterparts, semiconductor gates. Nevertheless, the brain can do very fast processing for tasks like vision, motor control, and decisions on the basis of incomplete and noisy data, tasks that are far beyond the capacity of a Cray supercomputer. This is obviously possible only because billions of neurons operate simultaneously. Imagine the capabilities of a

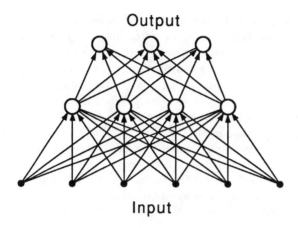

Output

Input

FIGURE 1.3 A two-layer
perceptron.

system which could operate in parallel like this but with switching times of current
semiconductor devices!

1.2 History

The history of these sorts of ideas in psychology originates with Aristotle. Yet as
a basis for computational or neural modelling we can trace them to the paper of
McCulloch and Pitts [1943], which introduced the model described above.

During the next fifteen years there was considerable work on the detailed logic of
threshold networks. They were realized to be capable of universal computation and
were analyzed as finite-state machines; see Minsky [1967]. The problem of making
a reliable network with unreliable parts was solved by the use of redundancy [von
Neumann, 1956], leading later to *distributed* redundant representations [Winograd
and Cowan, 1963].

At the opposite extreme to detailed logic, continuum theories were also developed. Known as **neurodynamics** or **neural field theory**, this approach used
differential equations to describe activity patterns in bulk neural matter [Rashevsky,
1938; Wiener, 1948; Beurle, 1956; Wilson and Cowan, 1973; Amari, 1977].

Around 1960 there was a wave of activity centered around the group of Frank
Rosenblatt, focusing on the problem of how to find appropriate weights w_{ij} for particular computational tasks. They concentrated on networks called **perceptrons**, in
which the units were organized into layers with feed-forward connections between
one layer and the next. An example is shown in Fig. 1.3. Very similar networks
called **adalines** were invented around the same time by Widrow and Hoff [1960;
Widrow, 1962].

For the simplest class of perceptrons without any intermediate layers, Rosenblatt [1962] was able to prove the convergence of a **learning algorithm**, a way to
change the weights iteratively so that a desired computation was performed. Many

people expressed a great deal of enthusiasm and hope that such machines could be a basis for artificial intelligence.

There was however a catch to the learning theorem, forcefully pointed out by Minsky and Papert [1969] in their book *Perceptrons*: the theorem obviously applies only to those problems which the structure *is capable of computing*. Minsky and Papert showed that some rather elementary computations could *not* be done by Rosenblatt's one-layer[2] perceptron. The simplest example is the **exclusive or** (XOR) problem: a single output unit is required to turn on ($n = +1$) if one or the other of two input lines is on, but not when neither or both inputs are on.

Rosenblatt had also studied structures with more layers of units and believed that they could overcome the limitations of the simple perceptrons. However, there was no learning algorithm known which could determine the weights necessary to implement a given calculation. Minsky and Papert doubted that one could be found and thought it more profitable to explore other approaches to artificial intelligence. With this most of the computer science community left the neural network paradigm for almost 20 years.

Still, there were a number of people who continued to develop neural network theory in the 1970's. A major theme was **associative content-addressable memory**, in which different input patterns become associated with one another (i.e., trigger the same response) if sufficiently similar. These had actually been proposed much earlier [Taylor, 1956; Steinbuch, 1961], and were later revived or rediscovered by Anderson [1968, 1970; Anderson and Mozer, 1981], Willshaw et al. [1969], Marr [1969, 1971] and Kohonen [1974–1989]. Grossberg [1967–1987] made a comprehensive reformulation of the general problem of learning in networks. Marr [1969, 1970, 1971] developed network theories of the cerebellum, cerebral neocortex, and hippocampus, assigning specific functions to each type of neuron. A number of people, including Marr [1982], von der Malsburg [1973], and Cooper [1973; Nass and Cooper, 1975], studied the development and functioning of the visual system.

Another thread of development can be traced to Cragg and Temperley [1954, 1955]. They reformulated the McCulloch-Pitts network as a spin (magnetic) system of the sort familiar in physics. Memory was believed to reside in the hysteresis of the domain patterns expected for such a system. Caianiello [1961] then constructed a *statistical* theory, using ideas from statistical mechanics, and incorporated learning in a way which drew on the ideas of Hebb [1949] about learning in the brain. The same theme was taken up in the 1970's by Little [1974; Little and Shaw, 1975, 1978] and again in 1981 by Hopfield [1982]. Hopfield was able to add some helpful physical insight by introducing an **energy function**, and by emphasizing the notion of memories as dynamically stable attractors. Hinton and Sejnowski [1983, 1986] and Peretto [1984] constructed formulations using **stochastic units** which follow the dynamics (1.1) or (1.3) only approximately, making "mistakes" with a certain probability analogous to temperature in statistical mechanics. The real power of

---

[2] We never count input lines as units in numbering layers. Figure 1.3 is thus a two-layer network. Until recently it would often have been called a three-layer network, but the convention is changing.

statistical mechanics was then brought to bear on the stochastic network problem by Amit et al. [1985a, b; Amit, 1989], using methods developed in the theory of random magnetic systems called **spin glasses**.

Perhaps the most influential development in this decade, however, takes up the old thread of Rosenblatt's perceptrons where it was cut 20 years ago. Various people have developed an algorithm which works quite well for adjusting the weights connecting units in successive layers of multi-layer perceptrons. Known as **back-propagation**, it appears to have been found first by Werbos [1974] in the mid-70's, and then independently rediscovered around 1985 by Rumelhart, Hinton, and Williams [1986a, b], and by Parker [1985]. Le Cun [1985] also proposed a related algorithm. Though not yet the holy grail of a completely general algorithm able to teach an arbitrary computational task to a network, it can solve many problems (such as XOR) which the simple one-layer perceptrons could not. Much current activity is centered on back-propagation and its extensions.

Many of the important early papers have been collected in Anderson and Rosen-feld [1988], including many of those mentioned here. This is an excellent collection for those interested in the history of neural networks. We also recommend the review article by Cowan and Sharp [1988a, b], which we drew on for this section.

1.3 The Issues

Massive parallelism in computational networks is extremely attractive in principle. But in practice there are many issues to be decided before a successful implementation can be achieved for a given problem:

- What is the best architecture? Should the units be divided into layers, or not? How many connections should be made between units, and how should they be organized? What sort of activation functions $g(x)$ should be used? What type of updating should be used: synchronous or asynchronous, deterministic or stochastic? How many units are needed for a given task?

- How can a network be programmed? Can it learn a task or must it be pre-designed? If it can learn a task, how many examples are needed for good performance? How many times must it go through the examples? Does it need the right answers during training, or can it learn from correct/incorrect reinforcement? Can it learn in real-time while functioning, or must the training phase be separated from the performance phase?

- What can the various types of network do? How many different tasks can they learn? How well? How fast? How robust are they to missing information, incorrect data, and unit removal or malfunction? Can they *generalize* from known tasks or examples to unknown ones? What classes of input-to-output functions can they represent?

- How can a network be built in hardware? What are the advantages and disadvantages of different hardware implementations, and how do they compare to simulation in software?

These questions are obviously coupled and cannot be answered independently. The architecture, for instance, strongly influences what the network can do, and what hardware options are available.

Much of this book will be concerned with refining and answering the above questions. However we will generally approach them from a theoretical point of view, rather than from a design one. That is, we will attempt to understand the behavior of networks as a function of their architecture, and only rarely raise the question of designing networks to fulfill particular goals. But of course the two viewpoints are not independent, and a strong understanding of principles is invaluable for good design.

Three of the issues raised above deserve a little more comment here, as general background before we become involved in details.

Hardware

Almost everything in the field of neural computation has been done by simulating the networks on serial computers, or by theoretical analysis. Neural network VLSI chips are far behind the models, as is natural at this point. The main problem with making neural network chips is that one needs a *lot* of connections, often some fraction of the square of the number of units. The space taken up by the connections is usually the limiting factor for the size of a network. The neural chips made so far contain of the order of 100 units, which is too few for most practical applications.

Potential alternatives to integrated circuit chips include optical computers. The field is very young, but electro-optical and optical associative memories have already been proposed or built.

Efficient hardware is crucially important in the long term if we are going to take full advantage of the capabilities of neural networks, and there is growing activity in this area. However, it is largely beyond the scope of this book; we return to hardware issues only briefly in Section 3.4.

Generalization

The reason for much of the excitement about neural networks is their ability to generalize to new situations. After being trained on a number of examples of a relationship, they can often induce a complete relationship that interpolates and extrapolates from the examples in a sensible way. But what is meant by *sensible* generalization is often not clear. In many problems there are almost infinitely many possible generalizations. How does a neural network—or a human for that matter—choose the "right" one? As an example one could train a neural network on three of the four XOR relations mentioned earlier, and it would be very unlikely that any of the known types of networks would actually generalize to the full XOR. Nevertheless neural networks commonly make very useful generalizations that would be judged sensible in human terms.

Programming

Like most of the work done in neural networks, much of this book is concerned with the problem of programming or learning: how do we choose the connection weights so the network can do a specific task?

We will encounter some examples where we can choose the weights *a priori* if we are a little clever. This **embeds** some information into the network by design. But such problems are the exception rather than the rule. In other cases we can often "teach" the network to perform the desired computation by iterative adjustments of the w_{ij} strengths. This may be done in two main ways:

- **Supervised learning.** Here the learning is done on the basis of direct comparison of the output of the network with known correct answers. This is sometimes called **learning with a teacher**. It includes the special case of **reinforcement learning**, where the only feedback is whether each output is correct or incorrect, not what the correct answer is.

- **Unsupervised learning.** Sometimes the learning goal is not defined at all in terms of specific correct examples. The only available information is in the correlations of the input data or signals. The network is expected to create categories from these correlations, and to produce output signals corresponding to the input category.

There are many exciting implications of the possibility of *training* a network to do a computation. Instead of having to specify every detail of a calculation, we simply have to compile a training set of representative examples. This means that we can hope to treat problems where appropriate rules are very hard to know in advance, as in expert systems and robotics. It may also spare us a lot of tedious (and expensive) software design and programming even when we do have explicit rules. John Denker has remarked that "neural networks are the second best way of doing just about anything." The *best* way is to find and use the right rules or the optimum algorithm for each particular problem, but this can be inordinately expensive and time consuming. There is plenty of scope for a second best approach based on learning by example.

TWO

The Hopfield Model

2.1 The Associative Memory Problem

Associative memory is the "fruit fly" or "Bohr atom" problem of this field. It illustrates in about the simplest possible manner the way that collective computation can work. The basic problem is this:

Store a set of p patterns ξ_i^μ in such a way that when presented with a new pattern ζ_i, the network responds by producing whichever one of the stored patterns most closely resembles ζ_i.

The patterns are labelled by $\mu = 1, 2, \ldots, p$, while the units in the network are labelled by $i = 1, 2, \ldots, N$. Both the stored patterns ξ_i^μ and the test patterns ζ_i can be taken to be either 0 or 1 on each site i, though we will adopt a different convention shortly.

We could of course do this serially in a conventional computer simply by storing a list of the patterns ξ_i^μ, writing a program which computed the Hamming distance[1]

$$\sum_i [\xi_i^\mu (1 - \zeta_i) + (1 - \xi_i^\mu)\zeta_i] \qquad (2.1)$$

between the test pattern ζ_i and each of the stored patterns, finding which of them was smallest, and printing the corresponding stored pattern out.

Here we want to see how to get a McCulloch-Pitts network to do it. That is, if we *start* in the configuration $n_i = \zeta_i$, we want to know what (if any) set of w_{ij}'s

---

[1] The Hamming distance between two binary numbers means the number of bits that are different in the two numbers.

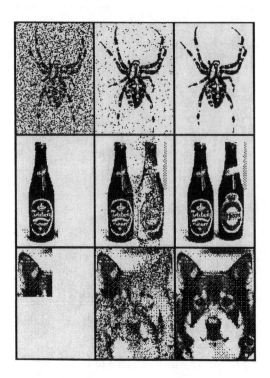

FIGURE 2.1 Example of how an associative memory can reconstruct images. These are binary images with 130 × 180 pixels. The images on the right were recalled by the memory after presentation of the corrupted images shown on the left. The middle column shows some intermediate states. A sparsely connected Hopfield network with seven stored images was used.

will make the network go to the state with $n_i = \xi_i^{\mu_0}$, where it is pattern number μ_0 that is the smallest distance (2.1) from ζ_i. Thus we want the memory to be **content-addressable** and insensitive to small errors in the input pattern.

A content-addressable memory can be quite powerful. Suppose, for example, we store coded information about many famous scientists in a network. Then the starting pattern "evolution" should be sufficient to recall everything about Darwin, and "$E = mc^3$" should recall Einstein, despite the error in the input pattern. Note that *some* pattern will always be retrieved for any clue (unless we invent a "don't know" pattern); the network will never retrieve a linear combination of, say, Darwin and Wallace in response to "evolution" but will pick the best match according to what has been stored. This depends on the nonlinearity of the network, and obviously has advantages for many practical applications.

Other common examples of applications for an associative memory are recognition and reconstruction of images (see Fig. 2.1), and retrieval of bibliographic information from partial references (such as from an incomplete title of a paper).

Figure 2.2 shows schematically the function of the dynamic associative (or content-addressable) memories that we construct in this chapter. The space of all possible states of the network—the **configuration space**—is represented by the region drawn. Within that space the stored patterns ξ_i^μ are **attractors**. The dynamics of the system carries starting points into one of the attractors, as shown by the trajectories sketched. The whole configuration space is thus divided up into

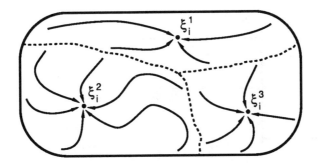

FIGURE 2.2 Schematic configuration space of a model with three attractors.

basins of attraction of the different attractors. This picture is very idealized, and in particular the space should really be a discrete set of points (on a hypercube), not a continuous region. But it is nevertheless a very useful image to keep in mind.

In the next section we treat the basic Hopfield [1982] model of associative memory. In Section 2.3 we turn to statistical mechanics, studying some magnetic systems that are analogous to our networks. Then in Section 2.4 we define a stochastic version of the original model, and analyze it using statistical mechanics methods. Finally, Section 2.5 presents a heuristic derivation of the famous $0.138N$ capacity of the Hopfield model. Various embellishments and generalizations of the basic model are discussed in the next chapter.

2.2 The Model

For mathematical convenience we now transform to a formulation where the activation values of the units are $+1$ (firing) and -1 (not firing) instead of 1 and 0. We denote[2] them by S_i rather than n_i. Conversion to and from the $n_i = 0$ or 1 notation is easy via $S_i = 2n_i - 1$. The dynamics of the network corresponding to (1.1) or (1.3) now reads

$$S_i := \mathrm{sgn}\left(\sum_j w_{ij} S_j - \theta_i\right) \tag{2.2}$$

where we take the sign function $\mathrm{sgn}(x)$ (illustrated in Fig. 2.3) to be

$$\mathrm{sgn}(x) = \begin{cases} 1 & \text{if } x \geq 0; \\ -1 & \text{if } x < 0; \end{cases} \tag{2.3}$$

---

[2] We reserve the symbol S_i for ± 1 units throughout this book.

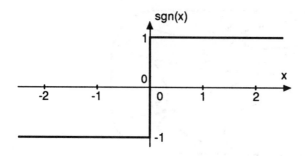

FIGURE 2.3 The function $\text{sgn}(x)$.

and the threshold θ_i is related to the μ_i in (1.1) by $\theta_i = 2\mu_i - \sum_j w_{ij}$. In the rest of this chapter we drop these threshold terms, taking $\theta_i = 0$, because they are not useful with the random patterns that we will consider. Thus we use

$$S_i := \text{sgn}\left(\sum_j w_{ij} S_j\right). \tag{2.4}$$

There are at least two ways in which we might carry out the updating specified by (2.4). We could do it *synchronously*, updating all units simultaneously at each time step. Or we could do it *asynchronously*, updating them one at a time. Both kinds of models are interesting, but the asynchronous choice is more natural for both brains and artificial networks. The synchronous choice requires a central clock or pacemaker, and is potentially sensitive to timing errors. In the asynchronous case, which we adopt henceforth, we can proceed in either of two ways:

- At each time step, select at random a unit i to be updated, and apply the rule (2.4).
- Let each unit independently choose to update itself according to (2.4), with some constant probability per unit time.

These choices are equivalent (except for the distribution of update intervals) because the second gives a random sequence; there is vanishingly small probability of two units choosing to update at exactly the same moment. The first choice is appropriate for simulation, with central control, while the second is appropriate for autonomous hardware units.

We also have to specify for how long (for how many updatings) we will allow the network to evolve before demanding that its units' values give the desired stored pattern. One possibility in the case of *synchronous* updating is to require that the network go to the correct memorized pattern right away on the first iteration. In the present discussion (using asynchronous updating) we demand only that the network settle eventually into a stable configuration—one for which no S_i changes any more.

Rather than study a specific problem such as memorizing a particular set of pictures, we examine the more generic problem of a *random* set of patterns drawn from a distribution. For convenience we will usually take the patterns to be made up of independent bits ξ_i which can each take on the values $+1$ and -1 with equal probability. More general situations are discussed in Section 3.2.

Our procedure for testing whether a proposed form of w_{ij} is acceptable is first to see whether the patterns to be memorized are themselves stable, and then to check whether small deviations from these patterns are corrected as the network evolves.

One Pattern

To motivate our choice for the connection weights, we consider first the simple case where there is just one pattern ξ_i that we want to memorize. The condition for this pattern to be stable is just

$$\text{sgn}\left(\sum_j w_{ij}\xi_j\right) = \xi_i \qquad \text{(for all } i\text{)} \tag{2.5}$$

because then the rule (2.4) produces no changes. It is easy to see that this is true if we take

$$w_{ij} \propto \xi_i\xi_j \tag{2.6}$$

since $\xi_j^2 = 1$. For later convenience we take the constant of proportionality to be $1/N$, where N is the number of units in the network, giving

$$w_{ij} = \frac{1}{N}\xi_i\xi_j \ . \tag{2.7}$$

Furthermore, it is also obvious that even if a number (fewer than half) of the bits of the starting pattern S_i are wrong (i.e., not equal to ξ_i), they will be overwhelmed in the sum for the net input

$$h_i = \sum_j w_{ij}S_j \tag{2.8}$$

by the majority that are right, and $\text{sgn}(h_i)$ will still give ξ_i. An initial configuration near (in Hamming distance) to ξ_i will therefore quickly relax to ξ_i. This means that the network will correct errors as desired, and we can say that the pattern ξ_i is an **attractor**.

Actually there are two attractors in this simple case; the other one is at $-\xi_i$. This is called a **reversed state**. All starting configurations with *more* than half the bits different from the original pattern will end up in the reversed state. The configuration space is symmetrically divided into two basins of attraction, as shown in Fig. 2.4.

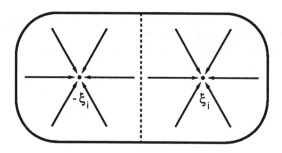

FIGURE 2.4 Schematic configuration space for the one pattern case, including the reversed state.

Many Patterns

This is fine for one pattern, but how do we get the system to recall the most similar of many patterns? The simplest answer is just to make w_{ij} a superposition of terms like (2.7), one for each pattern:

$$w_{ij} = \frac{1}{N} \sum_{\mu=1}^{p} \xi_i^\mu \xi_j^\mu . \tag{2.9}$$

Here p is the total number of stored patterns labelled by μ.

This is usually called the "Hebb rule" or the "generalized Hebb rule" because of the similarity between (2.9) and a hypothesis made by Hebb [1949] about the way in which synaptic strengths in the brain change in response to experience: Hebb suggested changes proportional to the correlation between the firing of the pre- and post-synaptic neurons. If we apply our set of patterns ξ_i^μ to the network during a **training phase**, and adjust the w_{ij} strengths according to such pre/post correlations, we arrive directly at (2.9). Technically, however, (2.9) goes beyond Hebb's original hypothesis because it changes the weights positively when neither of the units is firing ($\xi_i^\mu = \xi_j^\mu = -1$). This is probably not physiologically reasonable. Equation (2.9) can even cause a particular connection to change from excitatory to inhibitory or vice versa as more patterns are added, which is never believed to occur at real synapses. It is possible to modify the equation in various ways to remedy these defects [Toulouse et al., 1986], but here we use the simple form (2.9) unchanged.

An associative memory model using the Hebb rule (2.9) for all possible pairs ij, with binary units and asynchronous updating, is usually called a **Hopfield model**. The term is also applied to various generalizations discussed in the next chapter. Although most of the ingredients of the model were known earlier, Hopfield's influential paper [Hopfield, 1982] brought them together, introduced an energy function, and emphasized the idea of stored memories as dynamical attractors. Earlier related models, often also using the Hebb rule, are reviewed by Cowan and Sharp [1988a, b]. Particularly important is the Little model [Little, 1974; Little and Shaw, 1975, 1978], which is based however on *synchronous* updating.

Let us examine the stability of a particular pattern ξ_i^ν. The stability condition (2.5) generalizes to

$$\text{sgn}(h_i^\nu) = \xi_i^\nu \qquad \text{(for all } i\text{)} \tag{2.10}$$

where the net input h_i^ν to unit i in pattern ν is

$$h_i^\nu \equiv \sum_j w_{ij}\xi_j^\nu = \frac{1}{N}\sum_j\sum_\mu \xi_i^\mu \xi_j^\mu \xi_j^\nu . \tag{2.11}$$

We now separate the sum on μ into the special term $\mu = \nu$ and all the rest:

$$h_i^\nu = \xi_i^\nu + \frac{1}{N}\sum_j\sum_{\mu\neq\nu} \xi_i^\mu \xi_j^\mu \xi_j^\nu . \tag{2.12}$$

If the second term were zero, we could immediately conclude that pattern number ν was stable according to (2.10). This is still true if the second term is small enough: *if its magnitude is smaller than 1 it cannot change the sign of h_i^ν, and (2.10) will still be satisfied.*

It turns out that the second term, which we call the **crosstalk term**, *is* less than 1 in many cases of interest if p (the number of patterns) is small enough. We will discuss the details shortly; let us assume for now that the crosstalk term is small enough for all i and ν. Then the stored patterns are all stable—if we start the system from one of them it will stay there. Furthermore, a small fraction of bits different from a stored pattern will be corrected in the same way as in the single-pattern case; they are overwhelmed in the sum $\sum_j w_{ij}S_j$ by the vast majority of correct bits. A configuration near (in Hamming distance) to ξ_i^ν thus relaxes to ξ_i^ν. This shows that the chosen patterns are truly attractors of the system, as already anticipated in Fig. 2.2. The system works as expected as a content-addressable memory.

Storage Capacity

Consider the quantity

$$C_i^\nu \equiv -\xi_i^\nu \frac{1}{N}\sum_j\sum_{\mu\neq\nu} \xi_i^\mu \xi_j^\mu \xi_j^\nu . \tag{2.13}$$

This is just $-\xi_i^\nu$ times the crosstalk term in (2.12). If C_i^ν is negative the crosstalk term has the same sign as the desired ξ_i^ν term, and thus does no harm. But if C_i^ν is positive and larger than 1, it changes the sign of h_i^ν and makes bit (or unit) i of pattern ν unstable; if we start the system in the desired memory state ξ_i^ν, it will *not* stay there.

The C_i^ν's just depend on the patterns ξ_j^μ that we attempt to store. For now we consider purely *random* patterns, with equal probability for $\xi_j^\mu = +1$ and for

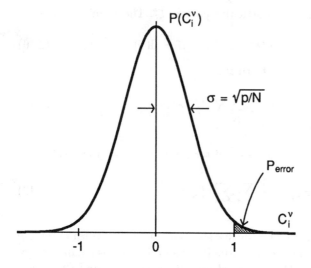

FIGURE 2.5 The distribution of values for the crosstalk C_i^ν given by (2.13). For p random patterns and N units this is a Gaussian with variance $\sigma^2 = p/N$. The shaded area is P_{error}, the probability of error per bit.

$\xi_j^\mu = -1$, independently for each j and μ. Then we can estimate the probability P_{error} that any chosen bit is unstable:

$$P_{\text{error}} = \text{Prob}(C_i^\nu > 1). \qquad (2.14)$$

Clearly P_{error} increases as we increase the number p of patterns that we try to store. Choosing a criterion for acceptable performance (e.g., $P_{\text{error}} < 0.01$) lets us find the storage **capacity** p_{max} of the network: the maximum number of patterns that can be stored without unacceptable errors. As we will see, there are actually several different expressions for p_{max}, depending on the type of criterion we use for P_{error}.

Let us first calculate P_{error}. It depends on the number of units N and the number of patterns p. We assume that both N and p are large compared to 1, because this is typically the case and because it makes the mathematics easier. Now C_i^ν is $1/N$ times the sum of about Np independent random numbers,[3] each of which is $+1$ or -1. From the theory of random coin tosses [Feller, 1968] it has therefore a **binomial distribution** with mean zero and variance $\sigma^2 = p/N$. But since Np is assumed large this can be approximated by a Gaussian distribution with the same mean and variance, as shown in Fig. 2.5.

The probability P_{error} that C_i^ν exceeds 1 is just the shaded area in Fig. 2.5. Thus

$$P_{\text{error}} = \frac{1}{\sqrt{2\pi}\sigma} \int_1^\infty e^{-x^2/2\sigma^2} \, dx$$
$$= \tfrac{1}{2}\left[1 - \text{erf}\left(1/\sqrt{2\sigma^2}\right)\right] = \tfrac{1}{2}\left[1 - \text{erf}\left(\sqrt{N/2p}\right)\right] \qquad (2.15)$$

---

[3]There are actually $N(p-1)$ terms if we include the $i = j$ terms, or $(N-1)(p-1)$ terms if we don't, but these are both approximately Np for large N and p.

TABLE 2.1 Capacities

P_{error}	p_{max}/N
0.001	0.105
0.0036	0.138
0.01	0.185
0.05	0.37
0.1	0.61

where the **error function** erf(x) is defined by

$$\text{erf}(x) \equiv \frac{2}{\sqrt{\pi}} \int_0^x \exp(-u^2)\, du\,. \tag{2.16}$$

Table 2.1 shows the values of p/N required to obtain various values of P_{error}. Thus if we choose the criterion $P_{error} < 0.01$ for example, we arrive at $p_{max} = 0.15N$.

This calculation only tells us about the *initial* stability of the patterns. If we choose $p < 0.185N$ for example, it tells us that no more than 1% of the pattern bits will be unstable initially. But if we start the system in a particular pattern ξ_i^ν and about 1% of the bits flip, what happens next? It may be that the first few flips will cause more bits to flip. In the worst case there could be an avalanche phenomenon in which more and more bits flip until the final state bears little or no resemblance to the original pattern. So our estimates for p_{max} are upper bounds; smaller values may be required to keep the final attractors close to the desired patterns. The more sophisticated calculation given in Section 2.5 deals with this problem, and shows that an avalanche occurs if $p > 0.138N$, making the whole memory useless. Thus $p_{max} = 0.138N$ if we are willing to accept the errors that occur up to that point. At $p = 0.138N$ table 2.1 shows that only 0.37% of the bits will be unstable initially, though it turns out that about 1.6% of them flip before a stable attractor is reached.

An alternative definition of the capacity insists that most of the memories be recalled *perfectly*. Since each pattern contains N bits, we need $P_{error} < 0.01/N$ to get all N bits right with 99% probability.[4] This clearly implies $p/N \to 0$ as $N \to \infty$, so we can use the asymptotic expansion of the error function

$$1 - \text{erf}(x) \to e^{-x^2}/\sqrt{\pi}x \qquad (\text{as } x \to \infty) \tag{2.17}$$

to obtain

$$\log(P_{error}) \approx -\log 2 - N/2p - \tfrac{1}{2}\log \pi - \tfrac{1}{2}\log(N/2p)\,. \tag{2.18}$$

This turns the condition $P_{error} < 0.01/N$ into

$$-\log 2 - N/2p - \tfrac{1}{2}\log \pi - \tfrac{1}{2}\log(N/2p) < \log 0.01 - \log N \tag{2.19}$$

---

[4]Strictly speaking we should write $(1 - P_{error})^N > 0.99$ here, but $P_{error} < 0.01/N$ is a good approximation from the binomial expansion.

or, taking only the leading terms for large N,

$$N/2p > \log N \qquad (2.20)$$

giving the capacity $p_{\text{max}} = N/2 \log N$ for this case.

Even more stringently, we could ask that *all* the patterns be recalled perfectly. This requires us to get Np bits right with, say, 99% probability, and so needs $P_{\text{error}} < 0.01/pN$. It is easy to see that this changes (2.20) to

$$N/2p > \log(Np) \qquad (2.21)$$

which gives $p_{\text{max}} = N/4 \log N$ because $\log(Np) \sim \log N^2 = 2 \log N$ in leading order.

Note that we have assumed in the perfect recall cases that the C_i^{ν}'s are independent of one another. Closer examination shows that this is justified. More detailed derivations of the $N/\log N$ results are available in Weisbuch and Fogelman-Soulié [1985] and McEliece et al. [1987].

In summary, the capacity p_{max} is proportional to N (but never higher than $0.138N$) if we are willing to accept a small percentage of errors in each pattern, but is proportional to $N/\log N$ if we insist that most or all patterns be recalled perfectly.

Realistic patterns will *not* in general be random, though some precoding can make them more so. The Hopfield model is usually studied with random patterns for mathematical convenience, though the effect of correlated patterns has also been examined (see Section 3.2). At the other extreme, if the different patterns are strictly *orthogonal*, i.e.,

$$\sum_j \xi_j^{\mu} \xi_j^{\nu} = 0 \qquad \text{for all } \mu \neq \nu \qquad (2.22)$$

then there is no crosstalk at all; $C_i^{\nu} = 0$ for all i and ν.

In this orthogonal case the memory capacity p_{max} is apparently N patterns, because at most N mutually orthogonal bit strings of length N can be constructed. But the *useful* capacity is somewhat smaller. Trying to embed N orthogonal patterns with the Hebb rule actually makes *all* states stable; the system stays wherever it starts, and is useless as a memory. This occurs because the orthogonality conditions (2.22) lead necessarily to[5]

$$w_{ij} = \begin{cases} 1 & \text{if } i = j; \\ 0 & \text{otherwise.} \end{cases} \qquad (2.23)$$

so each unit is connected only to itself. To define a useful measure of capacity for such a case it is clearly necessary to insist on a finite basin of attraction around each desired pattern. This leads to a useful capacity slightly less than N.

---

[5] Consider the matrix X with components $X_{\mu i} = \xi_i^{\mu}$. Equation (2.22) implies $XX^T = N\mathbf{1}$, where $\mathbf{1}$ is the unit matrix, while the Hebb rule (2.9) may be written $w = (1/N)X^T X$. Using $(AB)^T = B^T A^T$ leads immediately to $w = \mathbf{1}$.

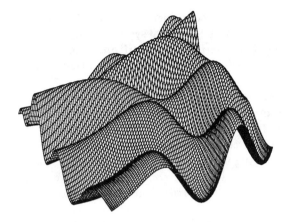

FIGURE 2.6 It is often useful (but sometimes dangerous) to think of the energy as something like this landscape. The z-axis is the energy and the 2^N corners of the hypercube (the possible states of the system) are formally represented by the x–y plane.

The Energy Function

One of the most important contributions of the Hopfield [1982] paper was to introduce the idea of an *energy function* into neural network theory. For the networks we are considering, the energy function H is

$$H = -\frac{1}{2} \sum_{ij} w_{ij} S_i S_j \, . \tag{2.24}$$

The double sum is over all i and all j. The $i = j$ terms are of no consequence because $S_i^2 = 1$; they just contribute a constant to H, and in any case we could choose $w_{ii} = 0$. The energy function is a function of the configuration $\{S_i\}$ of the system, where $\{S_i\}$ means the set of all the S_i's. We can thus imagine an **energy landscape** "above" the configuration space of Fig. 2.2. Typically this surface is quite hilly. Figure 2.6 illustrates the idea.

The central property of an energy function is that *it always decreases (or remains constant) as the system evolves according to its dynamical rule.* We will show this in a moment for (2.24). Thus the attractors (memorized patterns) in Fig. 2.2 are at local minima of the energy surface. The dynamics can be thought of as similar to the motion of a particle on the energy surface under the influence of gravity (pulling it down) and friction (so that it does not overshoot). From any starting point the particle (representing the whole state $\{S_i\}$ of the system) slides downhill until it comes to rest at one of these local minima—at one of the attractors. The basins of attraction correspond to the valleys or catchment areas around each minimum. Starting the system in a particular valley leads to the lowest point of that valley.

The term **energy function** comes from a physical analogy to magnetic systems that we will discuss in the next section. But the concept is of much wider applicability; in many fields there is a state function that always decreases during dynamical evolution, or that must be minimized to find a stable or optimum state.

In some fields the convention is reversed; the function increases or must be maximized. The most general name, from the theory of dynamical systems, is **Lyapunov function** [Cohen and Grossberg, 1983]. Other terms are **Hamiltonian** in statistical mechanics, **cost function** or **objective function** in optimization theory, and **fitness function** in evolutionary biology.

For neural networks in general an energy function exists if the connection strengths are *symmetric*, i.e., $w_{ij} = w_{ji}$. In real networks of neurons this is an unreasonable assumption, but it is useful to study the symmetric case because of the extra insight that the existence of an energy function affords us. The Hebb prescription (2.9) which we are now studying automatically yields symmetric w_{ij}'s. Gérard Toulouse has called Hopfield's use of symmetric connections a "clever step backwards from biological realism." The cleverness arises from the existence of an energy function.

For symmetric connections we can write (2.24) in the alternative form

$$H = C - \sum_{(ij)} w_{ij} S_i S_j \qquad (2.25)$$

where (ij) means all the distinct pairs ij, counting for example 12 as the same pair as 21. We exclude the ii terms from (ij); they give the constant C.

It now is easy to show that the dynamical rule (2.4) can only decrease the energy. Let S_i' be the new value of S_i given by (2.4) for some particular unit i:

$$S_i' = \text{sgn}(\sum_j w_{ij} S_j). \qquad (2.26)$$

Obviously if $S_i' = S_i$ the energy is unchanged. In the other case $S_i' = -S_i$ so, picking out the terms that involve S_i,

$$
\begin{aligned}
H' - H &= -\sum_{j \neq i} w_{ij} S_i' S_j + \sum_{j \neq i} w_{ij} S_i S_j \\
&= 2S_i \sum_{j \neq i} w_{ij} S_j \\
&= 2S_i \sum_j w_{ij} S_j - 2w_{ii}.
\end{aligned}
\qquad (2.27)
$$

Now the first term is negative from (2.26), and the second term is negative because the Hebb rule (2.9) gives $w_{ii} = p/N$ for all i. Thus the energy decreases every time an S_i changes, as claimed.

The **self-coupling terms** w_{ii} may actually be omitted altogether, both from the Hebb rule (where we can simply define $w_{ii} = 0$) and from the energy function. It is straightforward to check that they make no appreciable difference to the stability of the ξ_i^ν patterns in the large N limit. But they *do* affect the dynamics and the number of spurious states, and it turns out to be *better* to omit them [Kanter and

Sompolinsky, 1987]. We can see why simply by separating the self-coupling term out of the dynamical rule (2.4):

$$S_i := \text{sgn}\left(w_{ii}S_i + \sum_{j \neq i} w_{ij}S_j\right). \tag{2.28}$$

If w_{ii} were larger than $\sum_{j \neq i} w_{ij}S_j$ in some state, then $S_i = +1$ and $S_i = -1$ could *both* be stable.[6] This can produce additional stable **spurious states** in the neighborhood of a desired attractor, reducing the size of the basin of attraction. If $w_{ii} = 0$, then this problem does not arise; for a given configuration of the other spins S_i will always pick one of its states over the other.

Starting from an Energy Function

The idea of the energy function as something to be minimized in the stable states gives us an alternate way to derive the Hebb prescription (2.9). Let us start again with the single-pattern case. We want the energy to be minimized when the overlap between the network configuration and the stored pattern ξ_i is largest. So we choose

$$H = -\frac{1}{2N}\left(\sum_i S_i\xi_i\right)^2 \tag{2.29}$$

where the factor $1/2N$ is the product of inspired hindsight. For the many-pattern case, we can try to make each of the ξ_i^μ into local minima of H just by summing (2.29) over all the patterns:

$$H = -\frac{1}{2N}\sum_{\mu=1}^p\left(\sum_i S_i\xi_i^\mu\right)^2. \tag{2.30}$$

Multiplying this out gives

$$H = -\frac{1}{2N}\sum_{\mu=1}^p\left(\sum_i S_i\xi_i^\mu\right)\left(\sum_j S_j\xi_j^\mu\right) = -\frac{1}{2}\sum_{ij}\left(\frac{1}{N}\sum_{\mu=1}^p \xi_i^\mu\xi_j^\mu\right)S_iS_j \tag{2.31}$$

which is exactly the same as our original energy function (2.24) if w_{ij} is given by the Hebb rule (2.9).

This approach to finding appropriate w_{ij}'s is generally useful. If we can write down an energy function whose minimum satisfies a problem of interest, then we can multiply it out and identify the appropriate connection strength w_{ij} from the coefficient of S_iS_j. We will encounter several applications in Chapter 4. Of course we may find other terms, not of the S_iS_j form. Constants are no problem, and terms linear in a single S_i can be represented by thresholds or by a connection to a clamped S_0 unit. But terms like $S_iS_jS_k$ take us outside the present framework of pairwise connections.

---

[6]We assume that w_{ii} is positive or zero. The energy is no longer a Lyapunov function if negative self-couplings are allowed.

Spurious States

We have shown that the Hebb prescription (2.9) gives us (for small enough p) a dynamical system that has attractors—local minima of the energy function—at the desired points ξ_i^μ. These are sometimes called the **retrieval states**. But we have *not* shown that these are the only attractors. And indeed there are others.

First of all, the reversed states $-\xi_i^\mu$ are minima and have the same energy as the original patterns. The dynamics and the energy function both have a perfect symmetry, $S_i \leftrightarrow -S_i$ for all i. This is not too troublesome for the retrieved patterns; we could agree to reverse all the remaining bits when a particular "sign bit" is -1 for example.

Second, there are stable **mixture states** ξ_i^{mix}, which are not equal to any single pattern, but instead correspond to linear combinations of an odd number of patterns [Amit et al., 1985a]. The simplest of these are symmetric combinations of three stored patterns:

$$\xi_i^{\text{mix}} = \text{sgn}(\pm \xi_i^{\mu_1} \pm \xi_i^{\mu_2} \pm \xi_i^{\mu_3}). \qquad (2.32)$$

All eight sign combinations are possible, but we consider for definiteness the case where all the signs are chosen as $+$'s. The other cases are similar. Observe that on average ξ_i^{mix} has the same sign as $\xi_i^{\mu_1}$ three times out of four; only if $\xi_i^{\mu_2}$ and $\xi_i^{\mu_3}$ both have the opposite sign can the overall sign be reversed. So ξ_i^{mix} is Hamming distance $N/4$ from $\xi_i^{\mu_1}$, and of course from $\xi_i^{\mu_2}$ and $\xi_i^{\mu_3}$ too; the mixture states lie at points equidistant from their components. This also implies that $\sum_i \xi_i^{\mu_1} \xi_i^{\text{mix}} = N/2$ on average. Now to check the stability of (2.32), still with all $+$ signs, we can repeat the calculation of (2.11) and (2.12), but this time pick out the three special μ's:

$$h_i^{\text{mix}} = \frac{1}{N} \sum_{j\mu} \xi_i^\mu \xi_j^\mu \xi_j^{\text{mix}} = \frac{1}{2}\xi_i^{\mu_1} + \frac{1}{2}\xi_i^{\mu_2} + \frac{1}{2}\xi_i^{\mu_3} + \text{cross-terms}. \qquad (2.33)$$

Thus the stability condition (2.10) is indeed satisfied for the mixture state (2.32). Similarly 5, 7, ... patterns may be combined. The system does not choose an *even* number of patterns because they can add up to zero on some sites, whereas the units have to be ± 1.

Third, for large p there are local minima that are not correlated with any finite number of the original patterns ξ_i^μ [Amit et al., 1985b]. These are sometimes called **spin glass states** because of a close correspondence to spin glass models in statistical mechanics. We will meet them again in Section 2.5.

So the memory does not work perfectly; there are all these additional minima in addition to the ones we want. The second and third classes are generally called **spurious minima**. Of course we only fall into one of them if we start close to it, and they tend to have rather small basins of attraction compared to the retrieval states. There are also various tricks that we will consider later, including finite temperature and biased patterns, that can reduce or remove the spurious minima.

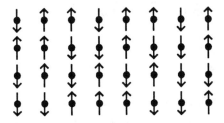

FIGURE 2.7 A very simplified picture of a magnetic material described by an Ising model.

2.3 Statistical Mechanics of Magnetic Systems

There is a close analogy between Hopfield networks and some simple models of **magnetic materials** in statistical physics. The analogy becomes particularly useful when we generalize the networks to use **stochastic units**, which brings the idea of **temperature** into network theory. We will explore this development in the next section, after providing here the necessary background in the statistical mechanics of magnetic systems. The Appendix goes further into statistical mechanics.

A simple description of a magnetic material consists of a set of atomic magnets arranged on a regular lattice that represents the crystal structure of the material (metals are crystals in this sense). We will use the term **spins** for these atomic magnets. The name comes from the quantum mechanical origin of the magnetic moments. The spins can each point in various directions, the number of possibilities depending on the type of atom considered. Particularly simple is the case of "spin 1/2" atoms, in which only two distinct directions are possible. This is represented in an **Ising model** by a variable S_i for each lattice site i, with allowed values ± 1. The spin is oriented "up" if $S_i = +1$ and "down" if $S_i = -1$. Figure 2.7 illustrates a possible configuration with spins shown by arrows pointing up or down.

The analogy of the Ising model spins to the activation of units in a neural network is obvious, and indeed we have used the same symbol S_i for both. An active unit (firing cell) in the network corresponds to "spin up" in the magnet and an inactive one to "spin down". Ising models are in fact used widely [e.g., Ma, 1985; Huang, 1987], not only for spin 1/2 magnetic materials, but also for many physical systems which can be described by binary (i.e., two-valued) variables. In many cases the description is very idealized. One might for instance simplify a continuous variable to an Ising one (as we are doing with our McCulloch-Pitts assumption), or one might describe a gas by specifying an Ising variable (for *filled* or *empty*) in each of a fine grid of cells covering the system. There is a great deal of accumulated knowledge about Ising models, some of which can be applied to neural networks.

An Ising model is not fully specified until the interactions and dynamics of the spins are given. In a magnetic material each of the spins is influenced by the **magnetic field** h at its location. This magnetic field consists of any **external field** h^{ext} applied by the experimenter, plus an **internal field** produced by the other spins. The contribution of each atom to the internal field at a given location is proportional to its own spin. Thus, adding up the contributions from all the neighboring atoms, we have a magnetic field

$$h_i = \sum_j w_{ij} S_j + h^{\text{ext}} \tag{2.34}$$

influencing S_i. The coefficients w_{ij} measure the strength of the influence of spin S_j on the field at S_i, and are called **exchange interaction strengths**. In a magnet these interactions are necessarily symmetric; it is always true that $w_{ij} = w_{ji}$. Depending on the microscopic details and on the relative location of sites i and j, the w_{ij}'s can be either positive or negative, and may vary considerably in strength.

The magnetic field h_i at spin S_i controls the dynamics of S_i. *At low temperature,* a spin tends to line up parallel to the local field h_i acting on it, so as to make $S_i = \text{sgn}(h_i)$. This can be taken to happen asynchronously in random order. We will discuss higher temperature in a moment.

Another way of specifying the interactions is to specify a potential energy corresponding to the interaction. The appropriate energy corresponding to (2.34) is

$$H = -\frac{1}{2} \sum_{ij} w_{ij} S_i S_j - h^{\text{ext}} \sum_i S_i . \tag{2.35}$$

The Appendix discusses further the importance of the energy function in statistical mechanics.

The match with the Hopfield model is thus complete. The connection strengths in the network correspond to the exchange interaction strengths of the magnet. The net input to a unit corresponds to the field acting on a spin, with the external field (if any) representing a threshold. The energy function for the network *is* just the energy of the magnet (with $h^{\text{ext}} = 0$), whence the name. The dynamics of the spins aligning with their local fields is equivalent to the McCulloch-Pitts rule (2.4).

Finite Temperature Dynamics

If the temperature is *not* very low, there is a complication in the magnetic problem. **Thermal fluctuations** tend to flip the spins, from down to up or from up to down, and thus upset the tendency of each spin to align with its field. The two influences— field and thermal fluctuations—are always present, one trying to align the spins, the other disrupting the alignment. Their relative strength depends on the temperature, with thermal fluctuations becoming decreasingly important at low temperature and vanishing altogether at the absolute zero of temperature ($-273°C$). At high temperature the thermal fluctuations dominate and a spin is nearly as often opposite to its field as aligned with it.

There is no direct equivalent of thermal fluctuations in the original Hopfield model, though we will introduce the idea in the next section. Here, however, we focus

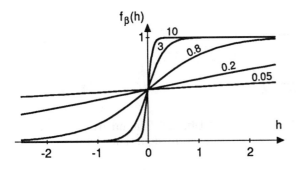

FIGURE 2.8 The sigmoid function $f_\beta(h)$ given by (2.37) for several values of β. This is sometimes called the *logistic* function.

on the magnetic problem and postpone drawing analogies in the finite[7] temperature case.

The conventional way to describe mathematically the effect of thermal fluctuations in an Ising model is with **Glauber dynamics** [Glauber, 1963]. We replace the previous deterministic dynamics by a *stochastic* rule:

$$S_i := \begin{cases} +1 & \text{with probability } g(h_i)\,; \\ -1 & \text{with probability } 1 - g(h_i)\,. \end{cases} \tag{2.36}$$

This is taken to be applied whenever spin S_i is updated. The function $g(h)$ depends on the temperature. There are several choices possible, as discussed in the Appendix. We adopt the usual "Glauber" choice of the sigmoid-shaped function

$$g(h) = f_\beta(h) \equiv \frac{1}{1 + \exp(-2\beta h)} \tag{2.37}$$

illustrated in Fig. 2.8. The parameter β is related to the *absolute* temperature[8] T by

$$\beta = \frac{1}{k_B T} \tag{2.38}$$

where k_B is **Boltzmann's constant**, with value 1.38×10^{-16} erg/K. Note that

$$1 - f_\beta(h) = f_\beta(-h) \tag{2.39}$$

so we can write the the dynamical rule in the symmetrical form[9]

$$\text{Prob}(S_i = \pm 1) = f_\beta(\pm h_i) = \frac{1}{1 + \exp(\mp 2\beta h_i)}. \tag{2.40}$$

---

[7]We use "finite" here (and frequently elsewhere) in the sense of non-zero, as opposed to the mathematical usage meaning non-infinite.

[8]Absolute temperature in Kelvin (K) is measured on the Celsius scale but from absolute zero upwards, so absolute zero is $T = 0\,\text{K}$ and $0°\text{C}$ is $273\,\text{K}$.

[9]An equation like this applies *either* with all the upper signs *or* with all the lower signs.

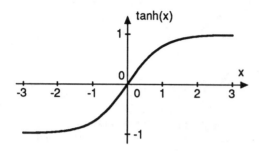

FIGURE 2.9 The function $\tanh(x)$ defined by equation (2.43).

The temperature controls the steepness of the sigmoid near $h = 0$. As we see in Fig. 2.8, at high temperature (small β) $f_\beta(h)$ goes very smoothly from 0 to 1 as h goes from $-\infty$ to $+\infty$. At low temperature (large β) it makes a rather sudden rise from near 0 to near 1 over a narrow range of h of width proportional to $k_B T$. In the limit $T \to 0$ or $\beta \to \infty$, $f_\beta(h)$ just reduces to a step function $\Theta(h)$, so (2.36) or (2.40) reduces to the deterministic rule $S_i := \text{sgn}(h_i)$. In the other limit, $T \to \infty$ or $\beta = 0$, every decision about the next value of an S_i is completely random.

A Single Spin in Equilibrium

We now apply this dynamical rule to the simplest problem in statistical mechanics, that of a single spin in a fixed external magnetic field. With only one spin we can drop the i subscripts and use $S = \pm 1$ for the spin and $h = h^{\text{ext}}$ for the field. Then using (2.40) we can calculate the average of S, usually called the **average magnetization**, which we denote by $\langle S \rangle$:

$$\langle S \rangle \equiv \text{Prob}(+1) \cdot (+1) + \text{Prob}(-1) \cdot (-1) \tag{2.41}$$

$$= \frac{1}{1 + e^{-2\beta h}} - \frac{1}{1 + e^{2\beta h}} = \frac{e^{\beta h}}{e^{\beta h} + e^{-\beta h}} - \frac{e^{-\beta h}}{e^{-\beta h} + e^{\beta h}}$$

$$= \tanh \beta h \tag{2.42}$$

where the **hyperbolic tangent function** $\tanh(x)$ is defined by

$$\tanh(x) \equiv \frac{e^x - e^{-x}}{e^x + e^{-x}} \, . \tag{2.43}$$

The $\tanh(x)$ function has the same kind of shape as $f_\beta(h)$ except that it goes from -1 to $+1$ instead of 0 to $+1$; see Fig. 2.9. In fact $\tanh(\beta h) = 2 f_\beta(h) - 1$. So as one sweeps the field from $-\infty$ to $+\infty$, the average magnetization goes from -1 to $+1$, with most of the increase happening for h within about $k_B T$ of zero. The sharp threshold behavior which the unit exhibited at zero temperature is thus smoothed out into a continuous transition in $\langle S \rangle$. But one must remember that at a given time, S itself is still either $+1$ or -1. It flips back and forth randomly between these two values, taking on one of them more frequently according to $f_\beta(h)$.

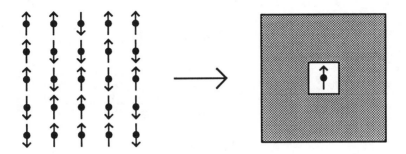

FIGURE 2.10 Mean field theory. All the spins but one are replaced by their average values.

It is worth remarking that our result (2.42) also applies to a whole collection of N spins if they all experience the same external field and have no influence on one another. Such a system is called a **paramagnet**. The famous Curie law for paramagnets, $\partial M/\partial h \propto 1/T$ at $h = 0$, follows immediately upon identifying the total magnetization M with $N\langle S\rangle$.

Mean Field Theory

In a problem of many interacting spins the problem is not so easily solved. The evolution of spin S_i depends on a local field $h_i = \sum_j w_{ij}S_j + h^{\text{ext}}$ which involves variables S_j that themselves fluctuate back and forth. There is in general no way to solve the many-spin problem exactly, but there is an approximation which is sometimes quite good. It will turn out to be very useful for the analysis of neural networks. Known as **mean field theory**, it consists of replacing the true fluctuating h_i by its average value

$$\langle h_i\rangle = \sum_j w_{ij}\langle S_j\rangle + h^{\text{ext}}. \qquad (2.44)$$

We can then compute the average $\langle S_i\rangle$ just as in (2.42) for the single-unit problem:

$$\langle S_i\rangle = \tanh(\beta\langle h_i\rangle) = \tanh\left(\beta\sum_j w_{ij}\langle S_j\rangle + \beta h^{\text{ext}}\right). \qquad (2.45)$$

These are still N nonlinear equations in N unknowns, but at least they no longer involve stochastic variables.

Figure 2.10 shows the essential idea. We focus on a single spin and replace all the other spins by an average background field. No fluctuations of the other spins are taken into account, not even in response to changing the state of the chosen spin.

This mean field approximation often becomes exact in the limit of **infinite range interactions**, where each spin interacts with all the others. Crudely speaking, this is because h_i is then the sum of very many terms, and a central limit

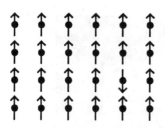

FIGURE 2.11 An Ising ferromagnet well below the critical temperature T_c. Almost all of the spins point in the same direction.

theorem can be applied. Even for short range interactions, for which $w_{ij} \approx 0$ if spins i and j are more than a few lattice sites apart, mean field theory can often give a good qualitative description of the phenomena.

The Ferromagnet

In a **ferromagnet**, all the w_{ij}'s are positive. Thus the interactions tend to make all the spins line up with each other, while thermal fluctuations tend to disrupt this ordering. It turns out that (in the absence of an external field) the thermal fluctuations win above a certain **critical temperature** T_c, making $\langle S \rangle = 0$, while the interactions win below T_c, making $\langle S \rangle \neq 0$, the same on all sites. The system thus exhibits a **phase transition** at T_c. This actually happens in materials like iron, for example, for which T_c is 770°C. Above this temperature (which is still well below the melting point of 1538°C) iron loses all its magnetization. Below T_c a majority of the spins are lined up in one direction, as shown in Fig. 2.11.

The simplest model of a ferromagnet is one in which all the w_{ij} are the same. We take

$$w_{ij} = \frac{J}{N} \qquad \text{(for all } ij) \tag{2.46}$$

for N spins where J is a constant. The $1/N$ dependence is needed to make the scaling with N sensible. These infinite range interactions are unrealistic but mathematically convenient.

Note that at *zero temperature* this infinite range ferromagnet corresponds precisely (for $J = 1$) to the one-pattern Hopfield model defined in (2.7) for a pattern with $\xi_i = 1$ for all i. A mathematical transformation also makes it equivalent to a one-pattern memory with any other pattern ξ_i; one defines new variables $\tilde{S}_i = S_i \xi_i$ and observes that the model is ferromagnetic in the new variables [Mattis, 1976]. The stability of the stored pattern in the one-pattern memory is equivalent to the stability of the fully magnetized state $S_i = 1$ for all i in the ferromagnet. This also gives us a concrete physical way to think about the *dynamics* of the recall process; it is just like the way the spins in a ferromagnet line up with the net field from their neighbors until they all point in the same direction. When we go to the many-pattern memory, the generalization is direct; the network is now like a magnetic system which has a number of locally stable magnetization patterns. For sufficiently

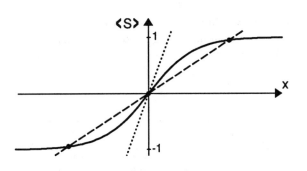

FIGURE 2.12 Solutions to equation (2.47) are intersections between the straight line $\langle S \rangle = x/\beta J$ and the sigmoid curve $\langle S \rangle = \tanh(x)$. When $\beta J \leq 1$ (dotted line) there is only one solution, at $\langle S \rangle = 0$, whereas for $\beta J > 1$ (dashed line) there are three solutions.

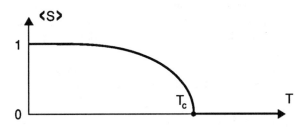

FIGURE 2.13 The positive solution of equation (2.47) as a function of temperature.

small p/N, the relaxation toward any one of them is just like the relaxation of a ferromagnet to a uniformly magnetized state.

At finite temperature we use mean field theory. Using (2.46) in (2.45) and *putting in what we know characterizes the ferromagnetic state*, i.e., that its magnetization is uniform, $\langle S_i \rangle = \langle S \rangle$, we obtain a single equation to be solved for $\langle S \rangle$:

$$\langle S \rangle = \tanh(\beta J \langle S \rangle). \tag{2.47}$$

Here we have set $h^{\text{ext}} = 0$ for convenience, although the generalization is obvious.

Equation (2.47) is easily solved graphically for $\langle S \rangle$ as a function of T, as shown in Fig. 2.12. The kinds of solutions depend on whether βJ is smaller or larger than one. This corresponds to the different behavior above and below the critical temperature T_c, so we can deduce that $T_c = J/k_B$. When $T \geq T_c$ (or $\beta J \leq 1$), there is only the trivial solution $\langle S \rangle = 0$; the spin on each site points up and down equally often. But for $T < T_c$, there are two other solutions with $\langle S \rangle \neq 0$, one the negative of the other. It turns out that the new solutions are stable against small disturbances in $\langle S \rangle$, while in this temperature range the trivial $\langle S \rangle = 0$ solution is unstable. This says that the ferromagnet can be found with its spins either predominantly up or predominantly down. If it is in one of these phases, it will not flip over to the other (in the limit of an infinite system). The magnitude of the average **spontaneous magnetization** $\langle S \rangle$ rises sharply (continuously, but with infinite derivative at $T = T_c$) as one goes below T_c; see Fig. 2.13. As T approaches 0, $\langle S \rangle$ approaches ± 1; all spins point in the same direction.

Thus this simple mean field theory approximation does give the right behavior for a ferromagnet. It is actually exact for the infinite range case defined by (2.46), but is qualitatively correct for any ferromagnet.

2.4 Stochastic Networks

We now apply much of the preceding section to neural networks, making the units stochastic and introducing the analogue of temperature. This will enable us to use mean field theory and hence ultimately to compute the storage capacity of the network. Taking the zero-temperature limit will always reduce our system to a deterministic Hopfield network, but the finite temperature extension will prove very useful for analysis.

We make our units behave stochastically exactly as for the spins in an Ising model [Hinton and Sejnowski, 1983; Peretto, 1984]. That is, we use (2.40),

$$\text{Prob}(S_i = \pm 1) \ = \ f_\beta(\pm h_i) \ = \ \frac{1}{1 + \exp(\mp 2\beta h_i)} \,. \tag{2.48}$$

for unit S_i whenever it is selected for updating, and select units in a random order as before. The function $f_\beta(h)$ is given by (2.37) and illustrated in Fig. 2.8. It is often called the **logistic function** in this context. We could actually make other choices for the activation function $g(h)$ in (2.36), but the choice $g(h) = f_\beta(h)$ allows the application of statistical mechanics. We can also reinterpret (2.48) as describing an ordinary deterministic threshold unit with a **random threshold** θ drawn from a probability density $f_\beta'(\theta)$.

What is the meaning of this stochastic behavior? In real neural networks, neurons fire with variable strength, and there are delays in synapses, random fluctuations from the release of transmitters in discrete vesicles, and so on. These are effects that we can loosely think of as **noise**, and crudely represent by thermal fluctuations as we have done in writing (2.48). Of course, the parameter β is *not* directly related to the physical temperature; it is simply a parameter controlling the noise level, or the likelihood that the deterministic rule (2.4) is violated. Nevertheless, it is useful to define a **pseudo-temperature** T for the network by

$$\beta \equiv \frac{1}{T} \,. \tag{2.49}$$

T is emphatically not the real temperature of a piece of brain, or that of a network of circuits. Note that we did not need to put a constant k_B into (2.49), since T is not a physical temperature. In effect we set $k_B = 1$ in applications of statistical mechanics formulae. Henceforth we use (2.49) constantly and without comment, converting freely between β and T as convenient. Moreover we generally refer to T as the "temperature" even though "pseudo-temperature" would be more appropriate.

As we noted for the magnetic system, and illustrated in Fig. 2.8, the temperature T controls the steepness of the sigmoid $f_\beta(h)$ near $h = 0$. At very low temperature the sigmoid becomes a step function and (2.48) reduces to the deterministic McCulloch-Pitts rule (2.4) for the original Hopfield network. As T is increased this sharp threshold is softened up in a stochastic way.

The use of stochastic units is not merely for mathematical convenience, nor simply to represent noise in our hardware or neural circuits. It is actually useful in many situations because it makes it possible to kick the system out of spurious local minima of the energy function. Generally the spurious minima will be less stable (higher in energy) than the desired retrieval patterns, and will not trap a stochastic system permanently.

The network will in general evolve differently every time it is run. Meaningful quantities to calculate are therefore averages over all possible evolutions, weighted by the probabilities of each particular history. This is just the type of calculation for which statistical mechanics is ideal. There is however an additional requirement for the use of most of statistical mechanics: we need to know that our network *comes to equilibrium*. This means that average quantities, such as the average value $\langle S_i \rangle$ of a particular S_i, eventually become *time-independent*. Luckily it can be proved that networks with an energy function—in the present context, networks with symmetric connections w_{ij}—do indeed come to equilibrium. So even though we can no longer talk about absolute stability of particular configurations $\{S_i\}$ of the network, we can still study stable **equilibrium states** $\{\langle S_i \rangle\}$ in which the average values do not change in time.

Mean Field Theory

We now apply the mean field approximation to the stochastic model we have just defined, with the Hebb rule (2.9) for the connection strengths w_{ij}. We restrict ourselves at present to the case of a relatively small number of patterns, $p \ll N$. The higher p case is much harder and will be treated in the next section. Technically the analysis here is valid for any fixed p in the $N \to \infty$ limit. The approach is due to Amit et al. [1985a]. A more formal derivation will be given in Chapter 10.

By direct analogy with (2.45), the mean field equations are

$$\langle S_i \rangle = \tanh\left(\frac{\beta}{N} \sum_{j,\mu} \xi_i^\mu \xi_j^\mu \langle S_j \rangle\right). \tag{2.50}$$

These are not obviously soluble, since they consist of N nonlinear equations. But we can follow the example of the ferromagnet and make an *ansatz* (hypothesis), taking $\langle S_i \rangle$ proportional to one of the stored patterns:

$$\langle S_i \rangle = m\xi_i^\nu . \tag{2.51}$$

We have already seen that states like this are stable in the deterministic limit $T = 0$ (with $m = 1$), so it is natural to look for similar average states in the stochastic case.

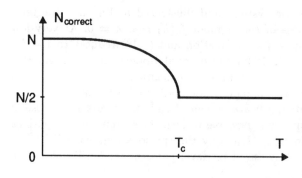

FIGURE 2.14 $\langle N_{\text{correct}} \rangle$ from (2.56) as a function of temperature T.

Using this *ansatz* in (2.50) produces

$$m\xi_i^\nu = \tanh\left(\frac{\beta}{N} \sum_{j,\mu} \xi_i^\mu \xi_j^\mu m \xi_j^\nu\right). \qquad (2.52)$$

Just as in the corresponding problem (2.11) for the deterministic network, the argument of the tanh can be split up into a term proportional to ξ_i^ν itself, plus a crosstalk term involving the overlap between ξ_i^ν and the other stored patterns. Again, in the limit where the number of stored patterns p is much smaller than N, the crosstalk term is negligible, so we find

$$m\xi_i^\nu = \tanh(\beta m \xi_i^\nu) \qquad (2.53)$$

or, since $\tanh(-x) = -\tanh(x)$,

$$m = \tanh(\beta m). \qquad (2.54)$$

This is just the same as (2.47) that we found for the spontaneous magnetization of the ferromagnet, and can be solved in the same way (Fig. 2.12). So we now know that the memory states will be stable for temperatures less than 1. Thus the critical temperature[10] T_c is 1 for the stochastic network with $p \ll N$.

Figure 2.13 immediately gives us m as a function of temperature T. Adapting (2.41), the number m may be written

$$m = \langle S_i \rangle / \xi_i^\nu = \text{Prob(bit } i \text{ is correct)} - \text{Prob(bit } i \text{ is incorrect)} \qquad (2.55)$$

(which is the same for all i in the mean field approximation), and thus the average number of correct bits in the retrieved pattern is

$$\langle N_{\text{correct}} \rangle = \frac{1}{2} N (1 + m). \qquad (2.56)$$

This is shown in Fig. 2.14 as a function of T. Note that above the critical temperature

---

[10]The value of the critical temperature depends of course on the choice of the particular coefficient $1/N$ in the expression (2.7) for the connection strengths; multiplying all w_{ij}'s by a constant factor will multiply the critical temperature by the same factor.

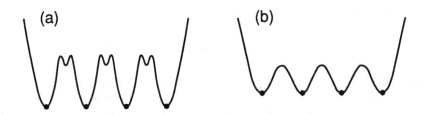

FIGURE 2.15 Schematic illustration of energy landscapes for $p \ll N$. (a) One can think of the mixture states as small dips between the desired pattern valleys. (b) At high enough temperature there are no mixture states.

$\langle N_{\text{correct}} \rangle$ is $N/2$, which is just the number of correct bits expected in a random pattern, whereas $\langle N_{\text{correct}} \rangle$ goes to N (all correct) at low temperature.

The fact that there is a *sharp* change in behavior at a particular noise level is another example of a **phase transition**. One might have assumed naively that the behavior would change smoothly as T was varied, but in a large system this is often not the case. It is in finding this kind of feature that the statistical mechanics approach makes an important contribution to the understanding of complex problems. In the present context it says that a large network abruptly ceases to function at all if a certain noise level is exceeded; this is obviously of great practical importance when it comes to designing devices.

Although we have shown that states with $\langle S_i \rangle$ proportional to a single pattern are stable at low temperatures, the system is not a perfect device. There are still the spurious states discussed earlier. The spin glass states are not relevant for $p \ll N$, but the reversed states and mixture states are both present. However each type of mixture state has its own critical temperature, above which it is no longer stable. Figure 2.15 illustrates the idea schematically. The *highest* of these critical temperatures is 0.46, for the combinations of three patterns given by (2.32). So for $0.46 < T < 1$ there are no mixture states, and only the desired patterns (and their reversals) are stable. This shows how noise (i.e., $T > 0.46$) can actually be useful for improving the performance of a network.

2.5 Capacity of the Stochastic Network

The preceding analysis was valid for $p \ll N$. A mean field analysis for p of the order of N is considerably harder and was first performed in a *tour de force* by Amit, Gutfreund, and Sompolinsky [1985b, 1987b; Amit, 1989]. Here we give a heuristic derivation due to Geszti [1990, chapter 4] (see also Peretto [1988]). The more formal derivation is provided in Chapter 10.

It is useful to define the **load parameter**

$$\alpha = \frac{p}{N} \tag{2.57}$$

i.e., the number of patterns we try to store expressed as a fraction of the number of units in the network. We now consider α of order unity, whereas previously $\alpha \sim 1/N$ since we considered p fixed and N large.[11] We will freely use the large N limit to drop lower order terms; we will, for instance, ignore any distinction between p and $p-1$.

The starting point is the same mean field equation (2.50) that we used above for the small p case, but now the crosstalk term in (2.52) can no longer be ignored. That is, we now have to pay attention to the overlaps between the state $\langle S_i \rangle$ and the patterns

$$m_\nu = \frac{1}{N} \sum_i \xi_i^\nu \langle S_i \rangle \tag{2.58}$$

for *all* the patterns, not just for the one being retrieved. We suppose that it is pattern number 1 whose retrieval we are studying. Then m_1 is of order unity, while each of the m_ν for $\nu \neq 1$ is small, of order $1/\sqrt{N}$ for our random patterns. Nevertheless the quantity

$$r = \frac{1}{\alpha} \sum_{\nu \neq 1} m_\nu^2 , \tag{2.59}$$

which is the mean square overlap of the system configuration with the nonretrieved patterns, is of order unity. The factor $1/\alpha = N/p$ in (2.59) makes r a true average over the p (or $p-1$) squared overlaps and cancels the expected $1/\sqrt{N}$ dependence of the m_ν's.

Capacity Calculation ⋆

Treatment of the network at finite α requires a self-consistent calculation of r and m_1. To do this, we use (2.58) to write the mean field equations (2.50) in the form

$$m_\nu = \frac{1}{N} \sum_i \xi_i^\nu \tanh\left(\beta \sum_\mu \xi_i^\mu m_\mu\right) \tag{2.60}$$

and, for $\nu \neq 1$, separate out explicitly the terms with $\mu = 1$ and with $\mu = \nu$:

$$m_\nu = \frac{1}{N} \sum_i \xi_i^\nu \xi_i^1 \tanh\left[\beta\left(m_1 + \xi_i^\nu \xi_i^1 m_\nu + \sum_{\mu \neq 1,\nu} \xi_i^\mu \xi_i^1 m_\mu\right)\right]. \tag{2.61}$$

Here we have taken advantage of the freedom to move factors like ξ_i^ν in and out of the tanh, because they are ± 1 and $\tanh(-x) = -\tanh(x)$.

[11] Actually the previous results are valid up to $p \sim \log N$ or $\alpha \sim (\log N)/N$.

The first term in the argument of the tanh is large (order 1) by hypothesis, because it is pattern 1 that we are retrieving, and the last term is large because there are about p terms in it. But the second term is small, order $1/\sqrt{N}$, so we can expand:

$$
m_\nu = \frac{1}{N} \sum_i \xi_i^\nu \xi_i^1 \tanh\left[\beta\left(m_1 + \sum_{\mu \neq 1,\nu} \xi_i^\mu \xi_i^1 m_\mu\right)\right]
$$
$$
+ \frac{\beta}{N} \sum_i \left\{1 - \tanh^2\left[\beta\left(m_1 + \sum_{\mu \neq 1,\nu} \xi_i^\mu \xi_i^1 m_\mu\right)\right]\right\} m_\nu \qquad (2.62)
$$

using $\frac{d}{dx}\tanh(x) = 1 - \tanh^2(x)$. We now *assume* that the small overlaps m_μ, $\mu \neq 1$, are independent random variables with mean zero and variance $\alpha r/p$, as suggested by (2.59). In the second line of (2.62), $\xi_i^\mu \xi_i^1$ is random and independent of m_μ, so by the central limit theorem the site average $N^{-1}\sum_i$ is effectively an average over a Gaussian "noise" $\sum_{\mu \neq 1,\nu} \xi_i^\mu \xi_i^1 m_\mu$ of variance αr. Thus (2.62) reduces to

$$
m_\nu = \frac{1}{N} \sum_i \xi_i^\nu \xi_i^1 \tanh\left[\beta\left(m_1 + \sum_{\mu \neq 1,\nu} \xi_i^\mu \xi_i^1 m_\mu\right)\right] + \beta m_\nu - \beta q m_\nu \qquad (2.63)
$$

or

$$
m_\nu = \frac{N^{-1} \sum_i \xi_i^\nu \xi_i^1 \tanh[\beta(m_1 + \sum_{\mu \neq 1,\nu} \xi_i^\mu \xi_i^1 m_\mu)]}{1 - \beta(1 - q)} \qquad (2.64)
$$

where

$$
q = \int \frac{dz}{\sqrt{2\pi}} e^{-\frac{1}{2}z^2} \tanh^2[\beta(m_1 + \sqrt{\alpha r}z)]. \qquad (2.65)
$$

Now we can calculate r. We just square (2.64)

$$
m_\nu^2 = \left[\frac{1}{1 - \beta(1 - q)}\right]^2 \frac{1}{N^2} \sum_{ij} \xi_i^\nu \xi_i^1 \xi_j^\nu \xi_j^1 \times \tanh\left[\beta\left(m_1 + \sum_{\mu \neq 1,\nu} \xi_i^\mu \xi_i^1 m_\mu\right)\right]
$$
$$
\times \tanh\left[\beta\left(m_1 + \sum_{\mu \neq 1,\nu} \xi_j^\mu \xi_j^1 m_\mu\right)\right] \qquad (2.66)
$$

and average the result over patterns. Since pattern ν does not occur inside the tanh's, the pattern factors $\xi_i^\nu \xi_i^1 \xi_j^\nu \xi_j^1$ outside the tanh's can be averaged separately, and only the $i = j$ term survives. Then the remaining average of the tanh's just gives a factor of q as in (2.62). The result is independent of ν, so from (2.59):

$$
r = \frac{q}{[1 - \beta(1 - q)]^2}. \qquad (2.67)
$$

We also need an equation for m_1. Using the same approach, starting again from (2.60) with $\nu = 1$, it is easy to obtain

$$
m_1 = \int \frac{dz}{\sqrt{2\pi}} e^{-\frac{1}{2}z^2} \tanh[\beta(m_1 + \sqrt{\alpha r}z)]. \qquad (2.68)
$$

The three equations (2.65), (2.67), and (2.68) can now be solved simultaneously for m_1, q, and r. In general this must be done numerically. We examine in detail only the $T \to 0$ (or $\beta \to \infty$) limit. In this limit it is clear from (2.65) that $q \to 1$, but the quantity $C \equiv \beta(1-q)$ remains finite. The $T \to 0$ limit lets us evaluate the integrals in (2.65) and (2.68), using the identities

$$\int \frac{dz}{\sqrt{2\pi}} e^{-z^2/2} (1 - \tanh^2 \beta[az+b])$$

$$\simeq \frac{1}{\sqrt{2\pi}} e^{-z^2/2}|_{\tanh^2 \beta[az+b]=0} \times \int dz \,(1 - \tanh^2 \beta[az+b])$$

$$= \frac{1}{\sqrt{2\pi}} e^{-b^2/2a^2} \frac{1}{a\beta} \int dz \,\frac{\partial}{\partial z} \tanh \beta[az+b]$$

$$= \sqrt{\frac{2}{\pi}} \frac{1}{a\beta} e^{-b^2/2a^2} \tag{2.69}$$

and

$$\int \frac{dz}{\sqrt{2\pi}} e^{-z^2/2} \tanh \beta[az+b]$$

$$\xrightarrow{T \to 0} \int \frac{dz}{\sqrt{2\pi}} e^{-z^2/2} \operatorname{sgn}[az+b]$$

$$= 2\int_{-b/a}^{\infty} \frac{dz}{\sqrt{2\pi}} e^{-z^2/2} - 1$$

$$= \operatorname{erf}\left(\frac{b}{\sqrt{2a}}\right) \tag{2.70}$$

where the error function erf(x) was defined in (2.16). Our three equations thus become:

$$C \equiv \beta(1-q) = \sqrt{\frac{2}{\pi\alpha r}} \exp\left(-\frac{m^2}{2\alpha r}\right) \tag{2.71}$$

$$r = \frac{1}{(1-C)^2} \tag{2.72}$$

$$m = \operatorname{erf}\left(\frac{m}{\sqrt{2\alpha r}}\right) \tag{2.73}$$

where we have written m for m_1. We can find the capacity of the network by solving these three equations. Setting $y = m/\sqrt{2\alpha r}$, we obtain the equation

$$y\left(\sqrt{2\alpha} + \frac{2}{\sqrt{\pi}} e^{-y^2}\right) = \operatorname{erf}(y) \tag{2.74}$$

which is easily solved graphically as shown in Fig. 2.16.

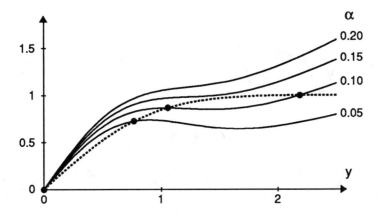

FIGURE 2.16 Graphical solution of equation (2.74). The solid lines show the left-hand side for several values of α, while the dotted line shows the right-hand side, $\mathrm{erf}(y)$. Nontrivial solutions with $m > 0$ are given by intersections away from the origin.

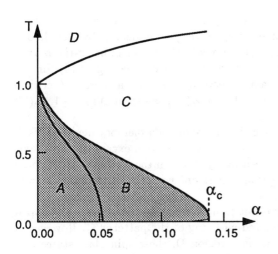

FIGURE 2.17 The phase diagram obtained by Amit et al.. The desired memory states are only stable in the shaded region.

The Phase Diagram of the Hopfield Model

By solving (2.74) we can see that there is a critical value α_c of α where the nontrivial solutions $(m \neq 0)$ disappear. A numerical evaluation gives

$$\alpha_c \approx 0.138. \tag{2.75}$$

The jump in m at this point is considerable: from $m \approx 0.97$ to zero. Recalling (2.56), this tells us that (at $T = 0$) we go discontinuously from a very good working memory with only a few bits in error for $\alpha < \alpha_c$ to a useless one for $\alpha > \alpha_c$.

Figure 2.17 shows the whole **phase diagram** for the Hopfield model, delineat-

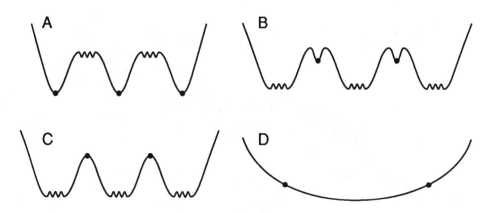

FIGURE 2.18 An attempt to visualize the energy landscape in different parts of the phase diagram. The dots show the desired memory states, while the small ripples represent spurious states. The four cases correspond to the four regions A-D of the phase diagram (figure 2.17).

ing different regimes of behavior in the T–α plane. There is a roughly triangular region where the network is a good memory device, as indicated by the shaded region of the figure. The result (2.75) corresponds to the upper limit on the α axis, while the critical temperature $T_c = 1$ derived previously (see Fig. 2.14) for the $p \ll N$ case sets the limit on the T axis. Between these limits there is a critical temperature $T_c(\alpha)$, or equivalently a critical load $\alpha_c(T)$, as shown. As $T \rightarrow 1$, $\alpha_c(T)$ goes to zero like $(1 - T)^2$.

Outside the shaded region the device is not useful as a memory device; m is 0. At the boundary m always jumps *discontinuously* down to 0, except on the T axis where the transition is continuous, as seen in Fig. 2.14. In the terminology of phase transitions this means that the transition is *first order* except at the point $\alpha = 0$, $T = 1$ where it is *second order*.

In region C the network still turns out to have many stable states, called **spin glass states**, but these are not correlated with any of the patterns ξ_i^μ. However, if T is raised to a sufficiently high value, into region D, these spin glass states also melt, and the only solution of the mean field equations is $\langle S_i \rangle = 0$.

Regions A and B both have the desired retrieval states, besides some percentage of wrong bits, but also have spin glass states. The spin glass states are the most stable states in region B, lower in energy[12] than the desired states, whereas in region A the desired states are the global minima. For small enough α and T there are also mixture states which are correlated with an odd number of the patterns as discussed earlier. These always have higher free energy than the desired states. Each type of mixture state is stable in a triangular region like AB, but with smaller

[12]Or, more correctly, *free energy*; see the Appendix.

intercepts on both axes. The most stable mixture states, given by (2.32), extend to 0.46 on the T axis and 0.03 on the α axis.

Figure 2.18 shows some highly idealized sketches of the form of the free energy landscape in regions A, B, C, and D of the phase diagram. Only in regions A and B are the desired retrieval states at (or near) energy minima, and only in region A are they the global minima.

THREE

Extensions of the Hopfield Model

The previous chapter dealt with the basic Hopfield model, including the finite-temperature stochastic version. Here we examine various extensions and generalizations of that model. In the first two sections we consider straightforward variations, including the modifications needed to cope with correlated patterns. Then, in Section 3.3, we treat networks of *continuous-valued units* with dynamics described by differential equations. A couple of hardware implementations of Hopfield-like networks are sketched in Section 3.4, one electrical and one optical. Finally in Section 3.5 we discuss models that recall *sequences* of states, so that the usual point attractors are replaced by limit cycles or by more complicated dynamical trajectories.

3.1 Variations on the Hopfield Model

In this section we discuss how the associative memory properties of the basic Hopfield model carry over to various more complicated situations. In many cases the results are qualitatively similar to the simple case, but changes occur in quantitative values such as the capacity $\alpha_c = p_{\max}/N$.

Connection Strength Inaccuracy and Clipping

We first consider the perturbation of the connection strengths away from those given by the Hebb prescription (2.9). This may be of practical importance when trying to build a network (in silicon, say) with connections of limited precision or range.

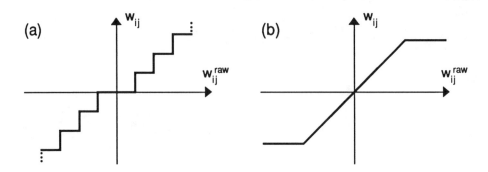

FIGURE 3.1 Examples of (a) discretization and (b) clipping. The raw connection weight w_{ij}^{raw} calculated from the Hebb rule is transformed to the actual w_{ij} by the function shown.

The addition of a small random number to each w_{ij} has only a qualitative effect, reducing α_c [Sompolinsky, 1987]. Similarly with **discretizing** or **clipping** the allowed values. Discretizing—allowing only a discrete set of values—may be useful when building circuits using a fixed number of standard resistors. Clipping means restricting all connections w_{ij} to some fixed range, say $|w_{ij}| \leq A$, and is clearly also useful (if not essential) in practical implementations. Figure 3.1 shows examples of both processes.

In the most extreme case of discretization and clipping we allow only two values for w_{ij}, sometimes referred to as **binarizing** the connections:

$$w_{ij} = \text{sgn}(w_{ij}^{\text{raw}}) = \text{sgn}\left(\sum_{\mu=1}^{p} \xi_i^{\mu} \xi_j^{\mu}\right). \tag{3.1}$$

This model can be solved exactly [van Hemmen and Kühn, 1986]; the result is that α_c is reduced from 0.138 to about 0.1. This represents a rather efficient use of the single bit of information retained per connection, compared to the $\log_2 p$ bits necessary to specify one of the p possible values of each w_{ij} with the full Hebb rule.

Clipping may also be viewed in the context of successively learning new patterns. We can imagine using the Hebb rule *incrementally* to continue adding new terms to each w_{ij}, so that

$$w_{ij}^{\text{new}} = w_{ij}^{\text{old}} + \eta \xi_i^{\mu} \xi_j^{\mu} \tag{3.2}$$

to add pattern μ. Here η is an **acquisition rate**. Applying clipping to this means restricting w_{ij}^{new} to a range $[-A, A]$ at all times; values outside these limits are immediately replaced by the appropriate limit value. This is called **learning within bounds** [Parisi, 1986; Nadal et al., 1986]. The most recently added memory patterns are then always recalled well, while older ones gradually decay away. The number of patterns that can be remembered depends on the value of η compared to A; if η is very large only the most recent pattern can be recalled, while for very small η the system can become overloaded ($\alpha > \alpha_c$) before the bounds are reached, and then not work at all. In between there is an optimum value for η. Then there is no sharp threshold α_c; adding more memories simply erases earlier ones, so the memory is termed a **palimpsest**.[1] This observation has been related to the limitations of short-term memory [Nadal et al., 1986] and to interpretations of dream sleep [Geszti and Pázmándi, 1987].

Synchronous Dynamics

As mentioned early in the discussion, we could choose to update all the units simultaneously instead of one at a time [Little, 1974; Amit et al., 1985a]. How does this affect the results we have found? Again, there is no significant change. In particular, the memory states (those with $\langle S_i \rangle \propto \xi_i^\mu$ for some μ) for $p \ll N$ are exactly the same as in the asynchronously updated network, because the mean field equations for time-independent states are the same in the two dynamics.

Weak Dilution

Another complication which can be easily dealt with is dilution of the connections. We start with **weak dilution**, where a finite fraction of the connections is removed randomly. Thus we set

$$w_{ij} = \begin{cases} \text{Hebb value} & \text{with probability } c\,; \\ 0 & \text{with probability } 1-c\,. \end{cases} \tag{3.3}$$

We can either enforce the $w_{ij} = w_{ji}$ symmetry or treat w_{ij} and w_{ji} independently. c is equal to the relative concentration of bonds remaining after dilution, and is taken to be independent of N.

Again, the effect is only quantitative in general, as is simple to see in the limit of small p/N. Let $C_{ij} = 1$ if there is a non-zero connection w_{ij} from unit j to unit i, and $C_{ij} = 0$ otherwise. Then

$$w_{ij} = C_{ij} w_{ij}^{\text{Hebb}} \tag{3.4}$$

where w_{ij}^{Hebb} is the Hebb connection strength (2.9), and the net input to unit i is

$$h_i = \sum_j C_{ij} w_{ij}^{\text{Hebb}} S_j\,. \tag{3.5}$$

As long as it is only a fixed fraction of the connections that are removed, there is still an infinite number of terms in the sum as $N \to \infty$. So we can apply mean field

---

[1]A writing material (as a parchment or tablet) used one or more times after earlier writing has been erased [Webster, 1988].

theory as before, replacing h_i by

$$\langle h_i \rangle = c \sum_j w_{ij}^{\text{Hebb}} \langle S_j \rangle . \tag{3.6}$$

Thus the previous mean field results apply (for p/N small), with a simple scaling of the temperature by a factor of c.

At larger p/N the situation is more complicated, but the qualitative feature of a capacity p_{max} of order N persists for the case of symmetric dilution $C_{ij} = C_{ji}$ [Sompolinsky, 1987; Canning and Gardner, 1988].

Strong Dilution

There is another limit of the dilution problem which can be solved exactly and rather simply. This is the case of **strong dilution**, where only an infinitesimal fraction of the original number of connections remain. Defining K as the average number of connections to and from each unit, the precise condition is that K not exceed something proportional to $\log N$ as N goes to infinity. The exact solution also requires another twist: the dilution must be performed independently on w_{ij} and w_{ji}, so that the factors C_{ij} and C_{ji} in (3.4) are independent random variables. The w_{ij} matrix is then no longer symmetric.

This model, first studied by Derrida, Gardner, and Zippelius [1987], can then be solved. We use

$$w_{ij} = \frac{1}{K} C_{ij} \sum_\mu \xi_i^\mu \xi_j^\mu \tag{3.7}$$

for the connection strengths, with $1/K$ rather than $1/N$ for the normalization so as to give sensible values of order unity. For any state $\{S_j\}$ of the network, we now break up the field h_i in (3.5) into a term coming from a particular pattern ν and a remaining crosstalk term:

$$
\begin{aligned}
h_i &= \sum_j w_{ij} S_j = \frac{1}{K} \sum_j C_{ij} \sum_\mu \xi_i^\mu \xi_j^\mu S_j \\
&= \frac{1}{K} \xi_i^\nu \sum_j C_{ij} \xi_j^\nu S_j + \eta_i^\nu
\end{aligned}
\tag{3.8}
$$

where

$$\eta_i^\nu = \frac{1}{K} \sum_{\mu \neq \nu} \xi_i^\mu \sum_j C_{ij} \xi_j^\mu S_j . \tag{3.9}$$

Note that the crosstalk term η_i^ν depends on the state $\{S_j\}$.

If we set $S_i = \xi_i^\nu$ in (3.8) and (3.9) we can see that ξ_i^ν is stable for small enough p. The first term on the right-hand side of (3.8) gives just ξ_i^ν on average, since $\langle \sum_j C_{ij} \rangle = K$. Meanwhile the second term η_i^ν, given by (3.9), becomes $1/K$

times a sum of about Kp independent random ± 1's. In the limit where $Kp \gg 1$ it therefore has a Gaussian distribution

$$P(\eta) = \frac{1}{(2\pi p/K)^{\frac{1}{2}}} \exp\left(-\frac{K\eta^2}{2p}\right) \tag{3.10}$$

just as for C_i^ν in Fig. 2.5. Apart from the replacement of N by K we are thus back in the same situation as the undiluted case; the memory works well if the Gaussian is narrow enough.

The capacity $p_{\max}$ is clearly of order K rather than order N. Just as we found the parameter $\alpha = p/N$ useful in the fully connected network, here the important parameter is $\alpha' = p/K$. The maximum capacity $p_{\max}$ is always proportional to the average number of other units that a given unit is connected to, corresponding to a particular value of α', not α.

In the very dilute case we can easily go further and calculate $p_{\max}$ (or α'_c) exactly. Instead of testing the stability of one of the original patterns ξ_i^ν, we focus on an actual stable configuration $S_i = \text{sgn}(h_i)$ of the network and ask how close it is to ξ_i^ν. We therefore examine the overlap

$$m_\nu = \frac{1}{N} \sum_i \xi_i^\nu S_i = \frac{1}{K} \sum_j C_{ij} \xi_j^\nu S_j . \tag{3.11}$$

These two definitions of m_ν are equivalent for large K (with any i in the last term). Using $S_i = \text{sgn}(h_i)$ and (3.8) we can now write a self-consistent equation for m_ν:

$$m_\nu = \frac{1}{N} \sum_i \text{sgn}(\xi_i^\nu h_i) = \frac{1}{N} \sum_i \text{sgn}(m_\nu + \xi_i^\nu \eta_i^\nu)$$

$$= \int d\eta \, P(\eta) \, \text{sgn}(m_\nu + \eta) . \tag{3.12}$$

To replace the sum by the integral we had to assume the statistical independence of ξ_i^ν and η_i^ν, so that the symmetry $P(\eta) = P(-\eta)$ could be used. Moreover we are assuming that η_i^ν has the distribution $P(\eta)$ given by (3.10), even though that was strictly valid only for $S_i = \xi_i^\nu$. It can be shown that these assumptions *are* valid if (a) dilution is independent for ij and ji, and (b) we are in the strongly diluted regime $K \ll N$.

Using (3.10) we can reduce (3.12) to

$$m_\nu = \text{erf}\left(\frac{m_\nu}{\sqrt{2\alpha'}}\right) \tag{3.13}$$

where $\text{erf}(x)$ is defined by (2.16). Equation (3.13) is essentially the same as (2.47), except that the $\text{erf}(x)$ replaces the $\tanh(x)$, and the control parameter is α' rather than the temperature. As shown in Fig. 3.2, the $\text{erf}(x)$ function is very similar to a $\tanh(x)$ function, but with a slope of $2/\sqrt{\pi}$ instead of 1 at the origin. Thus (3.13) can be solved graphically as in Fig. 2.12.

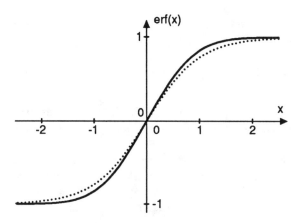

FIGURE 3.2 The function erf(x). The dotted line shows tanh(x) for comparison.

There is a critical value $\alpha'_c = 2/\pi$ of α' beyond which the only solution is $m_\nu = 0$, but below which there are solutions with $m_\nu \neq 0$. Thus the crosstalk acts in a way similar to thermal noise. Note however that m_ν goes continuously to zero as α' approaches α'_c, in contrast to what happened in the fully connected case as α approached α_c, where the jump down to zero was discontinuous. The origin of this difference can be understood by comparing (3.13) with the corresponding equation (2.73) for the fully connected case. The latter had an extra factor of $1/\sqrt{r}$ in the argument of the erf, and r (and q) had to be determined self-consistently with m.

We can generalize the treatment to finite temperature simply by replacing the sgn(x) in (3.12) by a tanh(x):

$$m_\nu = \int d\eta \, P(\eta) \tanh[\beta(m_\nu + \eta)] \,. \tag{3.14}$$

As in the full connectivity case, the effect of finite temperature is to reduce the capacity from its zero-temperature value.

The model can be solved for both synchronous and asynchronous updating (with the same result for the capacity), but is apparently not so easy to solve if the connection matrix is constrained to be symmetric.

Random Asymmetric Connections

The densely connected model may also be studied when the connections are allowed to be asymmetric, $w_{ij} \neq w_{ji}$. If the asymmetry is systematic, or very strong, it can produce limit cycles or chaotic behavior, as we will study later. But if it is random and not too strong it mainly plays the role of noise. Random asymmetry can be introduced by random dilution or by adding a random number to each connection, independently for ij and ji in both cases.

For $p \ll N$ there seems to be no difference from the symmetric case. As in the case of weak dilution, the argument relies on the fact that there are still of order N

terms in the weighted input sum h_i, which can therefore be replaced by its average value without approximation in the large N limit.

At finite α, the asymmetric fully connected model differs from the symmetric one in that it does not have any of the spurious spin glass states at any $T > 0$ [Hertz et al., 1986, 1987; Crisanti and Sompolinsky, 1987]. But retrieval of the stored patterns is not qualitatively different from that in the symmetric model. Thus random asymmetry may improve performance by removing the spin glass states. On the other hand, the asymmetry introduces some fluctuations and slows down the approach to an attractor [Parisi, 1986].

Unipolar Connections

For some applications it is inconvenient to require both positive and negative connection weights w_{ij}. This is particularly true when implementing a network in electronic or optical hardware, as discussed in Section 3.4. It is however easy to modify the design so that all the weights are positive; i.e., we can replace *bipolar* connections by *unipolar* ones [Denker, 1986].

The trick is simply to add a constant κ to every connection w_{ij},

$$w'_{ij} = w_{ij} + \kappa \tag{3.15}$$

and compensate for this with an extra term $-\kappa \sum_j S_j$ in the input h_i at every unit. Then the total input h'_i is given by

$$h'_i = \sum_j (w_{ij} + \kappa) S_j - \kappa \sum_j S_j = \sum_j w_{ij} S_j = h_i \tag{3.16}$$

exactly as before. Thus there is *no* overall effect on the network's behavior.

We choose κ large enough to make w'_{ij} positive (or perhaps zero) for all ij. For the usual Hebb rule (2.9) the value $\kappa = 1$ suffices. The compensating term $-\kappa \sum_j S_j$ is the same for all units, and may be calculated by one extra unit. It is sometimes referred to as an **adaptive threshold** because in effect it changes the threshold of every unit by an amount depending on the total activity $\sum_j S_j$.

3.2 Correlated Patterns

The Pseudo-Inverse

The crosstalk term in h_i which sets the fundamental limit on the network capacity comes from the overlap between random patterns. This overlap is much more of a problem when the different patterns are correlated. Then a standard network may not even recall patterns reliably in the limit $p \ll N$. There is however a general solution to this problem called the **pseudo-inverse** approach. It works for any set of $p < N$ linearly independent patterns [Kanter and Sompolinsky, 1987; Personnaz et al., 1985, 1986; Kohonen, 1974, 1989].

It is simplest just to write the prescription down and then show that it works. If we define the $p \times p$ overlap matrix $\mathbf{Q}$ by

$$\mathbf{Q}_{\mu\nu} = \frac{1}{N} \sum_i \xi_i^\mu \xi_i^\nu \tag{3.17}$$

then (if $\mathbf{Q}$ is non-singular) the appropriate connection matrix is

$$w_{ij} = \frac{1}{N} \sum_{\mu\nu} \xi_i^\mu (\mathbf{Q}^{-1})_{\mu\nu} \xi_j^\nu . \tag{3.18}$$

To verify that this will recall any pattern ξ_i^λ correctly we just calculate[2] h_i when $S_i = \xi_i^\lambda$:

$$\begin{aligned}
h_i^\lambda &\equiv \sum_j w_{ij} \xi_j^\lambda = \frac{1}{N} \sum_{j\mu\nu} \xi_i^\mu (\mathbf{Q}^{-1})_{\mu\nu} \xi_j^\nu \xi_j^\lambda \\
&= \sum_{\mu\nu} \xi_i^\mu (\mathbf{Q}^{-1})_{\mu\nu} \mathbf{Q}_{\nu\lambda} \\
&= \sum_\mu \xi_i^\mu \delta_{\mu\lambda} \\
&= \xi_i^\lambda .
\end{aligned} \tag{3.19}$$

Thus (just as for the Hebb prescription applied to orthogonal patterns), the application of this modified prescription for correlated patterns gives a net input to each unit equal to its output. Therefore all the patterns are stable.

Another way to picture how this prescription works is obtained by defining a set of patterns

$$\eta_i^\mu = \sum_\nu (\mathbf{Q}^{-1})_{\mu\nu} \xi_i^\nu \tag{3.20}$$

which have the property that

$$\frac{1}{N} \sum_i \eta_i^\mu \xi_i^\nu = \delta_{\mu\nu} . \tag{3.21}$$

This is where the name *pseudo-inverse* arises; if we regard ξ_i^ν and η_i^μ as $N \times p$ matrices, then (3.21) says that $\boldsymbol{\eta}^T \boldsymbol{\xi} = N\mathbf{1}$, and $N^{-1}\boldsymbol{\eta}^T$ is called the pseudo-inverse of $\boldsymbol{\xi}$. More generally for any $m \times n$ matrix $\mathbf{M}$ with $m \geq n$ the pseudo-inverse (or

---

[2] The Kronecker delta symbol δ_{rs} is defined to be 1 if $r = s$ and 0 otherwise, so $\delta_{\mu\lambda}$ means the identity matrix.

generalized inverse, or Moore-Penrose inverse) $M^{\ddagger}$ can be defined as $(M^T M)^{-1} M^T$ whenever $M^T M$ is non-singular. It obeys $M^{\ddagger} M = 1$ but not normally $M M^{\ddagger} = 1$. See Rao and Mitra [1971] for details.

In terms of η_i^{μ} the connection matrix can be written

$$w_{ij} = \frac{1}{N} \sum_{\mu} \xi_i^{\mu} \eta_j^{\mu} . \tag{3.22}$$

Thinking now of the η_i^{μ} and the ξ_i^{μ} as components of vectors $\boldsymbol{\eta}^{\mu}$ and $\boldsymbol{\xi}^{\mu}$, these equations tell us that each $\boldsymbol{\eta}^{\mu}$ is orthogonal to all $\boldsymbol{\xi}^{\nu}$'s (for $\nu \neq \mu$), and that the matrix $\mathbf{w}$ is a sum of outer products of $\boldsymbol{\xi}^{\mu}$ and $\boldsymbol{\eta}^{\mu}$ vectors:

$$\mathbf{w} = \frac{1}{N} \sum_{\mu} \boldsymbol{\xi}^{\mu} (\boldsymbol{\eta}^{\mu})^T . \tag{3.23}$$

It therefore naturally takes every $\boldsymbol{\xi}^{\mu}$ into itself:

$$\mathbf{w} \boldsymbol{\xi}^{\mu} = \boldsymbol{\xi}^{\mu} \qquad \text{for all } \mu . \tag{3.24}$$

We can also regard (3.24) as a set of equations to be solved for the $N \times N$ weight matrix $\mathbf{w}$, given the p pattern vectors $\boldsymbol{\xi}^{\mu}$. If $p < N$ there are many different solutions to these equations, including the obvious $N \times N$ identity matrix. The solution chosen by the pseudo-inverse prescription (3.23) is the unique one where, for any vector $\mathbf{y}$ orthogonal to all the pattern vectors, we have

$$\mathbf{w} \mathbf{y} = 0 . \tag{3.25}$$

From (3.24) and (3.25) it follows that $\mathbf{w}$ projects *onto* the p-dimensional subspace spanned by the pattern vectors. This is why this method is also called the **projection method**. Note that *any* vector in this pattern subspace is stable in the sense that (3.24) is satisfied, though of course not all such vectors have purely ± 1 components.

It is easy to see that the overlap matrix $\mathbf{Q}$ cannot be inverted if there are any linear dependencies among the patterns. But then a suitable $\mathbf{w}$ matrix can be found by restricting attention in (3.17) and (3.18) to a linearly independent subset of pattern vectors that spans the pattern subspace. If, however, the patterns span the whole N-dimensional space the recipe makes *all* patterns stable, and the memory is useless; $\mathbf{w}$ becomes an identity matrix. The same problem was encountered in (2.23) for the Hebb rule with N orthogonal patterns. So up to $N - 1$ linearly independent patterns can be usefully stored by this method. That is the capacity $p_{\max}$ in most practical situations.[3]

---

[3] In principle more linearly *dependent* patterns can be usefully stored if the size of any linearly independent subset is $N - 1$ or less. But for *random* ± 1 patterns the probability of linear dependency is small (becoming 0 as $N \to \infty$) for $p \leq N$, and 1 for $p > N$.

For all this linear algebra to hold, one has to retain the self-coupling terms—the diagonal terms w_{ii} of **w**. But it can be shown that they can usually be set equal to zero without problems [Kanter and Sompolinsky, 1987]. Indeed, making them zero usually *improves* the performance, as for the Hebb rule in (2.28).

The original Hebb rule had a biological appeal, in that such connections could develop in response to correlations of firings between only the pre- and post-synaptic cells. With (3.18) no such interpretation is possible; the pseudo-inverse prescription is **nonlocal**, because computation of w_{ij} requires knowledge of ξ_k^μ for *all* k. There is however an iterative scheme which converges to the same set of w_{ij}'s using only local information (including h_i^λ); see Diederich and Opper [1987] for details. We will encounter similar schemes for layered networks in Chapter 5.

Special Cases

There are many special cases of correlated patterns which can be treated with the general prescription (3.18) or by direct analysis. An important one is patterns which have a certain "average pattern" in common [Amit et al., 1987b]. The simplest case is of **biased patterns**, in which the probabilities for $\xi_i^\mu = +1$ and $\xi_i^\mu = -1$ are unequal (like a biased coin), but still independent for all i and μ. A generalization of this is a hierarchical correlation of patterns, where they can be grouped into families, families within families, and so on, on the basis of their mutual overlaps [Cortes et al., 1987; Krogh and Hertz, 1988; Parga and Virasoro, 1986; Gutfreund, 1988]. In both of these cases the result of the prescription (3.18) is a fairly simple generalization of the Hebb rule. One simply has to imprint both the average features (e.g., the mean value for biased patterns) and the deviations from them in a Hebb-like fashion, with suitable relative weights which can be calculated from (3.18).

Another interesting case is that of **sparse patterns**, in which almost all the ξ_i^μ are the same. This would be the case, for example, in patterns which are mostly "off", such as a small fraction of black pixels on a predominantly white screen. In this limit it is convenient to go over to units n_i and patterns ξ_i^μ which both take values 0 or 1 instead of ± 1. For random biased patterns with a fraction a of ones, the appropriate distribution is

$$\xi_i^\mu = \begin{cases} 1 & \text{with probability } a\,; \\ 0 & \text{with probability } 1 - a\,; \end{cases} \tag{3.26}$$

and we consider the case $a \ll 1$. In this limit a very simple Hebb-like rule

$$w_{ij} = \frac{1}{N} \sum_\mu (\xi_i^\mu - a)(\xi_j^\mu - a)\,, \tag{3.27}$$

together with an optimized choice of threshold (the same for all units), will store a very large number of patterns. The capacity is approximately [Tsodyks and Feigel'man, 1988]

$$\alpha_c \approx \frac{1}{2a|\log a|} \tag{3.28}$$

which is in fact of the same order as the maximum possible capacity that could be achieved by any choice of w_{ij}'s, as we will see in Chapter 10. Thus the simple Hebb-like guess (3.27) is a very good choice for the connection weights in this limit.

An even simpler rule for the sparse 0/1 case is the **Willshaw model** [Willshaw et al., 1969] in which one takes

$$w_{ij} = \begin{cases} 1 & \text{if } \xi_i^\mu = \xi_j^\mu = 1 \text{ for any } \mu ; \\ 0 & \text{otherwise;} \end{cases} \qquad (3.29)$$

and again uses an optimized threshold. This rule is extremely simple, and easy to implement in hardware [Thakoor et al., 1987], though its capacity (which does not scale with N) is not nearly as good as (3.28). The addition of a uniform inhibition between all pairs of units gives rise to an interesting model which has only **partial ordering** [Golomb et al., 1990]; the attractors are dynamic, with some units frozen in the 0 state and some fluctuating between 0 and 1. Only the time average $\langle n_i \rangle$ gives the correct stored pattern. The average activity $\langle n_i \rangle$ can be quite low even on the fluctuating units, which brings the model closer to biological realism.

3.3 Continuous-Valued Units

In this section we discuss a different way to generalize the original McCulloch-Pitts model, by making the output of a unit a continuous variable instead of a binary 0/1 or ± 1. This is more realistic for real neurons, is sometimes more convenient for analog hardware implementation, and in some contexts makes analysis easier.

We consider only the case where the output V_i of unit i is (in equilibrium, as discussed below) a continuous function of its net input u_i:

$$V_i = g(u_i) = g\left(\sum_j w_{ij} V_j\right). \qquad (3.30)$$

V_i and u_i correspond directly to the S_i and h_i used in the case of binary units, but the new notation is convenient; in particular we never use S_i for anything but a ± 1 unit, in agreement with the convention in statistical mechanics.

The activation function $g(u)$ is usually nonlinear. In most cases we want it to have a **saturation nonlinearity** so that $g(u)$ levels off and approaches fixed limits for large negative and positive u. Then V_i will always remain between those limits. $g(u)$ is often called a **squashing function** in this context. Commonly used functions include $\tanh(\beta u)$ for a $[-1, +1]$ range (Fig. 2.9, Eq. (2.43)) and the sigmoid function $f_\beta(u)$ for a $[0, 1]$ range (Fig. 2.8, Eq. (2.37)).

There are several possible choices for updating the units:

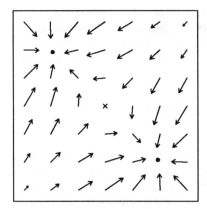

FIGURE 3.3 Motion towards attractors in a two-unit network. There are two attractors shown by dots. The center point is a saddle point of the energy, not an attractor. The system moves in the direction of the arrows to one of the attractors; *which* one depends on the starting point.

- **Asynchronous updating.** One unit at a time is selected to have its output set according to (3.30).

- **Synchronous updating.** At each time step all units have their outputs set according to (3.30).

- **Continuous updating.** All units continuously and simultaneously change their outputs towards the values given by (3.30). The u_i's change continuously too, according to $u_i = \sum_j w_{ij} V_j$.

The third possibility is the new one [Cohen and Grossberg, 1983; Hopfield, 1984], and is of particular interest for the circuit implementations discussed in Section 3.4. It can be represented by the set of differential equations

$$\tau_i \frac{dV_i}{dt} = -V_i + g(u_i) = -V_i + g\left(\sum_j w_{ij} V_j\right) \tag{3.31}$$

where τ_i are suitable time constants.

If $g(u)$ has a saturation nonlinearity and the w_{ij}'s are symmetric, the solution $V_i(t)$ to these equations always settles down to a stable equilibrium solution, as we will prove in the next subsection. At an equilibrium $V_i(t)$ ceases to change, so $dV_i/dt = 0$ for all i. Then the right-hand side of (3.31) shows that (3.30) is obeyed on all units. Thus the desired state satisfying (3.30) is an **attractor** of the dynamical rule (3.31).

Figure 3.3 shows a simple example for a system with two units. A state of the system corresponds to a point in the V_1–V_2 plane illustrated. At any such point, the equations (3.31) (one for $i = 1$, one for $i = 2$) give a velocity vector $d\mathbf{V}/dt$, shown in the figure by an arrow. The state of the system moves from its initial point in the direction of the arrows, faster for larger arrows. Thus it ends up at one of the two attractors shown, where $d\mathbf{V}/dt = 0$.

A very similar dynamical rule with the same end result arises from letting the inputs u_i continuously approach their correct values $\sum_j w_{ij} V_j$, with $V_i = g(u_i)$

always true:

$$\tau_i \frac{du_i}{dt} = -u_i + \sum_j w_{ij} V_j = -u_i + \sum_j w_{ij} g(u_j). \qquad (3.32)$$

This equation has the same equilibrium solution (3.30) as (3.31). Indeed the whole attractor structure of the two equations is identical if the matrix **w** is invertible [Pineda, 1988].

If the β parameter in $g(u) = \tanh(\beta u)$ is made large, then most of the outputs V_i will come close to the limiting values ± 1. This applies with either dynamical equation, and a similar result holds for the 0/1 case. Thus we can obtain binary yes/no "answers" from a network that uses continuous-valued units with continuous differential updating. Even without β large we could take any positive V_i as a *yes* and any negative V_i as a *no*. In other cases we may actually be interested in continuous-valued answers; examples will appear in Chapter 4.

Alternatively we can note that the equations (3.30) are just like (2.45) for the *average* values $\langle S_i \rangle$ in the stochastic model. Indeed, if we use $g(u) = \tanh(\beta u)$, then the equations are exactly the same; u_i here plays the role of the average internal field $\langle h_i \rangle$ and V_i is equivalent to the average unit value $\langle S_i \rangle$. Thus we can use a continuous-valued network to solve the mean field problem for a stochastic network at finite temperature.

Continuous-valued units obeying (3.31) or (3.32) are normally implemented as electrical circuits, as discussed in Section 3.4. The equations may alternatively be integrated numerically by computer, in which case an adaptive step-size technique is well worth trying. The Bulirsch-Stoer method [Press et al., 1986] should be suitable.

We have so far assumed that the connections w_{ij} are symmetric; this guarantees the existence of attractors. If on the other hand the connections are *not* symmetric, then an oscillatory or chaotic $V_i(t)$ is possible. In the extreme case of purely random connections, with mean 0 and variance σ^2, there is a transition from stable to chaotic behavior as σ^2 is increased [Sompolinsky et al., 1988].

The Energy Function

Just as for the binary units discussed in Chapter 2 we can define an energy function that always decreases during the dynamical evolution (3.31) or (3.32). The equilibrium solutions are then energy minima, and we can view the dynamics as sliding downhill on the energy surface and coming to rest at a minimum.

We consider only version (3.32) of the dynamics. The appropriate energy function is then [Hopfield, 1984]:

$$H = -\frac{1}{2} \sum_{ij} w_{ij} V_i V_j + \sum_i \int_0^{V_i} g^{-1}(V)\, dV \qquad (3.33)$$

where V_i is always equal to $g(u_i)$. To show that H decreases, we differentiate (3.33) with respect to time, which enters implicitly through V_i:

$$
\begin{aligned}
\frac{dH}{dt} &= -\tfrac{1}{2}\sum_{ij} w_{ij}\frac{dV_i}{dt}V_j - \tfrac{1}{2}\sum_{ij} w_{ij}V_i\frac{dV_j}{dt} + \sum_i g^{-1}(V_i)\frac{dV_i}{dt} \\
&= -\sum_i \frac{dV_i}{dt}\left(\sum_j w_{ij}V_j - u_i\right) \\
&= -\sum_i \tau_i \frac{dV_i}{dt}\frac{du_i}{dt} \\
&= -\sum_i \tau_i g'(u_i)\left(\frac{du_i}{dt}\right)^2 \leq 0 .
\end{aligned}
\tag{3.34}
$$

Here we used the $w_{ij} = w_{ji}$ symmetry in obtaining the second line, and used (3.32) for the third line. The result is negative or zero because τ_i is positive, $g(u)$ is monotonic, and du_i/dt appears squared. It is only zero if we are at an equilibrium point, where $du_i/dt = 0$ for all i.

Equation (3.34) shows that the dynamical equation (3.32) continuously decreases the energy function until it reaches its lower bound of 0 at an equilibrium point. This proves that the equilibrium points are attractors of the system, and that they are the *only* attractors. Limit cycles, for example, are *not* possible; the energy cannot decrease continuously around a closed curve.

We can imagine an energy landscape given by (3.33) "above" a velocity field such as that shown in Fig. 3.3. The attractors are at energy minima, and the motion is always downhill. The velocity $d\mathbf{u}/dt$ is not strictly in the negative gradient direction $-\nabla_{\mathbf{u}}H$ (in fact $\tau_i du_i/dt = -\partial H/\partial V_i$) but always has a positive projection onto it.

Terminal Attractors $\star$

The dynamical system described by (3.31) or (3.32) approaches its attractors in an exponential way, as is readily seen by linearizing about an attractor. Taking (3.32) for example, if we assume that u_i^* is an attractor (i.e., satisfies (3.30)) and expand about it,

$$
u_i = u_i^* + \varepsilon_i ,
\tag{3.35}
$$

we obtain

$$
\tau_i\frac{d\varepsilon_i}{dt} = -\varepsilon_i - u_i^* + \sum_j w_{ij}g(u_j^* + \varepsilon_j) = -\varepsilon_i + \sum_j w_{ij}g'(u_j^*)\varepsilon_j
\tag{3.36}
$$

to order ε_i, since $u_i^* = \sum_j w_{ij}g(u_j^*)$ by hypothesis. Equation (3.36) is linear, and has the general solution

$$
\varepsilon_i(t) = \sum_k a_k \varepsilon_i^{(k)} e^{-\lambda^{(k)}t}
\tag{3.37}
$$

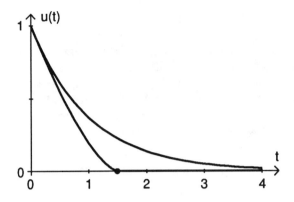

FIGURE 3.4 Approach to attractors. The upper curve is for an ordinary attractor, given by (3.39). The lower curve is for a terminal attractor, given by (3.41). Parameter values are $u(0) = 1$, $u^* = 0$, $\tau = 1$.

where $\lambda^{(k)}$ and $\varepsilon_i^{(k)}$ are the eigenvalues and eigenvectors of the matrix $\mathbf{T}^{-1}(\mathbf{1} - \mathbf{w}\mathbf{G})$ and a_k are coefficients chosen to satisfy the initial conditions. Here $\mathbf{T}$ and $\mathbf{G}$ are diagonal matrices constructed respectively from τ_i and $g'(u_j^*)$. All the eigenvalues $\lambda^{(k)}$ must be positive if u_i^* is an attractor (they are necessarily real if $\mathbf{w}$ is symmetric). So, if we start u_i close enough to the attractor u_i^* for the linearization to be valid, the displacement ε_i from the attractor decays away exponentially.

This exponential approach to the attractor in principle takes forever. Although this may not be of importance in practice, it is interesting that the dynamical equations may be modified so that the attractor is reached after a finite time [Zak, 1988, 1989]. It is easiest to examine this first for a simple one-dimensional problem.

An ordinary attractor at u^* would be governed by the equation

$$\tau \frac{du}{dt} = -(u - u^*) \tag{3.38}$$

with the solution

$$u(t) = u^* + [u(0) - u^*]e^{-t/\tau} \tag{3.39}$$

that shows the usual exponential approach to the attractor. But now suppose that we replace the right-hand side of (3.38) by a singular term:

$$\tau \frac{du}{dt} = -(u - u^*)^{1/3} . \tag{3.40}$$

This has the solution

$$u(t) = \begin{cases} u^* + [(u(0) - u^*)^{2/3} - 2t/3\tau]^{3/2} & \text{if } t < t^* ; \\ u^* & \text{if } t \geq t^* \end{cases} \tag{3.41}$$

where

$$t^* = \tfrac{3}{2}\tau[u(0) - u^*]^{2/3} . \tag{3.42}$$

Thus the attractor u^* is reached at time t^*. Such an attractor is called a **terminal attractor**. $u = u^*$ is a singular solution of the dynamical system, and is infinitely stable (the eigenvalues found upon linearization are infinite). Figure 3.4 compares the approach to an ordinary attractor and to a terminal attractor.

The same effect is achieved if we *add* the singular term $-(u - u^*)^{1/3}$ to the right-hand side of the original equation (3.38) instead of simply replacing the linear term, because near the attractor the singular term dominates the behavior. The choice of $1/3$ for the exponent is also not essential; $1/k$ with k odd would also work.

To apply the same idea to the coupled equations (3.32) we add a singular term proportional to $[u_i - u_i^k]^{1/3}$ for each of the attractors u_i^k. So that these additions do not interfere with each other we also need to suppress them away from their attractors with an additional exponential factor. Thus overall we can use

$$\tau_i \frac{du_i}{dt} = -u_i + \sum_j w_{ij} g(u_j) - \alpha \sum_k [u_i - u_i^k]^{1/3} e^{-\gamma[u_i - u_i^k]^2} \qquad (3.43)$$

where α and γ are suitable constants. The value of γ (which can be allowed to depend on i and k) must be large enough to make $\gamma(u_i^k - u_i^l)^2 \gg 1$ for every pair of distinct attractors u_i^k and u_i^l.

A neural network circuit implemented using (3.43) would certainly approach its attractors more rapidly than with (3.32). Barhen et al. [1989] found a remarkable speedup in learning time in an application to inverse kinematics for robot manipulators. More control over spurious attractors and basin sizes also seems to be possible. In applications to networks with hidden units (discussed in Chapter 7) it can be shown that terminal attractors never become repellers, as can happen in some cases with normal attractors. Note however that the locations of the attractors must be known in advance, which, while still appropriate for content-addressable memory, limits the applicability to many other problems in neural computation.

3.4 Hardware Implementations

In this section we briefly discuss two hardware implementations of Hopfield models. One is an analog electrical circuit using continuous-valued units. The other is an optical implementation using binary units and clipped connections. This is the only place in the book where hardware is discussed, and even here we cover only some basic principles.

Electrical Circuit Implementation

Figure 3.5(a) shows an electrical circuit constructed to implement a unit obeying (3.32) [Hopfield, 1984; Hopfield and Tank, 1986]. A network of four such circuits is shown in Fig. 3.5(b), making it clear that the resistors R_{ij} play the role of the connections w_{ij}. Such circuits have actually been made in analog VLSI.

Each unit i is composed of the circuit shown in Fig. 3.5(a). u_i is the input voltage, V_i is the output voltage, and the operational amplifier has the transfer function $V_i = g(u_i)$. The input of each unit is connected to ground with a resistor

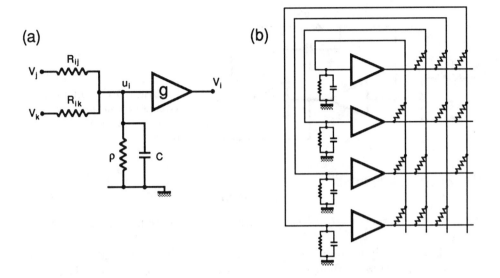

FIGURE 3.5 (a) The circuit described in the text. (b) A network of such circuits.

ρ in parallel with a capacitor C, and the output of unit j is connected to the input of unit i with a resistor R_{ij}. In terms of modelling a real neuron, we can regard ρ and C as the transmembrane resistance and input capacitance respectively.

The circuit equations are

$$C\frac{du_i}{dt} + \frac{u_i}{\rho} = \sum_j \frac{1}{R_{ij}}(V_j - u_i) \tag{3.44}$$

or, equivalently,

$$\tau_i\frac{du_i}{dt} = -u_i + \sum_j w_{ij}g(u_j) \tag{3.45}$$

where

$$\tau_i = R_iC \tag{3.46}$$

with

$$\frac{1}{R_i} = \frac{1}{\rho} + \sum_j \frac{1}{R_{ij}} \tag{3.47}$$

and

$$w_{ij} = R_i/R_{ij}. \tag{3.48}$$

Equation (3.45) is identical to (3.32). From any starting state the circuit therefore settles down to a steady state in which (3.30) is obeyed.

The circuit implementation of an abstract network involves choosing resistances R_{ij} to satisfy (3.48). If we choose ρ small enough we can make $R_i \approx \rho$ for all i and

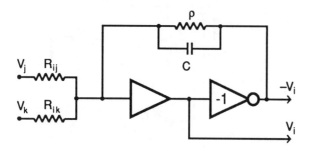

FIGURE 3.6 Circuit for a virtual ground unit, with normal and inverted outputs.

thus simplify (3.48) to

$$w_{ij} \approx \rho/R_{ij}. \tag{3.49}$$

Then the conductance $1/R_{ij}$ of each connection is chosen to be proportional to the desired w_{ij}.

This works only for positive w_{ij}'s. To deal with negative connection strengths we can add inverters to the circuit so that $-V_i$ is available as well as V_i at each unit. If a desired w_{ij} is negative, the term $w_{ij}V_j$ is simply replaced by $|w_{ij}|(-V_j)$; a positive connection $|w_{ij}|$ is made to the inverted output $-V_j$. Alternatively, we can transform to unipolar connections, as described on page 49; this requires one extra unit to compute $\sum_j V_j$, which is then fed to all other units.

It is possible to avoid the need for small ρ, and yet obtain $w_{ij} \propto 1/R_{ij}$ (and τ_i independent of i), by modifying the circuit slightly [Denker, 1986]. As shown in Fig. 3.6, we simply connect the resistor ρ and the capacitor C between the input and the inverted output, instead of between the input and ground. Then if we use a high gain amplifier the negative feedback loop keeps the input close to ground potential, giving rise to the name **virtual ground unit**. In the high gain limit it is easy to show that the circuit equations reduce to

$$\tau \frac{dV_i}{dt} = -V_i + \sum_j w_{ij}V_j \tag{3.50}$$

where $\tau = \rho C$ and $w_{ij} = \rho/R_ij$. The result is linear even if the operational amplifier is nonlinear. To regain a saturation property for V_i one can replace the resistor ρ by a nonlinear element such as a reversed pair of parallel diodes [Denker, 1986].

The hardest problem in creating networks of these circuits in VLSI is fabricating the connection resistors R_{ij}. We need $2N^2$ resistors for N fully connected units (using inverters), and the resistance values must be large to limit power dissipation. High resistance paths are difficult to make in conventional CMOS technology, and especially difficult to make small. Nevertheless Graf et al. [1986] were able to make custom chips with $N = 256$ fully connected units, using $2N^2 \approx 130,000$ resistor sites. Each resistor had to be added to the otherwise finished CMOS chip using electron-beam lithography, and all had approximately the same resistance. Resistors could not be changed once made.

Others have employed active elements (transistors) for the connections instead of passive resistive paths [Sivilotti et al., 1987; Alspector et al., 1988; Mead, 1989]. Typically the connection strengths have a discrete set of gray levels (e.g., $1/R_{ij} = 0, 1, \ldots, 16$). Most importantly, they can be programmed after fabrication, allowing general purpose chips and the possibility of learning. In most applications it is important to make the connection strengths non-volatile, so that they retain their values even when the device is turned off.

Optical Implementation

In silicon technology it is easy to construct the units but hard to make the many connections needed. Optical technology has the reverse problem; because of the inherent linearity of most optical media, it is relatively easy to make the connections but harder to construct the units. Different rays can cross or be superposed without interfering with one another, so many optical connections can be made by a set of crossing light rays. But this typical linearity makes it hard to construct units with appropriate nonlinearity such as threshold behavior. Many of the earlier optical implementations of neural networks were therefore hybrid **optoelectronic** systems, in which the nonlinearity was placed in the electronic part. But now there is more effort going into fully optical systems, using nonlinear optical media and devices.

We first describe a very simple optoelectronic system due to Farhat et al. [1985]. There were 32 units, each represented by a separate electronic circuit shown schematically in Fig. 3.7(a). Each unit had an LED (Light Emitting Diode) as its output and a pair of photodiodes (PDs) as its input, one for excitatory and one for inhibitory signals. The LED was on when the unit was firing, and not otherwise. The desired connections were binarized so that $w_{ij} = -1$ or 1 for each ij (and zero on the diagonal). Thus the optical problem was simply to make the light from the LED for unit j shine on unit i's excitatory photodiode, or inhibitory photodiode, or neither, according to the value of w_{ij}. The weight matrix was photographically coded onto a two-dimensional mask with two pixels for each of the 32×32 connections. The two pixels were opaque and transparent as required for the desired w_{ij} (black–white, white–black, or black–black corresponding to 1, −1, or 0).

Figure 3.7(b) shows how the optical interconnections were made. The transparency encoding the weights was placed in front of an array of 64 long photodiodes such that each column of the transparency exactly covered one diode.[4] Each of the LED's in the vertical output array illuminated one row of the transparency. This was done by vertically focusing the light and horizontally smearing it with a pair of lenses so that the light from one unit became a horizontal line on the appropriate row of the transparency. The input to each unit was accumulated in the corresponding pair of vertical photodiodes (inhibition in one, excitation in the other).

---

[4]Actually the transparency was cut into two halves, each covering 32 photodiodes, for technical convenience.

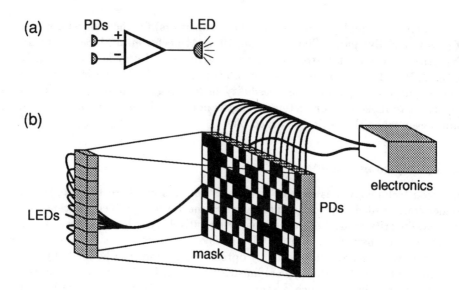

FIGURE 3.7 Optoelectronic Hopfield network of Farhat et al. [1985]. (a) Single unit. The signals from the excitatory and inhibitory photodiodes (PDs) are amplified and thresholded before driving the light emitting diode (LED). (b) Optical arrangement for the interconnections. Lenses are not shown, and the number of units has been scaled down by a factor of four.

Incoherent light was used, so the total intensity on each photodiode automatically gave the appropriate summation of total input.

The electronic circuitry allowed the designers to selectively turn units on and off, and then release them. When three or fewer random patterns were stored in the memory mask it was found that any could be recalled by starting the system close enough to it. Indeed the successes and failures corresponded almost exactly to those for the same Hopfield network simulated by computer.

In later versions of this scheme the need for excitatory and inhibitory inputs was eliminated by using unipolar connections, as described on page 49. The adaptive threshold term $\sum_j S_j$ was easily computed by focusing light from all the LEDs onto a single photodiode.

Most recent work has employed a hologram rather than a mask for the connection matrix [Cohen, 1986; Abu-Mostafa and Psaltis, 1987; Hsu et al., 1988; Peterson et al., 1990]. Using coherent light, a hologram can direct a light wave from any incoming direction independently to any outgoing direction, and thus connect many pairs of directions. A plane hologram on photographic film can implement about 10^8 such connections per square inch, so 10^4 sources can easily be fully connected to the same number of sensors. Volume holograms made from photorefractive crystals have an even greater potential; in principle they can hold more than 10^{12} connections per cm^3, corresponding to a million fully connected units. However, the readout from a volume hologram is partially destructive, so the stored patterns decay slowly unless refreshed.

Of course one does not want to *calculate* the appropriate hologram for a particular set of memories; one wants to produce it optically from, e.g., a set of pictures. The details of this are beyond the scope of this book, but it is worth distinguishing *angular* and *spatial* multiplexing of different stored patterns. In the angular case the patterns are stored at different input angles, or at different lateral locations on the hologram, while spatial multiplexing involves assigning them to different layers of a volume hologram. Spatial multiplexing appears capable of storing more patterns [Peterson et al., 1990].

Many optical implementations are all-optical rather than optoelectronic. This means that a nonlinear medium or device must be found to perform the thresholding function for the units. Various approaches have been tried, including the use of strongly pumped *phase-conjugate mirrors* [Soffer et al., 1986; Anderson, 1986], *etalon arrays* [Wagner and Psaltis, 1987], and two-dimensional *spatial light modulators* [Abu-Mostafa and Psaltis, 1987; Hsu et al., 1988]. The last of these seems to have the greatest potential.

Some recent optical work has gone beyond Hopfield networks with the construction of *adaptive* optical neural networks that are able to learn. Simple perceptrons, multilayer networks with back-propagation, and deterministic Boltzmann machines have all been implemented [Wagner and Psaltis, 1987; Hsu et al., 1988; Peterson et al., 1990].

3.5 Temporal Sequences of Patterns

We have so far been concerned with networks that evolve to a stable attractor and then stay there. They can thus act as content addressable memories for a set of static patterns. It is also interesting to investigate the possibility of storing, recalling, generating, and learning *sequences* of patterns. These obviously occur widely in biological systems.

In this section we mainly examine how to *generate* a sequence of patterns. Instead of settling down to an attractor we want the network to go through a predetermined sequence, usually in a closed **limit cycle**. The sequences will be embedded in the choice of connections, as elsewhere in Chapters 2 and 3. Networks that can *recognize* sequences, or *learn* sequences incrementally by example, are considered separately in Section 7.3.

Simple sequences can be generated by a set of units that are *synchronously* updated by connecting together a chain of units S_i with excitatory connections $w_{i+1,i}$ from each unit to its successor, and then turning the first unit on. A closed cycle can be made by joining the ends of the chain, as shown in Fig. 3.8. However this sort of scheme is very limited. It can only produce sequences of patterns related by shifting or permutation. It is far from robust; one lost bit and the whole scheme

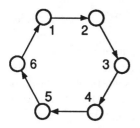

FIGURE 3.8 A very simple sequence generator. This relies on synchronous updating by a central clock, and must be initialized by turning on unit 1. Then it generates the cyclic sequence 1–2–3–4–5–6.

disintegrates. It needs a central pacemaker to control the synchronous updating, and is thus no good as an independent clock or oscillator.

Asymmetric Connections

How can we generate more arbitrary sequences of patterns, using only asynchronous updating? Suppose we have an associative network of units S_i ($i = 1, 2, \ldots, N$) and a sequence of patterns ξ_i^μ ($\mu = 1, 2, \ldots, p$) that we want it to produce *in order*, with pattern 1 following pattern p cyclically. Hopfield [1982] suggested that asymmetric connections,

$$w_{ij} = \frac{1}{N} \sum_\mu \xi_i^\mu \xi_j^\mu + \frac{\lambda}{N} \sum_\mu \xi_i^{\mu+1} \xi_j^\mu , \qquad (3.51)$$

might be able to achieve this. Here λ is a constant that governs the relative strength of symmetric and asymmetric terms. We take ξ_i^{p+1} to mean ξ_i^1. If such a system is in the state $S_i = \xi_i^\nu$ then the input h_i^ν to unit i is

$$
\begin{aligned}
h_i^\nu = \sum_j w_{ij} \xi_j^\nu &= \frac{1}{N} \sum_{\mu j} \xi_i^\mu \xi_j^\mu \xi_j^\nu + \frac{\lambda}{N} \sum_{\mu j} \xi_i^{\mu+1} \xi_j^\mu \xi_j^\nu \\
&= \xi_i^\nu + \lambda \xi_i^{\nu+1} + \text{cross-terms} .
\end{aligned}
\qquad (3.52)
$$

The cross-terms are small if the patterns are uncorrelated and there are not too many of them, as we have seen earlier. For $\lambda < 1$ the original pattern ξ_i^ν is stable because $\text{sgn}(\xi_i^\nu + \lambda \xi_i^{\nu+1}) = \text{sgn}(\xi_i^\nu)$. But for $\lambda > 1$ the second term dominates and tends to move the system to the next pattern, $\xi_i^{\nu+1}$.

For *correlated* patterns the Hebb rule (3.51) may be replaced by a pseudo-inverse rule [Personnaz et al., 1986] generalizing (3.18):

$$w_{ij} = \frac{1}{N} \sum_{\mu\nu} \xi_i^\mu (\mathbf{Q}^{-1})_{\mu\nu} \xi_j^\nu + \frac{\lambda}{N} \sum_{\mu\nu} \xi_i^{\mu+1} (\mathbf{Q}^{-1})_{\mu\nu} \xi_j^\nu \qquad (3.53)$$

where the matrix $\mathbf{Q}$ is given by (3.17) as before.

Unfortunately these schemes do not work very well in practice. The asynchronous updating tends to dephase the system so that one obtains states that overlap several consecutive patterns, and the sequence is soon lost. Only if the length of the sequence is very small ($p \ll N$) can limit cycles be embedded success-fully [Nishimori et al., 1990].

Buhmann and Schulten [1987, 1988] suggested a number of modifications that made stable sequence generation possible. They used sparse patterns (cf. (3.26)) with a stochastic network at $T > 0$, and added terms to (3.51) to inhibit transitions to pattern states that were *not* next in sequence. The parameter values were chosen so that the system tended to remain in a particular pattern state for some time before being driven to the next one by the thermal noise ($T > 0$); in effect λ in (3.51) was a little *less* than one. No transitions were made at all if the temperature T was too low. The time between transitions was rather unpredictable, but became sharper and sharper as the system size N was increased.

Fast and Slow Connections

Sompolinsky and Kanter [1986], Kleinfeld [1986], and Peretto and Niez [1986] found ways of controlling how long each state in the sequence stabilizes before moving on to the next. *In effect* the parameter λ is small when each new state is entered, and then grows steadily until it provokes the next transition. Peretto and Niez proposed doing this directly with connection strengths that changed dynamically. However, this is expensive to implement in hardware or software because there are so many connections. The same effect can be achieved with the dynamics located only in the units if we use two types of connections [Sompolinsky and Kanter, 1986; Kleinfeld, 1986]. Between units i and j we have the usual symmetric **short-time connections**,

$$w_{ij}^S = \frac{1}{N} \sum_\mu \xi_i^\mu \xi_j^\mu \tag{3.54}$$

that stabilize each pattern, and asymmetric **long-time connections**,

$$w_{ij}^L = \frac{1}{N} \sum_\mu \xi_i^{\mu+1} \xi_j^\mu \tag{3.55}$$

that tend to cause transitions in the sequence. The long-time connections represent slow synapses that have delayed or sluggish responses. Explicitly, the input $h_i(t)$ to unit i at time t is now given by

$$h_i(t) = \sum_j \left[w_{ij}^S S_j(t) + \lambda w_{ij}^L \overline{S_j}(t) \right] \tag{3.56}$$

where the delayed response $\overline{S_j}(t)$ is a weighted moving average (sometimes called a memory trace) over the past values of S_j:

$$\overline{S_j}(t) = \int_{-\infty}^t G(t - t') S_j(t') \, dt' . \tag{3.57}$$

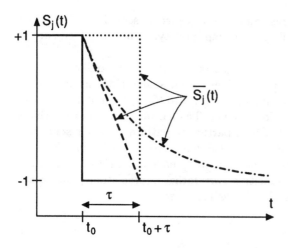

FIGURE 3.9 Illustration of (3.61). The solid curve shows $S_j(t)$ for some unit j. The other curves show $\overline{S_j}(t)$ using equations (3.58) [dotted], (3.59) [dashed], and (3.60) [dot-dashed].

Various forms can be used for the kernel $G(t)$, such as a delta function,[5]

$$G(t) = \delta(t - \tau) \tag{3.58}$$

a step function,

$$G(t) = \Theta(\tau - t)/\tau \tag{3.59}$$

or an exponential decay,

$$G(t) = \exp(-t/\tau)/\tau. \tag{3.60}$$

The last of these is easily implemented using decay units, as discussed in Section 7.3 (page 181). In all cases $\int_0^\infty G(t)\,dt = 1$.

If the network has been in pattern $\xi_i^{\nu-1}$ for a time long compared to τ and then changes to ξ_i^ν at time t_0, we obtain

$$\overline{S_j}(t) = \begin{cases} \xi_j^{\nu-1} & \text{for } t - t_0 \ll \tau; \\ \xi_j^\nu & \text{for } t - t_0 \gg \tau. \end{cases} \tag{3.61}$$

This is illustrated in Fig. 3.9. Thus

$$h_i = \begin{cases} (1+\lambda)\xi_i^\nu & \text{for } t - t_0 \ll \tau; \\ \xi_i^\nu + \lambda\xi_i^{\nu+1} & \text{for } t - t_0 \gg \tau \end{cases} \tag{3.62}$$

(besides cross-terms), which causes the next transition after a time of order τ if $\lambda > 1$.

---

[5] The Dirac delta function $\delta(x)$ is defined to be zero for $x \neq 0$, with an infinite peak of area 1 at $x = 0$. Thus $\int f(x)\delta(x)\,dx = f(0)$ for any continuous function $f(x)$ if the range of integration includes $x = 0$. $\delta(x)$ can also be regarded as the derivative of the step function $\Theta(x)$. The delta function is not strictly speaking a function, but can be defined as the limit of a sequence of functions.

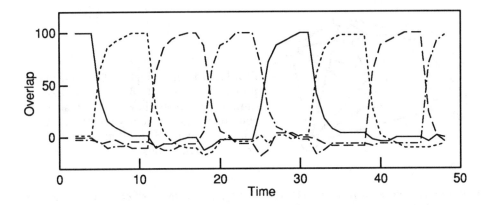

FIGURE 3.10 Example of sequence generation. The curves show the overlap of the state S_i at time t with each of the embedded patterns ξ_i^1, ξ_i^2, ξ_i^3, ξ_i^4. The overlaps were calculated using $\sum_i S_i \xi_i^\mu$. The parameters used were $N = 100$, $p = 4$, $\lambda = 2$, $\tau = 8$ using the step function kernel (3.59).

An intuitive picture of this process consists of an energy surface that has minima at each ξ_i^ν but which also tilts steadily while the system is in a particular state, so that a *downhill* move from ξ_i^ν to $\xi_i^{\nu+1}$ occurs eventually. But this picture is a little deceptive, because the dynamics cannot be represented simply as descent on an energy landscape unless the connections are symmetric.

In general the scheme works well and is quite robust. Figure 3.10 shows an example using the step function kernel (3.59). The parameters λ and τ must be chosen carefully however; in a detailed analysis Riedel et al. [1988] showed that some choices using (3.60) with large λ can lead to chaotic behavior instead of the desired sequence.

The essential idea can also be generalized in most of the ways considered in Section 3.1. For example Gutfreund and Mézard [1988] demonstrated the applicability to the strong dilution model of Derrida, Gardner, and Zippelius [1987] discussed on page 46.

Central Pattern Generators

Kleinfeld and Sompolinsky [1989] have applied the model just described to the **central pattern generator** for swimming in the mollusk *Tritonia diomedea*. Central pattern generators are groups of neurons, typically in the spinal cord, that collectively produce a cyclic sequence without either feedback from the controlled system or continuous control from the brain. There is no single pacemaker neuron. In some cases multiple patterns can be generated by the same set of neurons. Patterns can be started, stopped, and modulated in period by external control inputs. In the case of *Tritonia*, there are four neural groups that fire in a well-defined sequence. Modelling these groups by one unit each, Kleinfeld and Sompolinsky calculated the

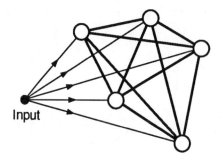

FIGURE 3.11 Architecture for counting pulses. The connections between the units (heavy lines) have fast and slow components, and are asymmetrical. The input is connected to *all* units with random ±1 connections.

required connections from (3.54) and (3.55), and then verified that the network generated the appropriate sequence. Particularly impressive was the comparison of their computed connection strengths with those measured experimentally—in *Tritonia* one can measure individual synapse strengths and identify delayed and non-delayed components. In every case where a synapse was found experimentally its sign was the same as that in the computed set.

Counting

Amit [1988] suggested a modification of the same sequence generation scheme to *count* input pulses in a network. For counting we want each transition $\nu \rightarrow \nu + 1$ to occur only when an input pulse is received. This can be achieved with the same architecture—using fast and slow connections—if we use $\lambda < 1$. Then no transitions occur in the absence of input (at $T = 0$). The input pulse is arranged to apply an additional input $\mu \phi_i$ to each unit. Here $\phi_i = \pm 1$, uncorrelated with any of the sequence patterns ξ_i^μ, and $1 + \lambda > \mu > 1 - \lambda$. Figure 3.11 shows the architecture.

If the system has been in state ξ_i^ν for a time long compared to τ, the total input to unit i is given by

$$h_i = \begin{cases} \xi_i^\nu + \lambda \xi_i^{\nu+1} & \text{without input pulse;} \\ \xi_i^\nu + \lambda \xi_i^{\nu+1} + \mu \phi_i & \text{with input pulse.} \end{cases} \qquad (3.63)$$

Without an input pulse the current state ξ_i^ν is stable because $\lambda < 1$. With an input pulse, there are two cases. If ξ_i^ν and $\xi_i^{\nu+1}$ have the same sign then they determine the sign of h_i. If they have the opposite sign, then ϕ_i determines the sign of h_i, favoring ξ_i^ν and $\xi_i^{\nu+1}$ equally. So the input pulse moves the state to a point approximately equidistant from ξ_i^ν and $\xi_i^{\nu+1}$, but in the subsequent dynamical motion the delayed connections break the tie and carry the state forward to $\xi_i^{\nu+1}$, as desired.

Delayed Synapses

In the case $G(t) = \delta(t - \tau)$ the long time connections simply correspond to **delayed** synapses that pass a given signal after a delay τ. Equation (3.62) makes it clear that τ is also the time between transitions, so the rule (3.55) for the delayed synapse strengths may be written

$$w_{ij}^L = \frac{1}{N} \sum_{t=0,\tau,2\tau,\dots} \xi_i(t+\tau)\xi_j(t) \qquad (3.64)$$

if we regard the desired sequence as a time-varying pattern $\xi_i(t)$. Equation (3.64) is now of the Hebb form; the synapse strength depends on the product of presynaptic potential $\xi_j(t)$ and postsynaptic potential $\xi_i(t+\tau)$ with the correct synapse delay. In general we would write

$$w_{ij}^L = \frac{1}{NT} \int_0^T \xi_i(t+\tau_{ij})\xi_j(t)\,dt \qquad (3.65)$$

allowing also for different delays τ_{ij} for different connections. Then we can compute the connection strengths by imposing the required sequence $\xi_i(t)$ on the network and letting the individual synapses "learn" according to (3.65). This is a more reasonable procedure for real neural circuits than is the abstract formulation (3.55).

The Hebbian form (3.64) or (3.65) rests on the assumption that the sequence transition time is the same as the synaptic delay time. But in fact more general sequences $\xi_i(t)$ that do not necessarily change at regular intervals may be learned using (3.65) [Coolen and Gielen, 1988; Herz et al., 1989; Kerszberg and Zippelius, 1990]. There must be a *range* of delay times τ_{ij} in the network, with a broad distribution that covers the relevant time scales of the sequence $\xi_i(t)$. Separate short-time connections w_{ij}^S are not needed. The connections that become strong are those with delays that approximately match the time course of the sequence itself. This can be seen as **learning by selection**, or **resonance** between the patterns to be learned and the natural modes inherent in the network [Toulouse et al., 1986; Dehaene et al., 1987].

FOUR

Optimization Problems

We have viewed the Hopfield associative memory model as performing a recognition or retrieval task. But it can also be seen as solving an optimization problem—the network is expected to find a configuration which minimizes an energy function. There are many other such problems which arise in general optimization theory, with applications to technological and economic systems. What we usually call the energy would in those cases be called the **cost function** or **objective function**. The problems which have been most intensively studied are, like our associative memory of random uncorrelated patterns, extreme idealizations of the real situations one wants to model, but one hopes to learn generic things from simple models amenable to mathematical analysis. Neural computational methods have recently been applied to some of these problems, and in this chapter we will examine a few of them. As in the preceding chapters, the idea will not be to solve specific problems explicitly, but just to exhibit how to formulate them in terms of neural networks.

Before we start we should emphasize that when we formulate these problems "in terms of neural networks" we are really constructing a particular kind of parallel algorithm for their solution. These algorithms lend themselves to a direct implementation in terms of the networks we have been studying, including the circuits discussed in Section 3.4. Even without actually building a circuit, however, the algorithms can be used as a basis for computing solutions on a conventional computer. The analogy between these algorithms and circuits (or other physical systems) helps to give insight into how to tune their parameters so that they work optimally, whether in hardware or software.

This chapter deals mainly with discrete systems where there is a large but finite set of possible solutions to the optimization problem. Typically if the problem is of "size" N, then there are order e^N or $N!$ possible solutions, of which we want the one that minimizes the cost function. This kind of problem is often referred to as a **combinatorial optimization problem**.

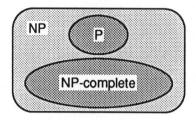

FIGURE 4.1 The class of NP problems assuming that P≠NP. Under this assumption it can be shown that there are NP problems that are neither NP-complete nor in P.

Optimization problems can be divided into classes according to the time it takes to solve them. If there exists an algorithm that solves the problem in a time that grows only polynomially (or slower) with the size N of the problem, then it is said to be polynomial and to belong to class P. P is a subclass of another class called NP (non-deterministic polynomial). NP problems are those for which one can test in polynomial time whether any "guess" of the solution is right.[1] An important subclass of NP is that of the NP-complete problems. They are the hardest NP-problems and are loosely characterized as follows: if one could find a deterministic algorithm that solves *one* NP-complete problem in polynomial time, then all other NP problems could be solved in polynomial time. In that case P, NP, and NP-complete would all be the same class, but it is most likely (though not proven) that P≠NP. The probable situation is sketched in Fig. 4.1. Empirically the time it takes to solve an NP-complete problem tends to scale exponentially with the size N.

A more extensive introduction to combinatorial optimization problems and NP-completeness at a mathematically higher level can be found in the books of Papadimitriou and Steiglitz [1982], Garey and Johnson [1979], and Mézard et al. [1987, pages 307–335].

4.1 The Weighted Matching Problem

We first examine the **weighted matching problem**. One has a set of N points with a known "distance" d_{ij} between each ij pair. The points may be taken as randomly located in some space, in which case d_{ij} can be the real [Euclidean] distance, or the points may represent abstract entities and their relations. For theoretical analysis one often takes the d_{ij} as independent random variables with some distribution $P(d_{ij})$. The problem is to link the points together in pairs, with each point linked to exactly one other point, so as to minimize the total length of the links. Figure 4.2 shows an example. Practical examples might include matching pairs of

---

[1] Strictly speaking, these definitions only hold for *decision* problems where the aim is a "yes" or "no" answer. But it is easy to generalize to optimization problems by posing decision problems of the form "is there a solution with cost less than C?".

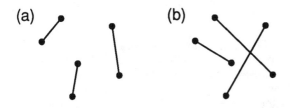

FIGURE 4.2 Illustration of the weighted matching problem, showing a good (a) and a bad (b) solution to the same problem.

electronic components (e.g., loudspeakers for stereo), optimal scheduling of jobs for two equivalent processors, and matching students to schools.[2]

Weighted matching is not actually a very hard problem computationally; it is not NP-complete, and fast polynomial algorithms are known for its solution [e.g., Lawler, 1976]. Special cases of the problem may also be solved analytically [Orland, 1985; Mézard and Parisi, 1985, 1988]. Nevertheless it serves as a simple example to demonstrate the neural network approach.

We formulate the problem in terms of a stochastic McCulloch-Pitts network like the one used for the associative memory problem in Section 2.4, except that we use 0/1 rather than ±1 units. We assign a unit n_{ij}, $i < j$, to each *pair* of sites in the problem, making $N(N-1)/2$ units in all. Each candidate solution of the problem corresponds to a state of the network with $n_{ij} = 1$ if there is a link connecting point i and point j and $n_{ij} = 0$ if there is not. Our problem is to find a way of specifying the values of the connection strengths $w_{ij,kl}$ between the units. In a symmetric network there are $N(N-1)[N(N-1)-2]/8$ such connections.

In the associative memory problem it was quite simple to choose an energy function (2.30), and thereby the appropriate connection weights (2.9). The situation here is one step more involved because we have to deal explicitly with constraints. We have to minimize the total length of the links

$$L = \sum_{i<j} d_{ij} n_{ij} \tag{4.1}$$

subject to the constraint that

$$\sum_j n_{ij} = 1 \qquad \text{for all } i. \tag{4.2}$$

This constraint says that each point should be connected to exactly one other point; we define $n_{ij} \equiv n_{ji}$ when $j < i$, and take $n_{ii} = 0$.

It is difficult to enforce the constraint rigidly from the beginning. Our strategy is instead to add a **penalty term** to the energy which is minimized when the

---

[2] In the last case, there are clearly two types of points, students and schools. Links d_{ij} (representing compatibility or preference) are possible only between one type and the other; the underlying graph is called *bipartite*. It is easy to see how to modify the neural network implementation for the bipartite case.

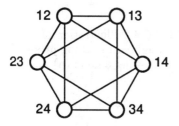

FIGURE 4.3 Stochastic network to solve a four-point weighted matching problem. There is one unit for each pair of points. The connections shown are all inhibitory with strength $-\gamma$. The distance information is in the thresholds (not shown); unit ij has threshold $d_{ij} - \gamma$.

constraint is satisfied. Thus we write down a total effective energy or cost function

$$H[\mathbf{n}] = \sum_{i<j} d_{ij} n_{ij} + \frac{\gamma}{2} \sum_{i} \left(1 - \sum_{j} n_{ij}\right)^2. \tag{4.3}$$

We can now build or simulate a network with this energy function. If we multiply out the summations in (4.3) we obtain constant terms, linear terms proportional to one n_{ij}, and quadratic terms with two n_{ij}'s. The quadratic terms can be represented by connections $w_{ij,kl}$ between the units, while the linear terms can be accounted for with thresholds.

To get the details correct in this 0/1 case, it is easiest to go back to the stochastic evolution rule, given by (2.48) for the ± 1 case. We reformulate this rule in terms of the **flip probability** that the state n_{ij} *changes* at a given step:

$$\text{Prob}(n_{ij} \to n'_{ij}) = \frac{1}{1 + \exp(\beta \Delta H_{ij})} \tag{4.4}$$

where

$$\Delta H_{ij} = H(n'_{ij}) - H(n_{ij}). \tag{4.5}$$

Here $n_{ij} = 0$ and $n'_{ij} = 1$ or vice versa, but the reformulation is also valid for the ± 1 case. It is easy to show that it is exactly equivalent to (2.48). Equation (4.4) makes clear that changes that decrease the energy are more likely than those that increase it. In the deterministic limit $\beta \to \infty$, only changes with $\Delta H_{ij} \leq 0$ are accepted.

For the present problem (4.3) may be used to derive

$$\Delta H_{ij} = \left(d_{ij} - \gamma + \gamma \sum_{k \neq j} n_{ik} + \gamma \sum_{k \neq i} n_{jk}\right) \Delta n_{ij} \tag{4.6}$$

where $\Delta n_{ij} = n'_{ij} - n_{ij}$. If the coefficient of Δn_{ij} is positive, then flips from $n_{ij} = 1$ to $n'_{ij} = 0$ are more likely than vice versa, while if the coefficient is negative the reverse is true. Thus the coefficient of Δn_{ij} plays the same role here (except for sign) that h_i did in the memory network. In the deterministic limit $\beta \to \infty$, unit ij will turn on if the coefficient is negative, and off if it is positive. Comparing with (1.1) we see that $d_{ij} - \gamma$ is the threshold for unit ij, and the remaining terms in (4.6) say that it receives input of strength $-\gamma$ (note the sign) from all other units that represent links to or from points i or j. In general $w_{ij,kl} = -\gamma$ whenever ij has an index in common with kl; otherwise $w_{ij,kl} = 0$. Figure 4.3 shows a simple example for $N = 4$.

We have thus found the necessary thresholds and connection strengths in the network. The values of these are rather easy to understand: the contribution d_{ij} to the threshold of unit ij just reflects the fact that without the constraint (4.2) the total length of links would be minimized by not having any links at all. The other contribution to the threshold, which has the opposite sign, counteracts this so that some positive number of links is allowed. Finally, the mutual inhibition represented by the $w_{ij,kl}$'s discourages configurations with more than one link coming in to any one point.

The magnetic analogy is interesting: the d_{ij}-dependent threshold term is like a random external field $-d_{ij}$, and the mutual inhibition terms are like **antiferromagnetic interactions** between all pairs of spins.[3]

To find a solution of the weighted matching problem, one has to run this network, following its stochastic dynamics and measuring the resulting average outputs $\langle n_{ij} \rangle$. This is a form of **Monte Carlo simulation** [Metropolis et al., 1953; Binder and Heerman, 1988]. Alternatively one could run a version of the network with continuous-valued units, as discussed in Section 3.3, and simply use the steady state value of n_{ij} (between 0 and 1) in place of $\langle n_{ij} \rangle$. In either case, if $\langle n_{ij} \rangle$ is near 1, it means the solution has a link between point i and point j. If it is near zero, there is no link. At low temperatures, the system will have $\langle n_{ij} \rangle$'s predominantly near 1 or 0.

There is the question of how big γ should be. A reasonable approach is to choose it about the same size as typical d_{ij}'s. Then there will be about the same priority on satisfying the constraint as on having short links. If it is chosen much smaller, we will tend to get solutions with very few links, violating the constraint (4.2), while in the opposite limit we will find configurations which satisfy the constraint but probably are not composed of especially short links. It might be useful to adjust γ while running the network, perhaps relaxing the constraint initially and then gradually enforcing it.

The system may take a long time to come to equilibrium if it is started out at the temperature at which we want to measure the average outputs. If we start, say, with an allowed configuration (i.e., one that satisfies the constraint), then any single change we make in the output state of a unit will take us to a configuration which violates the constraint. This tends to increase H (if γ is large enough), so the change will be unlikely at low T. Thus the system will tend to get stuck in *any* configuration which satisfies the constraint, even though there may be better configurations elsewhere.

A useful strategy may be to start the network at a relatively high temperature and gradually cool it down. When this is done in a simulation of the network, it is called **simulated annealing** [Kirkpatrick et al., 1983], by analogy to the real annealing an experimental physicist or metallurgist does in the laboratory: cooling

---

[3]Remember that in the ferromagnet (page 30) all the interactions w_{ij} were positive. Negative w_{ij}'s are called antiferromagnetic interactions.

(a) 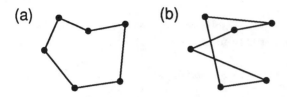 (b)

FIGURE 4.4 The travelling salesman problem, showing a good (a) and a bad (b) solution to the same problem.

the sample slowly so that it finds the true stable phase at low temperature where measurements are to be made. Simulated annealing may also be applied to many optimization problems without first constructing a network formulation; for recent reviews see Johnson et al. [1989], Romeo [1989], and the forthcoming special issue of *Algorithmica* on simulated annealing [1990 or 1991].

Another related approach is **mean field annealing** [Soukoulis et al., 1983; Peterson and Anderson, 1987; Bilbro et al., 1989] where we iteratively solve the coupled mean field equations for the averages $\langle n_{ij} \rangle$. This can be done by a continuous-valued network, as noted on page 55. It has proved to be better than simulated annealing in several optimization problems [Van den Bout and Miller, 1988, 1989; Cortes and Hertz, 1989; Bilbro and Snyder, 1989]. We discuss it further in Section 7.1 on page 171.

4.2 The Travelling Salesman Problem

The travelling salesman problem is similar to the weighted matching problem, but is much harder to solve by computer; it *is* an NP-complete problem. Again we have N points or **cities** in a space with distances d_{ij} between them. The task is to find the minimum-length closed tour that visits each city once and returns to its starting point. Figure 4.4 shows an example. Practical applications include scheduling of truck deliveries, of airline crews, and of the movements of an automatic drill-press or robot arm. In many applications the "distances" d_{ij} are abstract quantities not related to a Euclidean distance between points in a real space; they may not even satisfy the triangle inequality $d_{ij} \leq d_{ik} + d_{kj}$.

There is an enormous amount of literature about the travelling salesman problem, which has become a standard test-bed for methods of combinatorial optimization; see e.g., [Lawler et al., 1985]. A neural network approach was first suggested by Hopfield and Tank [1985, 1986]. Initial optimism was somewhat quenched by a paper of Wilson and Pawley [1988] showing that the original formulation didn't work very well. But then many were inspired to find improved formulations, and several effective approaches are now known. Similar approaches have been applied to many other combinatorial optimization problems [e.g., Tank and Hopfield, 1986; Ramanujam and Sadayappan, 1988; see also recent conference proceedings].

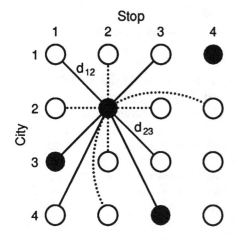

FIGURE 4.5 Network to solve a four-city travelling salesman problem. Solid and open circles denote units that are on and off respectively when the network is representing the tour 3–2–4–1. The connections are shown only for unit n_{22}; solid lines are inhibitory connections of strength $-d_{ij}$, and dotted lines are uniform inhibitory connections of strength $-\gamma$. All connections are symmetric. Thresholds are not shown.

We again choose stochastic binary units n_{ia} to represent possible solutions: $n_{ia} = 1$ if and only if city i is the ath stop on the tour. There are N^2 units in all. The total length of the tour is

$$L = \tfrac{1}{2} \sum_{ij,a} d_{ij} n_{ia}(n_{j,a+1} + n_{j,a-1}) \tag{4.7}$$

and there are two constraints:

$$\sum_a n_{ia} = 1 \qquad \text{(for every city } i) \tag{4.8}$$

and

$$\sum_i n_{ia} = 1 \qquad \text{(for every stop } a). \tag{4.9}$$

The first constraint says that each city appears only once on the tour, while the second says that each stop on the tour is at just one city.

Now we can construct the cost function by adding to the length L two penalty terms which are minimized when the constraints are satisfied:

$$H = \tfrac{1}{2} \sum_{ij,a} d_{ij} n_{ia}(n_{j,a+1} + n_{j,a-1})$$
$$+ \frac{\gamma}{2} \left[\sum_a \left(1 - \sum_i n_{ia}\right)^2 + \sum_i \left(1 - \sum_a n_{ia}\right)^2 \right]. \tag{4.10}$$

As in the weighted matching problem, calculating the ΔH resulting from a change in an n_{ia} now allows us to identify the connection parameters of the network. A network implementation is possible because multiplying (4.10) out yields terms no higher than quadratic in the n_{ia}'s. We leave the details for the reader, and simply summarize the result, illustrated in Fig. 4.5. Suppose we arrange our units into a square array in which each row represents a city and each column a stop (first, second, etc.) in the tour. We employ **periodic boundary conditions**, so that the top and bottom elements in a column are regarded as neighbors, as are the leftmost and rightmost elements in a row. Then from the first term of (4.10) we get connections of strength $-d_{ij}$ between all units in one column and all units in the columns to the left and right of it. The constraint terms proportional to γ give inhibitory connections between all pairs of units within a column and between all pairs of units within a row. They also provide an overall compensating positive input (or negative threshold), without which all units would simply turn off. Note that order N^3 connections are needed for an N city problem.

A solution for a given set of cities and distances can thus be found in the same way as in the weighted matching problem, either by running a physical network or by simulation. Using binary threshold units does *not* work well; we rapidly become stuck in a local minimum with a poor tour length. So we must use either stochastic units with simulated annealing, or continuous-valued units. The former generates better tours but is more time consuming. Hopfield and Tank [1985] used the electrical circuit implementation of continuous-valued units (Fig. 3.5 on page 58), simulating the equations of motion on a computer. This demonstrated nicely that parallel circuits of simple units could solve hard combinatorial optimization problems.

Note that there are $2N$ equivalent formal solutions for each optimal tour, since we don't care where we start the tour or which direction we go around it. However, the network represents these possibilities by different configurations. One or two cities can be clamped (e.g., so city 1 is always at stop 1) to break this degeneracy if desired, with little effect on the algorithm's overall performance.

The original formulation of Hopfield and Tank represented the constraints differently, using three independent parameters A, B, and C for the penalty functions instead of our single γ. It turned out that the success of the network was crucially dependent on the correct choice of these parameters; only in a small low-dimensional region of the A–B–C space does the method give valid tours satisfying the constraints [Hegde et al., 1988]. Thus most choices gave poor results, particularly as the network size is scaled up towards realistic problems [Wilson and Pawley, 1988]. Our formulation is more robust [Brandt et al., 1988]—in effect any reasonable choice of γ is equivalent to a compatible set of coefficients A, B, and C—but the tours produced are not quite so good. Various other modifications of the Hopfield-Tank architecture have also been explored; see e.g., Kahng [1989].

It is not even essential to use penalty functions to enforce the constraints. Van den Bout and Miller [1988, 1989] have suggested a modification *of the dynamics* for continuous units that keeps the N-city constraint (4.8) continuously satisfied. This idea has been developed further by Peterson and Söderberg [1989], and both groups get encouraging results.

Even with these improvements the solutions found by these networks are generally not as good as those found by the best conventional algorithms; the system still gets stuck in a local minimum. It is likely that the root of the problem lies in the underlying dynamics of the network. We discuss this for the stochastic network, but the point also applies to the continuous-valued network.

A good conventional algorithm [Lin and Kernighan, 1973] never makes the kind of moves between constraint-satisfying configurations and constraint-violating ones that we have in our stochastic dynamics. It operates strictly within the set of constraint-satisfying configurations and can get from one to another without going through others which have high energy values. It should be possible to formulate a better dynamical rule in which the elementary transitions do not involve just one unit at a time, but rather several units changing together in such a way as to describe, for example, interchanging the positions of two successive cities on the tour schedule. It is apparently this feature of such algorithms which is most crucial for good performance [Baum, 1986]. There is some recent work on neural network implementations in this direction [Lister, 1990].

Another algorithm called the **elastic net algorithm** for the travelling salesman problem has been developed by Durbin and Willshaw [1987]. It seems to work well at least for problems with cities in a low-dimensional Euclidean space. We discuss it further in Chapter 9.

Besides computer algorithms, there have been attempts to solve the travelling salesman problem analytically for particular distributions of the d_{ij} [Mézard and Parisi, 1986; Krauth and Mézard, 1989]; see Palmer [1988] for an overview. These approaches are based on a statistical mechanics formulation of the problem, using an energy function like (4.10). There are solutions which appear to give good agreement with the numerical simulations [Kirkpatrick and Toulouse, 1985], but it is not yet known whether they are exact.

4.3 Graph Bipartitioning

Our third example of a combinatorial optimization problem and its formulation in terms of a network is **graph bipartitioning** [Fu and Anderson, 1986]. As an example of the problem from chip design, consider N circuit elements that we want to put onto integrated circuit chips. If they do not all fit onto one chip we would like to have half on one chip and the rest on another, in such a way that the number of wires between the two chips is as small as possible. Choosing which circuit elements to put on which chip is then an optimization problem with the number of wires as the cost function.

We begin with a general **graph**, a set of N points or **vertices** and a set of **edges** which connect pairs of the vertices. N is taken to be even. A simple example of a graph would be a set of points on a regular lattice with edges connecting only nearest-neighbor vertices. Here we will focus instead on **random graphs** in which there is a fixed probability p of a given vertex being connected to any other. The average number pN of other vertices to which a given one is connected is called the **valency** of the graph. We can consider both *extensive* valency (valency of order N, or p of order unity) and *intensive* valency (valency of order unity, $p \propto 1/N$). We will be mostly concerned with the extensive case. Given such a graph, the task is

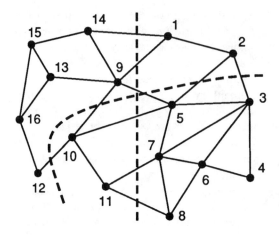

FIGURE 4.6 Example of a graph bipartitioning problem. The dashed lines show two different partitions, one with cost 6 and one with cost 5. An even better solution is possible.

to divide the vertices into two sets of equal size in such a way as to minimize the number of edges going between the sets. Fig. 4.6 shows an example.

Let us define $C_{ij} = 1$ if vertices i and j are connected and $C_{ij} = 0$ if they are not. At each vertex define a variable S_i which is $+1$ if the site is in one set and -1 if it is in the other. Then we want to minimize

$$L = -\sum_{\langle ij \rangle} C_{ij} S_i S_j \tag{4.11}$$

(where $\langle ij \rangle$ means a sum that counts each distinct pair once), subject to the constraint

$$\sum_i S_i = 0. \tag{4.12}$$

In magnetic terms this is a randomly diluted ferromagnet with a constraint of zero total magnetization.

A way to enforce the constraint softly (in analogy to (4.3) and (4.10)) is to add to the effective energy a term that penalizes total magnetizations different from zero. Thus we use a cost function

$$H = -\sum_{\langle ij \rangle} C_{ij} S_i S_j + \mu \left(\sum_i S_i \right)^2. \tag{4.13}$$

If we multiply out the last term, we obtain simply

$$H = N\mu - \sum_{\langle ij \rangle} w_{ij} S_i S_j \tag{4.14}$$

where

$$w_{ij} = C_{ij} - 2\mu. \tag{4.15}$$

This is *exactly of the Hopfield form* (2.24) (apart from the unimportant constant $N\mu$), and is thus easily minimized by a network.

Note that a choice of μ between 0 and 1/2 makes some of the connection strengths excitatory and some inhibitory. Viewed as a magnetic system, this produces a situation with random interactions of both signs between "spins" S_i, making the system into a **spin glass**. This kind of problem has been studied extensively in statistical mechanics; see, for example, Binder and Young [1986] and Fischer and Hertz [1990]. In the extensive-valency case, the interactions are effectively of infinite range (w_{ij} does not depend on the distance between vertices i and j), which turns the problem into the much studied **Sherrington-Kirkpatrick model** [Sherrington and Kirkpatrick, 1975; Mézard et al., 1987]. This model has been found to have many novel and interesting features such as *ultrametricity* and *replica symmetry breaking*, the details of which are beyond the compass of this book.

4.4 Optimization Problems in Image Processing

The previous three examples were of *combinatorial* optimization problems, where there are very many possible configurations arising from different combinations of the basic ingredients. The total number of possible configurations is typically exponentially large in the problem size N and a direct search is impractical for reasonable values of N. We now discuss a class of problems that do not all have the combinatorial aspect, but are nevertheless hard optimization problems amenable to solution by parallel networks. There are many aspects in common with the preceding problems. There may also be connections to parallel processing in real nervous systems (the visual cortex), though we do not discuss this in detail.

The general kind of problem is the **reconstruction** of an image from noisy or blurred data. The data are given as a set of values d_i for each of a 2-dimensional array of **pixels** (picture cells—small elements of an image). The actual values d_i might represent brightness, or a first or second spatial derivative of brightness, or a time derivative of brightness, or binocular disparity (the small displacement of an image point between two eyes), or some other spatially varying quantity. The data might be binary (e.g., bright/dark), or continuous-valued, say in the range 0 to 1. We take the continuous case, which is usually called **gray-level data** in this context.

We want to design a network that takes the noisy data d_i as inputs and produces outputs V_i that represent a possible reconstructed image. Figure 4.7 shows an example in one dimension. The problem is to find the architecture and parameters (connection strengths, thresholds, etc.) of that network to make the output give the best reconstruction.

We should emphasize that reconstruction always involves assumptions or *a priori* knowledge about the image, such as that it has smooth surfaces, straight edges, or whatever. The interesting problems occur when we do have reasons for such prejudices. If we have no such knowledge or expectation, then we can of course do no

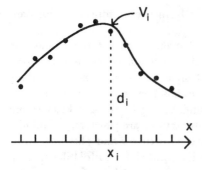

FIGURE 4.7 Fitting a smooth curve to noisy data. The data values d_i are given at a discrete set of locations x_i. The aim is to fit them with a *smooth* curve as shown, with value V_i at point x_i.

better than the dirty data themselves. Problems where there is insufficient information in the data are often called **ill-posed problems**. Techniques for transforming them into well-posed problems by introducing constraints (such as a smoothness requirement) are the subject of **regularization theory** [Poggio et al., 1985].

In the paper of Koch, Marroquin, and Yuille [1986], on which our presentation here is based, the specific problem arose in stereo vision. The data are binocular disparity values for elements of the image, given point by point over the field of view. The ultimate task is to reconstruct a depth map of the image. Here it should be reasonable to assume that surfaces in the image are fairly smooth except at edges, and that the edges themselves are straight or smoothly curved. We will make these assumptions below, although different assumptions might be necessary in another kind of task.

We could formulate this problem with either stochastic binary units (Section 2.4) or continuous-valued units (Section 3.3). Because of the continuous nature of the output we choose the latter possibility in most of the discussion here. If we were to do it with binary stochastic units, we would have to read off the solution as *average values* of the unit outputs. The temperature parameter would also have to be set fairly high in order not to get these averages concentrated near ± 1. Similarly, if we use analog units we have to choose the function $g(u)$ so that the units are not operating near saturation.

Reconstructing a Smooth Surface

Things are relatively simple if we know that our image is of a smooth surface without edges. The computation can be done by a network of *linear* elements, $g(u) = u$. We treat the one-dimensional case here, but (for the smooth surface problem, anyway) the generalization to higher dimensions is trivial. We approach the problem by finding a cost function whose minimum gives the best reconstruction. This allows us to identify the appropriate parameters for the neural network implementation.

Two considerations govern our choice of cost function: smoothness and fidelity to the data. We therefore write an H with two terms reflecting these priorities:

$$H = \tfrac{1}{2}\kappa \sum_i (V_i - V_{i+1})^2 + \tfrac{1}{2}\lambda \sum_i (V_i - d_i)^2 \tag{4.16}$$

where V_i is the activation of unit i. The relative weights placed on the two aspects will be reflected in the choice of the relative magnitudes of κ and λ. For a given choice of these parameters, the optimal solution is the one which minimizes H. Since it is just a quadratic form H has only one minimum, so here we do not have to worry about spurious stable states.

We want our network to have a dynamics which allows it to relax into the state with the minimum H. Picturing $H(V_1, V_2, ...V_N)$ as a surface in the N-dimensional V-space, we want to slide downhill until we settle into this valley. For now we do this in the simplest way, using sufficient frictional drag that we do not overshoot and go up the other side of the hill. This can be described by a **gradient descent rule**:

$$\kappa\tau\frac{dV_i}{dt} = -\frac{\partial H}{\partial V_i}\,. \tag{4.17}$$

That is, we slide downhill with a speed proportional to the slope of the hill.

In the present problem, (4.17) gives

$$\kappa\tau\frac{dV_i}{dt} = \kappa(V_{i+1} + V_{i-1} - 2V_i) + \lambda(d_i - V_i)\,. \tag{4.18}$$

The first term is just a discrete lattice version of a second derivative. So if there were no priority on the data at all (i.e., $\lambda = 0$), then (4.18) would just be a **diffusion equation** and the V_i would slowly smooth out until they all had the same value. With nonzero λ, this tendency is mitigated by the second term, which tends to pull the local value of V_i back toward the data value d_i. Thus we tend to get a locally smooth surface (or curve in one dimension) which runs as close as it can to the data points without becoming too rough.

Equation (4.18) is the same kind of equation as (3.31) or (3.32) that we used for associative memory, but with *linear* elements $g(u) = u$ and extra terms λd_i proportional to the data. We can construct an electrical circuit to solve the problem, as in Fig. 3.5. Comparing with (3.44), we identify $\kappa\tau$ with C, κ with the conductance $1/R_{i,i+1}$ between neighboring elements, λ with ρ^{-1}, and λd_i with an external current injected at node i. Thus this computation can be implemented on a simple linear network (Fig. 4.8) that will relax to its equilibrium configuration in a time of order τ.

We can also imagine a simple mechanical network (Fig. 4.9) which will do the same computation. The V_i are represented by masses connected by springs with spring constant λ to the data points d_i and by springs of spring constant κ to their neighboring masses. To get our equations of motion in exactly the form we wrote above, we then have to immerse the whole system in a viscous medium so that it moves in the **over-damped limit**, with *velocity* proportional to the applied force. The exact form is not important. In the opposite limit of weak friction the masses will wiggle back and forth many times before settling into their equilibrium positions, but the final positions will be the same. To get the fastest convergence possible, we could set the friction just strong enough so that the masses almost, but not quite, overshoot their final positions. This is called **critical damping**, and

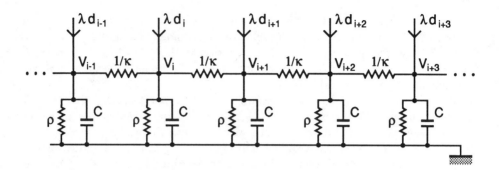

FIGURE 4.8 Electrical implementation of the network with cost function (4.16). The raw data d_i are represented by external currents λd_i injected at the nodes, while the smoothed outputs are given by the node voltages V_i. The capacitance C is $\kappa\tau$, the resistance ρ is $1/\lambda$.

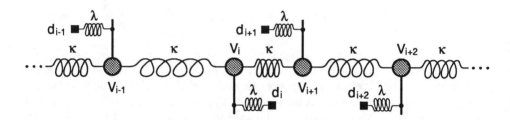

FIGURE 4.9 Mechanical implementation of the network with cost function (4.16). The raw data d_i are represented by springs with spring constant λ attached to anchors at locations d_i, while the smoothed outputs are given by the final positions V_i of the balls. An offset iL must be subtracted from both d_i and V_i, where L is the natural length of each spring (of spring constant κ) in the main chain. The data springs λ are shown displaced on alternate sides of the main chain for clarity. Damping is essential, but not shown.

avoids the very slow approach at the "bottom of the valley" that occurs in the over-damped limit (4.17): since the slope gets very small in the valley bottom, so does the velocity.

Critical damping can be achieved in the electrical circuit too, but requires inductive elements or additional amplifiers. Inductors are big and expensive, and are therefore rarely used. In a *software* implementation, it does help to add this kind of refinement. We will meet examples where people have added "inertia" in this fashion in later chapters.

Discontinuities in the Image

The foregoing network would not work very well for an image with discontinuities,

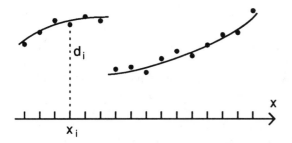

FIGURE 4.10 Fitting a piecewise smooth curve to noisy
data. Like Fig. 4.7 but with
discontinuity units that allow
a break in the fitted curve.

which it would try to fit with a smooth curve. The solution lies in adding to the
system another kind of unit that marks the discontinuities. Then the network can
find piecewise-smooth solutions, as shown in Fig. 4.10. Because of the all-or-nothing
character of discontinuities, it is appropriate that the units marking them be binary-
valued. We also make them stochastic, for reasons to be seen below. In the image-
processing literature the discontinuities are called **line processes**.

Again in one dimension, we put one discontinuity-marking unit between every
pair of adjacent V-units. We use S_i for the one lying between V_i and V_{i+1}. The
value $S_i = +1$ represents the hypothesis that there *is* a discontinuity in the fitted
curve between points i and $i+1$, whereas $S_i = -1$ means there is none. We modify
our energy function in two ways:

■ The energy cost for a change $V_{i+1} - V_i$ between adjacent pixels is removed if
the S_i between the pixels is $+1$.

■ In order to prevent the network from putting in discontinuities everywhere
(and thereby minimizing the data term in (4.16) absolutely), we assign a high
enough cost for each S_i which is $+1$ so that there aren't too many of them.

These changes lead to

$$H = \tfrac{1}{2}\kappa \sum_i \tfrac{1}{2}(1 - S_i)(V_i - V_{i+1})^2 + \tfrac{1}{2}\lambda \sum_i (V_i - d_i)^2 + \mu \sum_i S_i. \qquad (4.19)$$

In an electrical network implementation, the link $1/R_{i,i+1}$ between a neighboring pair of V-units must be cut when the binary S-unit between them is on $(+1)$.
Each binary unit S_i receives in turn an input h_i (the coefficient of S_i in (4.19))
which is a sum of two terms: a positive one proportional to the square of the difference between the neighboring V's, and a fixed negative one $-\mu$. If the first exceeds
the second, the unit will tend to fire $(S_i = +1)$. The first is simply proportional to
the power dissipated in the resistor R_{ij} when uncut.

We anticipate that there may be parts of the image which look initially as
though they have a discontinuity, but eventually turn out not to need one, or vice
versa. It is therefore useful to have the S-units spontaneously trying out hypotheses
about the presence or absence of discontinuities, which is why we chose stochastic
units. We might also want to tune the parameters κ, λ, and μ as the network relaxes.

For example we might strongly penalize discontinuities initially, and later become more tolerant.

It is obviously most natural in this problem that the V-units be analog and the S-units be binary. We can, however implement the network solely with stochastic binary units if we choose the parameters carefully. The temperature must be large compared to κ and λ so that the average values of these units respond more or less linearly to their inputs, while it must be small compared to μ so that the S-units are not indecisive. Alternatively one can perform the computation with all analog units if the S-units are gradually turned up into the large-gain (large β) regime [Cortes and Hertz, 1989].

Two Dimensions

We can generalize our treatment to two dimensions, and here another interesting complication shows up. We start with the obvious generalization

$$H = \tfrac{1}{2}\kappa \sum_{\langle ij \rangle} \tfrac{1}{2}(1 - S_{ij})(V_i - V_j)^2 + \tfrac{1}{2}\lambda \sum_i (V_i - d_i)^2 + \mu \sum_{\langle ij \rangle} S_{ij} \qquad (4.20)$$

where $\langle ij \rangle$ means a pair of sites which are **nearest neighbors** in the pixel lattice. Figure 4.11(a) shows the location of the units.

But in the presence of noise this can lead to *isolated* discontinuities, as in Fig. 4.11(a), whereas we expect that the image should break up into continuous regions bounded by discontinuity contours. The contours themselves should usually join up into continuous closed curves as in Fig. 4.11(b). This can be encouraged by adding a term [Hertz, 1989]

$$H_{\text{loop}} = -\gamma \sum_{\langle ijkl \rangle} S_{ij} S_{jk} S_{kl} S_{li} \qquad (4.21)$$

to the above energy. The notation $\langle ijkl \rangle$ means that the sites V_i, V_j, V_k, V_l are successive nearest neighbors, forming a square cell (or **plaquette**) in the pixel array. Figures 4.11(b) and (c) show by example that H_{loop} contributes $-\gamma$ for every cell except one in which a discontinuity contour terminates, where it gives $+\gamma$. Thus it puts a price on having too many such terminations (Geman and Geman [1984] suggest another way of doing this). Physicists may recognize H_{loop} as a gauge field energy, or as a penalty for frustrated plaquettes.

Notice that the network now has units which do more than compute a weighted linear combination of their inputs. The S_{ij} units, for example, must compute sums of products of two V_i's and of products of three other S's. So they are a little more complicated than the simple elements we used previously. On the other hand, this problem is simpler to implement than some others were, because the connections are all local—each unit only has input from nearby ones.

The approach to depth reconstruction discussed here can be applied to a number of other tasks in image processing. Problems such as surface interpolation,

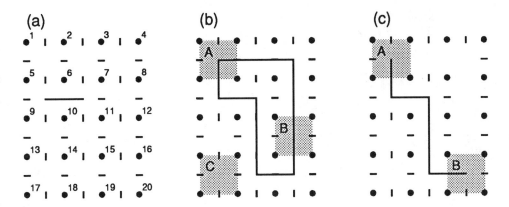

FIGURE 4.11 Two-dimensional image processing with discontinuity units. (a) The numbered dots are the pixel locations where the V_i units are situated. Between each neighboring pair of V_i units lies an S_{ij} unit, depicted by a horizontal or vertical line segment. When *on* ($S_{ij} = +1$), shown by a long line, it marks a discontinuity between those pixels. Only $S_{6,10}$ is shown on. (b) Now 10 S_{ij} units are shown on, forming a discontinuity contour that separates one image region from another. Note that every square cell (such as A, B, or C) has an even number of positive S_{ij}'s around it, because the discontinuity contour leaves each cell that it enters. Thus (4.21) gives $-\gamma$ for all cells. (c) Like (b) but with an open contour. The cells A and B at the ends of the contour have an odd number of positive S_{ij}'s, and therefore contribute $+\gamma$ to the energy.

edge detection, shape from shading, velocity field estimation, color determination, and structure from motion should all be amenable to such a neural network approach [Koch, Marroquin, and Yuille, 1986]. The computations can easily be done in parallel, and can be put onto VLSI chips [Koch et al., 1988]. A "silicon retina," incorporating both photoreceptors and network smoothing circuits (like Fig. 4.8) has also been fabricated in VLSI [Sivilotti et al., 1987; Mead, 1989].

FIVE

Simple Perceptrons

In all the problems we have studied so far, it has been possible with a modicum of cleverness to figure out *a priori* how to choose appropriate connection strengths for a network. This may not always be practical however, because it may involve a large optimization problem or a large matrix inversion. It is often easier to adopt an iterative approach, in which appropriate w_{ij}'s are found by successive improvement from an arbitrary starting point. We can then say that the network is **learning** the task.

This chapter and the next two are devoted to **supervised learning**. Recall from Chapter 1 that there are two general learning paradigms, supervised and unsupervised learning. In supervised learning the network has its output compared with known correct answers, and receives feedback about any errors. This is sometimes called **learning with a teacher**; the teacher tells the network what the right answers are, or at least (in reinforcement learning) whether or not its own answers are correct. In unsupervised learning there is no teacher and no right and wrong answers; the network must discover for itself interesting categories or features in the input data. Unsupervised learning is discussed in Chapters 8 and 9.

We usually consider networks with separate inputs and outputs, and assume that we have a list or **training set** of correct input-output pairs as examples. When we apply one of the training inputs to the network we can compare the network output to the correct output, and then change the connection strengths w_{ij} to minimize the difference. This is typically done incrementally, making small adjustments in response to each training pair, so that the w_{ij}'s converge—if it works—to a solution in which the training set is "known" with high fidelity. It is then interesting to try input patterns *not* in the training set, to see whether the network can successfully **generalize** what it has learned.

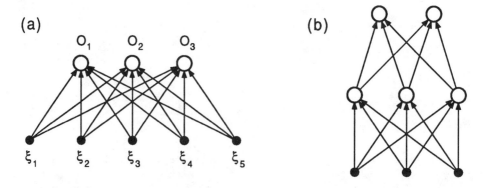

FIGURE 5.1 Perceptrons. (a) A simple perceptron which (by definition) has only one layer. (b) A two-layer perceptron. Inputs, shown as solid circles, perform no computation and are not included the count of layers.

5.1 Feed-Forward Networks

In this chapter and the next we study supervised learning in the context of a particular architecture: **layered feed-forward networks**. It is particularly simple to understand the learning process in this case. Note also that real neural structures in the brain are generally layered, and often largely (but not totally) feed-forward. We will turn to learning in more general networks in Chapter 7. Networks that are *not* strictly feed-forward, but include direct or indirect loops of connections, are often called **recurrent networks**. This includes the networks of Chapters 2 and 3 as well as those considered in Chapter 7.

Layered feed-forward networks were called **perceptrons** when first studied in detail by Rosenblatt and his coworkers 30 years ago [Rosenblatt, 1962]. Figure 5.1 shows two examples of perceptrons. There is a set of input terminals whose only role is to feed input patterns into the rest of the network. After this can come one or more intermediate layers of units, followed by a final output layer where the result of the computation is read off. In the restricted class of feed-forward networks under discussion, there are no connections leading from a unit to units in previous layers, nor to other units in the same layer, nor to units more than one layer ahead. Every unit (or input) feeds only the units in the next layer. The units in intermediate layers are often called **hidden units** because they have no direct connection to the outside world, neither input nor output.

There are two conventions in use for counting the number of layers in the network; some authors count the input terminals as a layer, some do not. We choose *not* to count them; we say for example that a network with one hidden layer is a two-layer network. This convention is becoming more frequently adopted, and seems more logical since the input "units" play no significant role. Note that an N–layer network has N layers of connections and $N - 1$ hidden layers.

Feed-forward networks have by definition *asymmetric* connection matrices w_{ij}; all connections are unidirectional.[1] In general this means that there is no energy function; only with symmetric connections can the existence of an energy function be guaranteed. Thus we cannot employ *equilibrium* statistical mechanical methods here, for those rely on an energy function. We can however do some simple statistical calculations if we use stochastic units.

In this chapter we restrict ourselves further to *one-layer* feed-forward networks. These are often known as **simple perceptrons**. There is a set of N inputs, and an output layer, but no hidden layers. Figure 5.1(a) shows an example and defines our notation; the inputs and outputs are called ξ_k and O_i respectively. Its computation is simply described by

$$O_i = g(h_i) = g\left(\sum_k w_{ik}\xi_k\right) \tag{5.1}$$

where $g(h)$ is the activation function computed by the units. $g(h)$ is usually taken to be nonlinear; we can use a threshold function, a continuous sigmoid, or a stochastically determined ± 1. Particular cases will be discussed later.

Note that the output is an explicit function of the input. This is true for all feed-forward networks; the input is propagated through the network and produces the output right away. In contrast, recurrent networks always need some kind of relaxation to reach an attractor.

We have omitted any thresholds from our description because they can always be treated as connections to an input terminal that is permanently clamped at -1. Specifically we can fix $\xi_0 = -1$ and choose connections strengths $w_{i0} = \theta_i$ to obtain

$$O_i = g\left(\sum_{k=0}^{N} w_{ik}\xi_k\right) = g\left(\sum_{k=1}^{N} w_{ik}\xi_k - \theta_i\right) \tag{5.2}$$

with thresholds θ_i.[2]

The general association task can always be cast in the form of asking for a particular output pattern ζ_i^μ in response to an input pattern ξ_k^μ. That is, we want the *actual* output pattern O_i^μ to be equal to the **target pattern** ζ_i^μ

$$O_i^\mu = \zeta_i^\mu \qquad \text{(desired)} \tag{5.3}$$

for each i and μ. For the simple perceptron the actual output O_i^μ is given by (5.1) when the input ξ_k is clamped to the pattern ξ_k^μ:

$$O_i^\mu = g(h_i^\mu) = g\left(\sum_k w_{ik}\xi_k^\mu\right). \tag{5.4}$$

---

[1] Feed-forward networks are in general characterized by the possibility of numbering the units so that the weights w_{ij} form a triangular matrix, in which all entries above (or below) the diagonal are zero.

[2] If $g(h)$ is a continuous function "threshold" is not a very good term for θ. Often $-\theta$ is called a *bias* instead.

We define p as the number of input-output pairs in the training set, so $\mu = 1, 2, \ldots, p$.

The inputs, outputs, and targets may be boolean (e.g., ± 1) or continuous-valued. For the outputs this depends of course on the nature of the activation function $g(h)$. Sometimes we will use continuous-valued output units but have boolean targets, in which case we can only expect the outputs O_i^μ to come within some margin of the targets.

The general association task (5.3) includes our old associative memory problem as a special case; there we wanted the memory patterns ξ_k^μ to reproduce themselves when used as inputs. That is sometimes called **auto-association**, in contrast to the **hetero-association** task in which the output patterns ζ_i^μ are distinct from the input patterns ξ_k^μ. In feed-forward networks we will focus on hetero-association and hardly ever consider auto-association (see pages 132 and 136). For hetero-association the number of output units may be larger or smaller than the number of input units. Hetero-association includes **classification problems** where the inputs must be divided into particular output categories—normally only one output is on for each input—though these usually require more than a simple perceptron.

For simple perceptrons we will see that *if* there is a set of weights w_{ik} which achieves a particular computation, then these weights can be found by a simple learning rule. The learning rule starts from a general first guess at the weight values and then makes successive improvements. It actually reaches an appropriate answer in a *finite* number of steps.

There are, however, some rather simple and conceptually important computations which a one-layer network cannot do. We will examine in this chapter just what a one-layer network can and cannot do. In the next chapter we will see that multi-layer networks can solve many problems that are impossible within the one-layer architecture.

5.2 Threshold Units

We start with the simplest case of deterministic threshold units, $g(h) = \text{sgn}(h)$, and assume that the targets ζ_i^μ also take ± 1 values. Then all that matters is the *sign* of the net input h_i^μ to output unit i; we want this sign to be the same as that of ζ_i^μ for each i and μ.

The output units are independent so it is often convenient to consider only one at a time and drop the i subscripts. Then the weights w_{ik} become a **weight vector** $\mathbf{w} = (w_1, w_2, \ldots, w_N)$ with one component for each input. Each input pattern ξ_k^μ can also be considered as a **pattern vector** $\boldsymbol{\xi}^\mu$ in this same N-dimensional space.

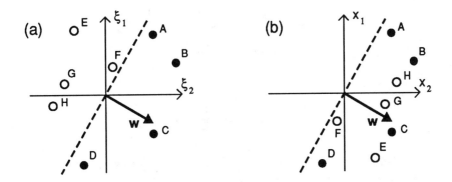

FIGURE 5.2 Geometrical illustration of the conditions (5.5) and (5.8). There are two continuous-valued inputs ξ_1 and ξ_2, and eight patterns ($\mu = 1 \ldots 8$) labelled A–H. Only one output is considered. The solid circles represent input patterns with $\zeta^\mu = +1$, whereas open circles mean $\zeta^\mu = -1$. In (a) the axes are ξ_1 and ξ_2 themselves, while in (b) they are x_1 and x_2; pattern μ has $\mathbf{x}^\mu = \zeta^\mu \boldsymbol{\xi}^\mu$. The condition for correct output is that the plane perpendicular to the weight vector $\mathbf{w}$ divides the points in (a), and lies on one side of all points in (b), as shown.

Then the condition (5.3) becomes[3]

$$\mathrm{sgn}(\mathbf{w} \cdot \boldsymbol{\xi}_\mu) = \zeta^\mu \qquad \text{(desired)} \qquad (5.5)$$

for every μ. This says that the weight vector $\mathbf{w}$ must be chosen so that the projection of pattern $\boldsymbol{\xi}^\mu$ onto it has the same sign as ζ^μ. But the boundary between positive and negative projections onto $\mathbf{w}$ is the plane[4] $\mathbf{w} \cdot \boldsymbol{\xi} = 0$ through the origin perpendicular to $\mathbf{w}$. So the condition for correct operation is that this plane should divide the inputs that have positive and negative targets, as illustrated in Fig. 5.2(a).

It is often convenient to use an alternate representation. By defining

$$x_k^\mu \equiv \zeta^\mu \xi_k^\mu \qquad (5.6)$$

or

$$\mathbf{x}^\mu \equiv \zeta^\mu \boldsymbol{\xi}^\mu \qquad (5.7)$$

we transform the condition (5.5) into

$$\mathbf{w} \cdot \mathbf{x}^\mu > 0 \qquad \text{(desired)} \qquad (5.8)$$

---

[3]The scalar product $\mathbf{A} \cdot \mathbf{B}$ or $\mathbf{B} \cdot \mathbf{A}$ of two vectors $\mathbf{A}$ and $\mathbf{B}$ means $\sum_k A_k B_k$. It is equal to $|\mathbf{A}||\mathbf{B}| \cos\phi$, where ϕ is the angle between $\mathbf{A}$ and $\mathbf{B}$, and can be thought of as $|\mathbf{B}|$ times the projection of $\mathbf{A}$ onto $\mathbf{B}$, or vice versa. It is also known as the dot product or inner product.

[4]Or hyperplane. We often use the word *plane* generically; it means a line in two dimensions, an ordinary plane in three dimensions, and a hyperplane in four or more dimensions.

TABLE 5.1 AND function

ξ_1	ξ_2	ζ
0	0	-1
0	1	-1
1	0	-1
1	1	$+1$

for every μ. This says that the **x** vectors (each of which depends on one input-output pair from the training set) must all lie on the *same* side of the plane perpendicular to **w**, as illustrated in Fig. 5.2(b).

Linear Separability

What happens if there is no such plane? Then the problem cannot be solved—the network cannot perform the task no matter how it is trained. So the condition for solvability of a problem by a simple perceptron with threshold units is whether or not that problem is **linearly separable**. A linearly separable problem is one in which a plane *can* be found in the ξ space separating the $\zeta^\mu = +1$ patterns from the $\zeta^\mu = -1$ ones. If there are several output units we must be able to find one such plane for *each* output.

If we have no threshold (or represent it implicitly by input ξ_0) the separating plane must go through the origin, as we have seen above. But it is interesting to reinstate an explicit threshold for a while. That turns the computation performed by the network into

$$O_i = \text{sgn}\left(\sum_{k>0} w_{ik}\xi_k - w_{i0}\right) \tag{5.9}$$

or, for one unit in vector notation,

$$O = \text{sgn}(\mathbf{w} \cdot \boldsymbol{\xi} - w_0). \tag{5.10}$$

Thus the regions in the N-dimensional input space $(\xi_1, \xi_2, \ldots, \xi_N)$ with different decisions (± 1) for O are separated by an $(N-1)$-dimensional plane

$$\mathbf{w} \cdot \boldsymbol{\xi} = w_0 \tag{5.11}$$

in this space, distance w_0 from the origin. So the effect of adding an explicit threshold is simply to allow the separating plane not to go through the origin. Again there is one such plane for each output unit.

Some examples are appropriate. We consider first a simple example that *is* linearly separable: the Boolean **AND function**. It is a function of two binary 0/1 variables, so a perceptron with two input units ξ_1 and ξ_2 is needed. Because we are using the sgn(h) function the output is ± 1 (instead of 0/1), and we want to get a 1

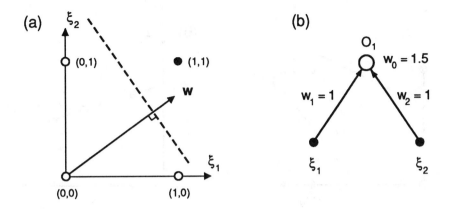

FIGURE 5.3 (a) The AND function is linearly separable. (b) A perceptron that implements AND.

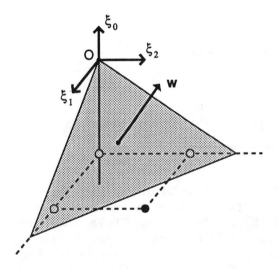

FIGURE 5.4 The AND problem with the threshold w_0 in an extra dimension ξ_0. Note that all the patterns have $\xi_0 = -1$ as in (5.2). The separating plane shown is perpendicular to the weight vector $\mathbf{w} = (1.5, 1, 1)$.

output only if both inputs are on. Table 5.1 lists the input-output pairs, or **truth table**, and Fig. 5.3(a) shows them in (ξ_1, ξ_2) space. It is easy to draw a line (a one-dimensional plane) to separate the "yes"-corner from the rest, so the problem is linearly separable and a simple perceptron can solve it. A suitable perceptron is shown in Fig. 5.3(b).

It is also interesting to consider the same problem with the threshold represented implicitly by weight w_0 to input $\xi_0 = -1$. Then the separating plane must go through the origin, but we have one extra dimension for ξ_0. Figure 5.4 shows the situation, with a separating plane drawn to correspond to the network of Fig. 5.3(b).

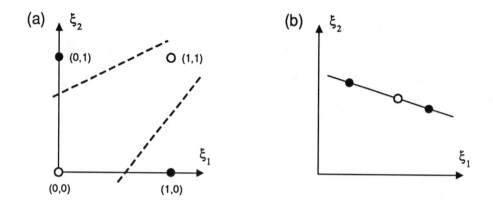

FIGURE 5.5 Problems that are *not* linearly separable. (a) The XOR problem. (b) Points that are not in "general position."

TABLE 5.2 XOR function

ξ_1	ξ_2	ζ
0	0	-1
0	1	$+1$
1	0	$+1$
1	1	-1

Now let us consider some examples that are *not* linearly separable. We return to representing thresholds explicitly, so separating planes need not be through the origin. Figure 5.5 shows two types of difficulties that can occur. In Fig. 5.5(a) there is clearly no plane (line) that can separate the two types of points. The same is true in (b), but only because the three points lie exactly in a straight line; if any of them were moved infinitesimally off the line a solution would be possible.

We discuss each case in a little more detail. The first, in Fig. 5.5(a), is the Boolean exclusive-OR or **XOR function**, with the truth table shown in table 5.2. The desired computation is a "yes" when one or the other of the inputs is on and a "no" when they are both off or both on. This is also the simplest case of the *N*-input **parity function** studied in detail by Minsky and Papert [1969].

The figure makes it clear that we cannot represent this function with a simple perceptron, but it is interesting to see the same thing algebraically. We just write out the equations (5.9) for the four patterns (suppressing the *i* index):

$$w_1 + w_2 \; < \; w_0 \tag{5.12}$$

$$-w_1 - w_2 \; < \; w_0 \tag{5.13}$$

$$w_1 - w_2 \; > \; w_0 \tag{5.14}$$

$$-w_1 + w_2 \; > \; w_0 \,. \tag{5.15}$$

Combining (5.12) and (5.15) we obtain $w_1 < 0$, while combining (5.13) and (5.14) we get the opposite, $w_1 > 0$. These cannot both be satisfied. Similarly, one finds impossible constraints on the other w_k's. Thus the network cannot do the computation.

Our other example, in Fig. 5.5(b), is very special. If, for example, we were choosing patterns from some continuous random distribution, there would be zero probability for three points to lie *exactly* on a line. Points are said to be **in general position** when this sort of special case does not occur. In an N-dimensional space a set of points is in general position if no more than $d + 1$ of them lie on any d-dimensional hyperplane, for any $d < N$. Without an explicit threshold the condition becomes that no more than d of them lie on any d-dimensional hyperplane *through the origin*, for any $d < N$; in two dimensions we'd need two points on a line through the origin to get into trouble. The no threshold condition is also equivalent to saying that all subsets of N or fewer points must be **linearly independent**.

We will show later that the first type of failure of linear separability, like XOR, can only occur when there are more patterns than inputs, $p > N$. On the other hand the second type can occur for any p.

A Simple Learning Algorithm

We now consider only linearly separable problems—so there is a solution—and ask how to find appropriate weights using a learning procedure. A simple procedure is to go through the input patterns one by one, and for each pattern go through the output units one by one, asking whether the output is the desired one ($O_i^\mu = \zeta_i^\mu$). If so, we leave the connections feeding into that unit alone. If not, then in the spirit of Hebb we add to each connection something proportional to the product of the input and the desired output. Specifically, we take

$$w_{ik}^{\text{new}} = w_{ik}^{\text{old}} + \Delta w_{ik} \tag{5.16}$$

where

$$\Delta w_{ik} = \begin{cases} 2\eta\zeta_i^\mu\xi_k^\mu & \text{if } \zeta_i^\mu \neq O_i^\mu \,; \\ 0 & \text{otherwise;} \end{cases} \tag{5.17}$$

or

$$\Delta w_{ik} = \eta(1 - \zeta_i^\mu O_i^\mu)\zeta_i^\mu\xi_k^\mu \tag{5.18}$$

or

$$\Delta w_{ik} = \eta(\zeta_i^\mu - O_i^\mu)\xi_k^\mu \,. \tag{5.19}$$

The parameter η is called the **learning rate**. In (5.19) one can think of the two terms as learning the desired association and "unlearning" the erroneous one.

Instead of just asking that the *sign* of the input h_i^μ to the output units is correct (i.e., equal to ζ_i^μ), it is sometimes a good idea also to require that its *size* be larger than some margin:

$$\zeta_i^\mu h_i^\mu \equiv \zeta_i^\mu \sum_k w_{ik}\xi_k^\mu > N\kappa \quad \text{(desired).} \quad (5.20)$$

The sum on k scales with N, so to keep the **margin size** κ fixed for any number of input units we include an N on the right-hand side. We can implement this new criterion by changing (5.18) to

$$\Delta w_{ik} = \eta\,\Theta(N\kappa - \zeta_i^\mu h_i^\mu)\zeta_i^\mu \xi_k^\mu \quad (5.21)$$

where Θ is the unit step function (1.2)—we add the weight increment whenever (5.20) is not satisfied. The simpler learning rule (5.18) is just the special case with $\kappa = 0$ (apart from a factor of 2 in η).

Equation (5.21) is called the **perceptron learning rule** [Rosenblatt, 1962]. It can be proved [Block, 1962; Minsky and Papert, 1969] to converge to weights which accomplish the desired association (5.20) in a finite number of steps (provided of course that the solution exists). We provide a proof in the next section.

For a single output unit introducing κ just changes (5.8) into

$$\mathbf{w}\cdot\mathbf{x}^\mu > N\kappa \quad \text{(desired).} \quad (5.22)$$

This says geometrically that all points in $\mathbf{x}$-space must be further than $N\kappa/|\mathbf{w}|$ from the plane perpendicular to $\mathbf{w}$. In Fig. 5.2, for example, we can see that pattern F would fail to satisfy the condition unless κ were very small or $|\mathbf{w}|$ were very large.

We can also give a geometrical picture of the learning process. In our single-output vector notation, (5.21) becomes

$$\Delta\mathbf{w} = \eta\,\Theta(N\kappa - \mathbf{w}\cdot\mathbf{x}^\mu)\mathbf{x}^\mu \quad (5.23)$$

which says that the weight vector $\mathbf{w}$ is changed a little in the direction of $\mathbf{x}^\mu$ if its projection $\mathbf{w}\cdot\mathbf{x}^\mu$ onto $\mathbf{x}^\mu$ is less than $N\kappa/|\mathbf{w}|$. This is done over and over again until all projections are large enough. An example for $\kappa = 0$ is sketched in Fig. 5.6. Observe that a final solution was found after three steps $\mathbf{w} \to \mathbf{w}' \to \mathbf{w}'' \to \mathbf{w}'''$; there are no patterns left in the bad region of $\mathbf{w}'''$ so no further updates will occur.

Whatever the value of κ, a *direction* for $\mathbf{w}$ in which all the projections are positive will give a solution if scaled up to a large enough magnitude $|\mathbf{w}|$. Depending on the pattern vectors $\mathbf{x}^\mu$, there may be a wide range of such directions, or only a narrow cone, or (if no solution exists) none at all (Fig. 5.7 shows two examples). We can use this observation to quantify how easy or hard a problem is. The quantity

$$D(\mathbf{w}) = \frac{1}{|\mathbf{w}|}\min_\mu \mathbf{w}\cdot\mathbf{x}^\mu \quad (5.24)$$

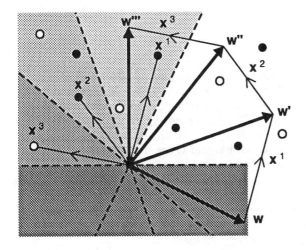

FIGURE 5.6 How the weight vector evolves during training for $\kappa = 0$, $\eta = 1$. Successive values of the weight vector are shown by $\mathbf{w}$, $\mathbf{w}'$, $\mathbf{w}''$, and $\mathbf{w}'''$. The darker and darker shading shows the "bad" region where $\mathbf{w} \cdot \mathbf{x} < 0$ for the successive $\mathbf{w}$ vectors. Each $\mathbf{w}$ is found from the previous one (e.g., $\mathbf{w}'$ from $\mathbf{w}$) by adding an $\mathbf{x}^\mu$ from the current bad region.

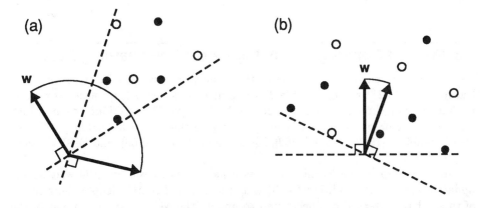

FIGURE 5.7 (a) An easy problem: any weight vector in the 135° angle shown will have positive pattern projections. (b) A harder problem where the weight vector must lie in a narrow cone.

depends on the *worst* of the projections. It is just the distance of the worst $\mathbf{x}^\mu$ to the plane perpendicular to $\mathbf{w}$, positive for the "good" side and negative for the "bad" side (see Fig. 5.8). The $1/|\mathbf{w}|$ factor makes it a function only of the direction of $\mathbf{w}$. If $D(\mathbf{w})$ is positive then all pattern points lie on the good side, and so a solution can be found for large enough $|\mathbf{w}|$.

If we maximize $D(\mathbf{w})$ over all possible weights we obtain the best direction for $\mathbf{w}$, along which a solution will be found for smallest $|\mathbf{w}|$. This solution is called the **optimal perceptron**. It can also be defined, equivalently, as the solution with largest margin size κ for fixed $|\mathbf{w}|$. The value

$$D_{\max} \equiv \max_{\mathbf{w}} D(\mathbf{w}) \tag{5.25}$$

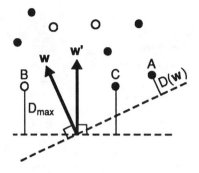

FIGURE 5.8 Definitions of $D(\mathbf{w})$ and D_{max}. Pattern A is nearest to the plane perpendicular to weight vector $\mathbf{w}$, so the distance to A gives $D(\mathbf{w})$. Maximizing $D(\mathbf{w})$ with respect to $\mathbf{w}$ gives $\mathbf{w}'$, with $D(\mathbf{w}') = D_{\mathrm{max}}$. Note that both B and C are distance D_{max} from the plane.

of $D(\mathbf{w})$ in this direction tells us how easy the problem is: the larger D_{max}, the easier the problem. If $D_{\mathrm{max}} < 0$, it cannot be solved. For example, D_{max} is $1/\sqrt{17}$ for the AND problem (from Fig. 5.4) and $-1/\sqrt{3}$ for XOR.

5.3 Proof of Convergence of the Perceptron Learning Rule ⋆

We assume that there *is* a solution to the problem and prove that the perceptron learning rule (5.21) reaches it in a *finite* number of steps. All we need to assume is that we can choose a weight vector $\mathbf{w}^*$ in a "good" direction; one in which $D(\mathbf{w}^*) > 0$. Our proof is related most closely to that given by Arbib [1987]; see also Rosenblatt [1962], Block [1962], Minsky and Papert [1969], and Diederich and Opper [1987].

At each step in the learning process a pattern is chosen and the weights are updated only if the condition (5.20) is *not* satisfied. Let M^μ denote the number of times that pattern μ has been used to update the weights at some point in the learning process. Then at that time

$$\mathbf{w} = \eta \sum_\mu M^\mu \mathbf{x}^\mu \tag{5.26}$$

if we assume that the initial weights are all zero.

The essence of the proof is to compute bounds on $|\mathbf{w}|$ and on the overlap $\mathbf{w}\cdot\mathbf{w}^*$ with our chosen "good" vector $\mathbf{w}^*$. These let us show that $\mathbf{w}\cdot\mathbf{w}^*/|\mathbf{w}|$ would get arbitrarily large if the total number of updates $M = \sum M^\mu$ kept on increasing. But this is impossible (since $\mathbf{w}^*$ is fixed), so the updating must cease at some finite M.

Consider $\mathbf{w}\cdot\mathbf{w}^*$ first. Using (5.26) and (5.24) we obtain

$$\mathbf{w}\cdot\mathbf{w}^* = \eta \sum_\mu M^\mu \mathbf{x}^\mu \cdot \mathbf{w}^* \geq \eta M \min_\mu \mathbf{x}^\mu \cdot \mathbf{w}^* = \eta M D(\mathbf{w}^*)|\mathbf{w}^*|. \tag{5.27}$$

Thus $\mathbf{w}\cdot\mathbf{w}^*$ grows like M. Now for an upper bound on $|\mathbf{w}|$, consider the change in length of $\mathbf{w}$ at a single update by pattern α:

$$
\begin{aligned}
\Delta|\mathbf{w}|^2 &= \left(\mathbf{w}+\eta\mathbf{x}^\alpha\right)^2 - \mathbf{w}^2 \\
&= \eta^2\left(\mathbf{x}^\alpha\right)^2 + 2\eta\mathbf{w}\cdot\mathbf{x}^\alpha \\
&\leq \eta^2 N + 2\eta N\kappa \\
&= N\eta(\eta+2\kappa).
\end{aligned}
\tag{5.28}
$$

The inequality comes directly from the condition $N\kappa \geq \mathbf{w}\cdot\mathbf{x}^\alpha$ for performing an update with pattern α. Note that we also used $x_k^\alpha = \pm 1$, so that $(\mathbf{x}^\alpha)^2 = N$, but the proof is easily generalized to other types of patterns. By summing the increments to $|\mathbf{w}|^2$ for M steps we obtain the desired bound

$$
|\mathbf{w}|^2 \leq MN\eta(\eta+2\kappa).
\tag{5.29}
$$

Thus $|\mathbf{w}|$ grows no faster than $\sqrt{M}$, and therefore from (5.27) the ratio $\mathbf{w}\cdot\mathbf{w}^*/|\mathbf{w}|$ grows at least as fast as $\sqrt{M}$. But this cannot continue, so M must stop growing.

More precisely, we can bound the normalized scalar product

$$
\phi = \frac{(\mathbf{w}\cdot\mathbf{w}^*)^2}{|\mathbf{w}|^2|\mathbf{w}^*|^2}
\tag{5.30}
$$

which is the squared cosine of the angle between $\mathbf{w}$ and $\mathbf{w}^*$. Because it is the squared cosine it is obviously less than or equal to 1 (as also follows from the Cauchy-Schwarz inequality). But with (5.27) and (5.29) we find

$$
1 \geq \phi \geq M\frac{D(\mathbf{w}^*)^2\eta}{N(\eta+2\kappa)}
\tag{5.31}
$$

which gives us an upper bound on the number of weight updates (using the best possible $\mathbf{w}^*$):

$$
M \leq N\frac{1+2\kappa/\eta}{D_{\max}^2}.
\tag{5.32}
$$

This bound is proportional to the number of input units, but interestingly enough it does *not* depend on the number of patterns p. Of course the real convergence time does depend on p because one typically has to continue checking all p patterns to find the ones for which a weight update is needed; the number of such checks increases with p even if the number of actual updates does not. Additionally, $D_{\max}$ typically decreases with increasing p, resulting in a growing M. Note also that the bound on M grows linearly with κ, because for larger κ the learning must reach a larger $|\mathbf{w}|$ along any given good direction.

5.4 Linear Units

So far in our study of simple perceptrons we have considered only threshold units, with $g(h) = \text{sgn}(h)$. We turn now to continuous-valued units, with $g(h)$ a continuous and *differentiable* function of u. The great advantage of such units is that they allow us to construct a cost function $E[\mathbf{w}]$ which measures the system's performance error as a differentiable function of the weights $\mathbf{w} = \{w_{ik}\}$. We can then use an optimization technique, such as gradient descent, to minimize this error measure.

We start in this section with **linear units**, for which $g(h) = h$. These are not as useful practically as the nonlinear networks considered next, but are simpler and allow more detailed analysis. The output of a linear simple perceptron subjected to an input pattern ξ_k^μ is given by

$$O_i^\mu = \sum_k w_{ik}\xi_k^\mu \tag{5.33}$$

and the desired association is $O_i^\mu = \zeta_i^\mu$ as usual, or

$$\zeta_i^\mu = \sum_k w_{ik}\xi_k^\mu \quad \text{(desired)}. \tag{5.34}$$

Note that the O_i^μ's are continuous-valued quantities now, although we could still restrict the desired values ζ_i^μ to ± 1.

Explicit Solution

For a linear network we can actually compute a suitable set of weights explicitly using the **pseudo-inverse** method. We saw on page 50 how to find the weights to satisfy (5.34) for the special case of auto-association, $\zeta_i^\mu = \xi_i^\mu$. The generalization to hetero-association is just

$$w_{ik} = \frac{1}{N}\sum_{\mu\nu} \zeta_i^\mu (\mathbf{Q}^{-1})_{\mu\nu}\xi_k^\nu \tag{5.35}$$

where

$$\mathbf{Q}_{\mu\nu} = \frac{1}{N}\sum_k \xi_k^\mu \xi_k^\nu \tag{5.36}$$

is the overlap matrix of the *input* patterns. It is straightforward to check that (5.35) solves (5.34), as in (3.19).

Observe that (5.35) only applies if $\mathbf{Q}^{-1}$ exists, and that this condition only depends on the *input* patterns. It requires the input patterns to be **linearly independent**. If on the contrary there exists a linear relationship

$$a_1\xi_k^1 + a_2\xi_k^2 + \cdots + a_p\xi_k^p = 0 \quad \text{(for all } k) \tag{5.37}$$

between them, then the outputs O_i^μ cannot be independently chosen and the problem is normally insoluble.

Linear independence is the general condition for solvability in the linear network. It is actually a sufficient, but not a necessary, condition; even with linearly dependent input patterns it *might* happen that the targets were such that a solution could be found, though not by the present method. But a very special choice would be needed.

A set of p input patterns can only be linearly independent if $p \leq N$, so we can store at most N arbitrary associations in a linear network. But a set of N or fewer patterns is not necessarily linearly independent. If a dependency exists, then the input pattern vectors ξ_k^1, ξ_k^2, ..., ξ_k^p only span a **pattern subspace** of the input space, and the solution (5.35) is not unique. If w_{ik} is one solution, and ξ_k^* is any vector orthogonal to the subspace of the input patterns, so $\sum_k \xi_k^\mu \xi_k^* = 0$ for all μ, then the weights w'_{ik} given by

$$w'_{ik} = w_{ik} + a_i \xi_k^* \tag{5.38}$$

for any a_i provide another solution.

The linear independence condition for linear (and nonlinear) units is quite distinct from the linear separability condition found for threshold units. Linear independence does imply linear separability, but the reverse is not true. In fact most of the problems of interest in threshold networks do *not* satisfy the linear independence condition, because they typically have $p > N$. This includes AND and XOR, and indeed all the other threshold network examples used or illustrated in this chapter.

Gradient Descent Learning

We could use (5.35) and (5.36) to compute a set of weights w_{ik} that produce exactly the desired outputs from each input pattern. But we are more interested here in finding a *learning* rule that allows us to find such a set of weights by successive improvement from an arbitrary starting point.

We define an error measure or **cost function** by

$$E[\mathbf{w}] = \frac{1}{2} \sum_{i\mu} (\zeta_i^\mu - O_i^\mu)^2 = \frac{1}{2} \sum_{i\mu} \left(\zeta_i^\mu - \sum_k w_{ik} \xi_k^\mu \right)^2. \tag{5.39}$$

This is smaller the better our w_{ik}'s are; E is normally positive, but goes to zero as we approach a solution satisfying (5.34). Note that this cost function depends only on the weights w_{ik} and the problem patterns. In contrast, the energy functions considered in Chapters 2 to 4 depended on the current state of the network, which evolved so as to minimize the energy. Here the evolution is of the weights (learning), not of the activations of the units themselves.

Given our error measure $E[\mathbf{w}]$, we can improve on a set of w_{ik}'s by sliding downhill on the surface it defines in $\mathbf{w}$ space. Specifically, the usual **gradient descent algorithm** suggests changing each w_{ik} by an amount Δw_{ik} proportional to the gradient of E at the present location:

$$\Delta w_{ik} = -\eta \frac{\partial E}{\partial w_{ik}}$$
$$= \eta \sum_{\mu} (\zeta_i^{\mu} - O_i^{\mu}) \xi_k^{\mu} . \qquad (5.40)$$

If we make these changes individually for each input pattern ξ_k^{μ} in turn, we have simply

$$\Delta w_{ik} = \eta (\zeta_i^{\mu} - O_i^{\mu}) \xi_k^{\mu} \qquad (5.41)$$

for the changes in response to pattern μ, or

$$\Delta w_{ik} = \eta \delta_i^{\mu} \xi_k^{\mu} \qquad (5.42)$$

if we define the errors (or **deltas**) δ_i^{μ} by

$$\delta_i^{\mu} = \zeta_i^{\mu} - O_i^{\mu} . \qquad (5.43)$$

This result—(5.41), or (5.42) plus (5.43)—is commonly referred to as the **delta rule**, or the **adaline rule**, or the **Widrow-Hoff rule**, or the **LMS** (least mean square) **rule** [Rumelhart, McClelland, et al., 1986; Widrow and Hoff, 1960]. It is also nearly identical to the Rescorla-Wagner model of classical conditioning in behavioral psychology [Rescorla and Wagner, 1972].

Equation (5.41) is identical to our simple rule (5.19) for the threshold unit network. Note however that (5.19) originated in an empirical Hebb assumption, but now we have derived it from gradient descent. As we will see, the new approach is easily generalized to more layers, whereas the Hebb rule is not. The new approach obviously requires continuous-valued units with differentiable activation functions.

Actually the equivalence of (5.41) and (5.19) is a little deceptive because O_i^{μ} is a different function of the inputs. In the threshold case the term $\zeta_i^{\mu} - O_i^{\mu}$ is either 0 or ± 2, whereas here it can take any value. As a result the final weight values are *not* the same in the two cases. Nor are the conditions for existence of a solution (linear separability versus linear independence). And in the threshold case the learning rule stops after a finite number of steps, while here it continues in principle for ever, converging only asymptotically to the solution.

The cost function (5.39) is just a quadratic form in the weights. In the subspace spanned by the patterns the surface is a parabolic bowl with a single minimum. Assuming that the pattern vectors are linearly independent, so that there *is* a solution to (5.34), the minimum is at $E = 0$. In the directions (if any) of $\mathbf{w}$-space orthogonal to all the pattern vectors the error is constant, as may be seen by inserting (5.38) into (5.39); the ξ_k^* term makes no difference because $\sum_k \xi_k^{\mu} \xi_k^* = 0$.

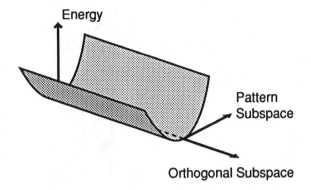

FIGURE 5.9 The "rain gutter" shape of the error surface for linear units. The error takes its minimum value of 0 along level valleys if the patterns ξ_k^μ do not span the whole of ξ-space.

In other words, the error surface in weight space is like a **rain gutter**, as pictured in Fig. 5.9, with infinite level valleys in the directions orthogonal to the pattern vectors.

The gradient descent rule (5.40) or (5.41) produces changes in the weight vectors $\mathbf{w}_i = (w_{i1}, \ldots, w_{iN})$ *only* in the directions of the pattern vectors ξ^μ. Thus any component of the weights orthogonal to the patterns is left unchanged by the learning, leaving an inconsequential uncertainty of exactly the form (5.38) in the final solution. Within the pattern subspace the gradient descent rule necessarily decreases the error if η is small enough, because it takes us in the downhill gradient direction. Thus with enough iterations we approach the bottom of the valley arbitrarily closely, from any starting point. And any point at the bottom of the valley solves the original problem (5.34) exactly.

Convergence of Gradient Descent ⋆

The argument just given for convergence to the bottom of the valley is intuitively reasonable but bears a little further analysis. The first step is to diagonalize the quadratic form (5.39). As long as the pattern vectors are linearly independent, this allows us to write $E[\mathbf{w}]$ in the form

$$E = \sum_{\lambda=1}^{M} a_\lambda \left(w_\lambda - w_\lambda^0\right)^2. \tag{5.44}$$

Here M is the total number of weights, equal to N times the number of output units, and the w_λ's are linear combinations of the w_{ik}'s. The a_λ's and w_λ^0's are constants depending only on the pattern vectors. The eigenvalues a_λ are necessarily positive or zero because of the sum-of-squares form of (5.39); the quadratic form is positive semi-definite.

Notice first that if some of the a_λ's are zero then E is independent of the corresponding w_λ's. This is equivalent to the rain gutters already described; the

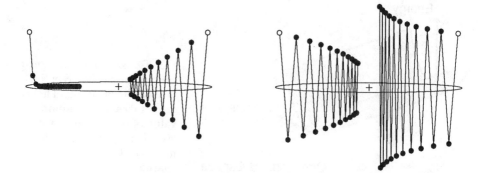

FIGURE 5.10 Gradient descent on a simple quadratic surface (the left and right parts are copies of the same surface). Four trajectories are shown, each for 20 steps from the open circle. The minimum is at the + and the ellipse shows a constant error contour. The only significant difference between the trajectories is the value of η, which was 0.02, 0.0476, 0.049, and 0.0505 from left to right.

corresponding eigenvector directions in **w** space are orthogonal to all the pattern vectors.

Now let us perform gradient descent on (5.44). Because the transformation from w_{ik} to w_λ is linear, this is entirely equivalent to gradient descent in the original basis. But the diagonal basis makes it much easier:

$$\Delta w_\lambda = -\eta \frac{\partial E}{\partial w_\lambda} = -2\eta a_\lambda \left(w_\lambda - w_\lambda^0 \right). \qquad (5.45)$$

Thus the distance $\delta w_\lambda = w_\lambda - w_\lambda^0$ from the optimum in the λ-direction is transformed according to

$$\delta w_\lambda^{\text{new}} = \delta w_\lambda^{\text{old}} + \Delta w_\lambda = (1 - 2\eta a_\lambda)\delta w_\lambda^{\text{old}}. \qquad (5.46)$$

In the directions for which $a_\lambda > 0$ we therefore get closer to the optimum as long as $|1 - 2\eta a_\lambda| < 1$. The approach is first order; each distance δw_λ gets multiplied by a fixed factor at each iteration. The value of η is limited by the largest eigenvalue a_λ^{max}, corresponding to the steepest curvature direction of the error surface; we must have $\eta < 1/a_\lambda^{\text{max}}$ or we will end up jumping too far, further up the other side of the valley than we are at present. But the rate of approach to the optimum is usually limited by the *smallest* non-zero eigenvalue a_λ^{min}, corresponding to the shallowest curvature direction. If $a_\lambda^{\text{max}}/a_\lambda^{\text{min}}$ is large progress along the shallow directions can be excruciatingly slow.

Figure 5.10 illustrates these points in a simple case. We show gradient descent on the surface $E = x^2 + 20y^2$ for 20 iterations at different values of η. This quadratic form is already diagonal, with $a_1 = 1$ and $a_2 = 20$. Notice the distance of the last point from the minimum. At $\eta = 0.02$ we reach $y \approx 0$ fairly quickly, but then make only slow progress in x. At the other extreme, if $\eta > 1/20 = 0.05$ the algorithm produces a divergent oscillation in y. The fastest approach is approximately when the x and y multipliers $|1 - 2\eta|$ and $|1 - 40\eta|$ are equal, giving $\eta = 1/21 = 0.476$ (second illustration).

Of course this analysis assumes that we actually take steps along the gradient. Progress will normally be somewhat slower if instead we change one component at a time, as in (5.41), though the incremental algorithm is often more convenient in practice. Another alternative sometimes used is to take steps of constant *size* (usually decreasing with time) in the gradient *direction*; this can help to speed up convergence when $a_\lambda^{\max}/a_\lambda^{\min}$ is large, but requires careful control.

We have also assumed of course that a perfect solution exists, which requires linear independence of the patterns. It is however interesting to ask what happens if the patterns are *not* linearly independent. Then we can only diagonalize (5.39) in the form

$$E = E_0 + \sum_{\lambda=1}^{M} a_\lambda \left(w_\lambda - w_\lambda^0\right)^2 \tag{5.47}$$

with $E_0 > 0$. There is still a single minimum (or gutter), but $E = E_0 > 0$ there, showing that the desired association has not been found. Trying gradient descent on the XOR problem, for example, produces a single minimum with $E = 2$ at $w_0 = w_1 = w_2 = 0$, which makes the output always 0. The XOR problem is obviously not linearly independent, since $p > N$.

5.5 Nonlinear Units

It is straightforward to generalize gradient descent learning from the linear $g(h) = h$ discussed in the previous section to networks with any differentiable $g(h)$. The sum-of-squares cost function becomes

$$E[\mathbf{w}] = \frac{1}{2} \sum_{i\mu} (\zeta_i^\mu - O_i^\mu)^2 = \frac{1}{2} \sum_{i\mu} \left[\zeta_i^\mu - g\left(\sum_k w_{ik}\xi_k^\mu\right)\right]^2. \tag{5.48}$$

We then find

$$\frac{\partial E}{\partial w_{ik}} = -\sum_\mu [\zeta_i^\mu - g(h_i^\mu)]g'(h_i^\mu)\xi_k^\mu \tag{5.49}$$

so the gradient descent correction $-\eta \partial E/\partial w_{ik}$ to w_{ik} after presentation with pattern number μ is of the same form as (5.42):

$$\Delta w_{ik} = \eta \delta_i^\mu \xi_k^\mu. \tag{5.50}$$

But now the quantity

$$\delta_i^\mu = [\zeta_i^\mu - O_i^\mu]g'(h_i^\mu) \tag{5.51}$$

has acquired an extra factor $g'(h_i^\mu)$ of the derivative of the activation function $g(h)$. With a sigmoid form for $g(h)$, such as $g(h) = \tanh(h)$, this derivative is largest when $|h_i^\mu|$ is small. Thus the changes are made most strongly on connections feeding into units with small $|h_i^\mu|$'s, those which are "in doubt" about their output.

We remark that a hyperbolic tangent function $g(h) = \tanh(\beta h)$ is particularly convenient because its derivative is given by $g'(h) = \beta(1-g^2)$. Thus one doesn't have to recompute the derivative of g in (5.51) once one has found $O_i^\mu = g(h_i^\mu)$ itself. The same applies of course to the corresponding choice of the sigmoid function $g(h) = f_\beta(h) = [1 + \exp(-2\beta h)]^{-1}$ for units with outputs between 0 and 1, for which $g'(h) = 2\beta g(1 - g)$.

The conditions for the existence of a solution are exactly the same as for the linear case: linear independence of the patterns. This is because the solution to our present problem is equivalent to the linear one with the targets ζ_i^μ replaced by $g^{-1}(\zeta_i^\mu)$.[5] But the question of whether gradient descent *finds* the solution, assuming it exists, is not the same as for the linear case. If the targets lie outside the range of $g(h)$ (e.g., ± 1 targets with $g(h) = \tanh h$), it is possible in the nonlinear case for the cost function to have **local minima** besides the global minimum at $E = 0$. The gradient descent algorithm can then become stuck in such a local minimum.

Nonlinear units do not allow a perfect solution of problems that are not linearly independent, but they may help by offering alternate **partial solutions**. If, for example, we try the XOR problem with a $\tanh(\beta h)$ activation function, we find five possible minima, each with $E = 2$. There is the one at $w_0 = w_1 = w_2 = 0$ (which makes the output always 0) that the linear network found. But now there are four more at $|w_0| = |w_1| = |w_2| \to \infty$ with 1 or 3 negative signs; these each produce the correct output for three out of the four input patterns, but get the wrong sign on the fourth. It is arguable that three out of four is better than four "don't know's". However the convergence is very slow for large β because most of the landscape is rather flat.

In simple perceptrons the main advantage of nonlinear activation functions is that they can keep the output between fixed bounds, such as ± 1 for a $\tanh(h)$ function. They are much more important in multi-layer networks where they make possible the solution of problems that are *not* possible with linear units. A multi-layer linear feed-forward network is exactly equivalent to a one-layer one in the computation it performs (since a linear transformation of a linear transformation is a linear transformation), so such a network has the same limitations as a one-layer one. In particular it can only work if the input patterns are linearly independent. But this restriction is *not* present for a multi-layer *nonlinear* feed-forward network.

Other Cost Functions

The quadratic cost function (5.48) is not the only possibility. Any differentiable

[5] We normally consider only monotonic activation functions, which are always invertible except at the endpoints.

function of ζ_i^μ and O_i^μ that is minimized by $O_i^\mu = \zeta_i^\mu$ could be used. The gradient descent rule $\Delta w_{ik} \propto -\partial E/\partial w_{ik}$ gives a prescription analogous to (5.50) for each such choice.

The choice

$$E = \sum_{i\mu} \left[\tfrac{1}{2}(1 + \zeta_i^\mu) \log \frac{1 + \zeta_i^\mu}{1 + O_i^\mu} + \tfrac{1}{2}(1 - \zeta_i^\mu) \log \frac{1 - \zeta_i^\mu}{1 - O_i^\mu} \right] \qquad (5.52)$$

has received particular attention [Baum and Wilczek, 1988; Hopfield, 1987; Solla et al., 1988]. It has a natural interpretation in terms of learning the correct probabilities of a set of hypotheses represented by the output units, using $\tfrac{1}{2}(1 + O_i^\mu)$ for the probability that the hypothesis represented by unit i is true: $O_i^\mu = -1$ means definitely false, and $O_i^\mu = +1$ means definitely true. Similarly $\tfrac{1}{2}(1 + \zeta_i^\mu)$ is interpreted as a target set of probabilities. Then information theory suggests the **relative entropy** (5.52) of these probability distributions as a natural measure of the difference between them [Kullback, 1959].

Like (5.48), (5.52) is always positive except when $O_i^\mu = \zeta_i^\mu$ for all i and μ, where $E = 0$. Its advantage is, qualitatively, that it diverges if the output of one unit saturates at the wrong extreme. The quadratic measure (5.48) just approaches a constant in that case, and therefore the learning can float around on a relatively flat plateau of E for a long time. The use of the entropy measure has been shown [Wittner and Denker, 1988] to solve some learning problems that cannot be solved using the quadratic E. It is also appropriate if the training set data are actually probabilistic or fuzzy as, for instance, in the association of symptoms with causes in medical diagnosis.

Differentiating (5.52) and taking g to be a tanh gives the same change of weights as (5.50) but with

$$\delta_i^\mu = \beta(\zeta_i^\mu - O_i^\mu). \qquad (5.53)$$

The only difference (besides a factor of β) from (5.51) is that the $g'(h_i^\mu)$ factor is missing. The result is essentially identical to the delta rules found in the linear network (5.43) and in the threshold network (5.19).

Gradient descent learning is sometimes used for **binary decision problems** where the network is trained with, say, ± 1 targets, but then used with any positive output $O_i > 0$ being taken as a "yes" and any negative output $O_i < 0$ as a "no". This can sometimes produce satisfactory results on problems which are linearly separable but not linearly independent. However gradient descent with the usual quadratic cost function does not necessarily find a viable solution in such cases, even though the perceptron learning rule *would* work. But Wittner and Denker [1988] have shown that a class of alternative **well-formed** cost functions do work in such situations, so that gradient descent always finds a solution if there is one. They define a well-formed cost function so that the magnitude of its gradient is always larger than a constant, not simply greater than zero, whenever an output has the wrong sign. The entropic measure (5.52) is well formed in this sense.

5.6 Stochastic Units

Another generalization is from our deterministic units to **stochastic units** S_i governed by (2.48):

$$\text{Prob}(S_i^\mu = \pm 1) = \frac{1}{1 + \exp(\mp 2\beta h_i^\mu)} \tag{5.54}$$

with

$$h_i^\mu = \sum_k w_{ik} \xi_k^\mu \tag{5.55}$$

as before. This leads to

$$\langle S_i^\mu \rangle = \tanh\left(\beta \sum_k w_{ik} \xi_k^\mu\right) \tag{5.56}$$

just as in (2.42). In the context of a simulation we can use (5.56) to *calculate* $\langle S_i^\mu \rangle$, whereas in a real stochastic network we would find it by averaging S_i for a while, updating randomly chosen units according to (5.54). Either way, we then use $\langle S_i^\mu \rangle$ as the basis of a weight change

$$\Delta w_{ik} = \eta \delta_i^\mu \xi_k^\mu \tag{5.57}$$

where

$$\delta_i^\mu = \zeta_i^\mu - \langle S_i^\mu \rangle . \tag{5.58}$$

This is just the average over outcomes of the changes we would have made on the basis of individual outcomes using the ordinary delta rule (5.43). We will find it particularly important when we discuss reinforcement learning in Section 7.4.

It is interesting to prove that this rule always decreases the average error given by the usual quadratic measure

$$E = \frac{1}{2} \sum_{i\mu} (\zeta_i^\mu - S_i^\mu)^2 . \tag{5.59}$$

Since we are assuming output units and patterns which are ± 1, this is just twice the total number of bits in error, and can also be written

$$E = \sum_{i\mu} (1 - \zeta_i^\mu S_i^\mu) . \tag{5.60}$$

Thus the *average* error in the stochastic network is

$$\langle E \rangle = \sum_{i\mu} (1 - \zeta_i^\mu \langle S_i^\mu \rangle)$$

$$= \sum_{i\mu} \left[1 - \zeta_i^\mu \tanh\left(\beta \sum_k w_{ik} \xi_k^\mu\right) \right] . \tag{5.61}$$

The change in $\langle E \rangle$ in one cycle of weight updatings is thus

$$\Delta \langle E \rangle = \sum_{ik} \frac{\partial \langle E \rangle}{\partial w_{ik}} \Delta w_{ik}$$

$$= -\sum_{i\mu k} \Delta w_{ik} \zeta_i^\mu \frac{\partial}{\partial w_{ik}} \tanh(\beta h_i^\mu)$$

$$= -\sum_{i\mu k} \eta[1 - \zeta_i^\mu \tanh(\beta h_i^\mu)]\beta \mathrm{sech}^2(\beta h_i^\mu) \qquad (5.62)$$

using[6] $d\tanh(x)/dx = \mathrm{sech}^2 x$. The result (5.62) is clearly always negative (recall $\tanh(x) < 1$), so the procedure always improves the average performance.

5.7 Capacity of the Simple Perceptron ⋆

In the case of the associative network in Chapter 2 we were able to find the **capacity** $p_{\max}$ of a network of N units; for random patterns we found $p_{\max} = 0.138N$ for large N if we used the standard Hebb rule. If we tried to store p patterns with $p > p_{\max}$ the performance became terrible.

Similar questions can be asked for simple perceptrons:

- How many *random* input-output pairs can we expect to store reliably in a network of given size?

- How many of these can we expect to *learn* using a particular learning rule?

The answer to the second question may well be smaller than the first (e.g., for nonlinear units), but is presently unknown in general. The first question, which this section deals with, gives the maximum capacity that *any* learning algorithm can hope to achieve.

For continuous-valued units (linear or nonlinear) we already know the answer, because the condition is simply linear independence. If we choose p *random* patterns, then they will be linearly independent if $p \leq N$ (except for cases with very small probability). So the capacity is $p_{\max} = N$.

The case of threshold units depends on linear separability, which is harder to deal with. The answer for random continuous-valued inputs was derived by Cover [1965] (see also Mitchison and Durbin [1989]) and is remarkably simple:

$$p_{\max} = 2N. \qquad (5.63)$$

As usual N is the number of *input* units, and is presumed large. The number of *output* units must be small and fixed (independent of N). Equation (5.63) is strictly true in the $N \to \infty$ limit.

---

[6]The function $\mathrm{sech}^2 x = 1 - \tanh^2 x$ is a bell-shaped curve with peak at $x = 0$.

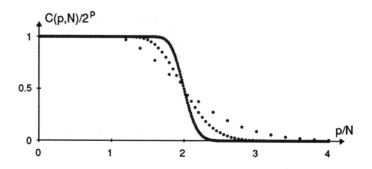

FIGURE 5.11 The function $C(p, N)/2^p$ given by (5.67) plotted versus p/N for $N = 5, 20,$ and $100.$

The rest of this section is concerned with proving (5.63), and may be omitted on first reading. We follow the approach of Cover [1965]. A more general (but much more difficult) method for answering this sort of question was given by Gardner [1987] and is discussed in Chapter 10.

We consider a perceptron with N continuous-valued inputs and one ± 1 output unit, using the deterministic threshold limit. The extension to several output units is trivial since output units and their connections are independent—the result (5.63) applies separately to each. For convenience we take the thresholds to be zero, but they could be reinserted at the expense of one extra input unit, as in (5.2).

In (5.11) we showed that the perceptron divides the N-dimensional input space into two regions separated by an $(N - 1)$-dimensional hyperplane. For the case of zero threshold this plane goes through the origin. All the points on one side give an output of $+1$ and all those on the other side give -1. Let us think of these as red $(+1)$ and black (-1) points respectively. Then the question we need to answer is: how many points can we expect to put randomly in an N-dimensional space, some red and some black, and then find a hyperplane through the origin that divides the red points from the black points?

Let us consider a slightly different question. For a given set of p randomly placed points in an N-dimensional space, for how many out of the 2^p possible red and black colorings of the points can we find a hyperplane dividing red from black? Call the answer $C(p, N)$. For p small we expect $C(p, N) = 2^p$, because we should be able to find a suitable hyperplane for *any* possible coloring; consider $N = p = 2$ for example. For p large we expect $C(p, N)$ to drop well below 2^p, so an arbitrarily chosen coloring will *not* possess a dividing hyperplane. The transition between these regimes turns out to be sharp for large N, and gives us $p_{\max}$.

We will calculate $C(p, N)$ shortly, but let us first examine the result. Figure 5.11 shows a graph of $C(p, N)/2^p$ against p/N for $N = 5, 20,$ and 100. Our expectations for small and large p are fulfilled, and we see that the transition occurs quite rapidly in the neighborhood of $p = 2N$, in agreement with (5.63). As N is made larger and

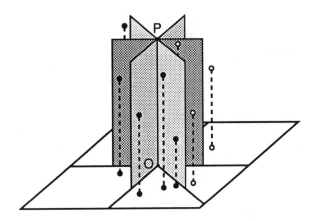

FIGURE 5.12 Finding separating hyperplanes constrained to go through a point P as well as the origin O is equivalent to projecting onto one lower dimension.

larger the transition becomes more and more sharp. Thus (5.63) is justified if we can demonstrate that Fig. 5.11 is correct.

The random placement of points is not actually necessary.[7] All that we need is that the points be in **general position**. As discussed on page 97, this means (for the no threshold case) that all subsets of N (or fewer) points must be linearly independent. As an example consider $N = 2$: a set of p points in a two-dimensional plane is in general position if no two lie on the same line through the origin. A set of points chosen from a continuous random distribution will obviously be in general position except for coincidences that have zero probability.

We can now calculate $C(p, N)$ by induction. Let us call a coloring that *can* be divided by a hyperplane a **dichotomy**. Suppose we start with p points and add a new point P. Then the $C(p, N)$ old dichotomies fall into two classes:

- For those previous dichotomies where the dividing hyperplane could have been drawn *through* point P, there'll be *two* new dichotomies, one with P red and one with it black. This is because when the points are in general position any hyperplane through P can be shifted infinitesimally to go either side of it, without changing the side of any of the other p points.

- For the remainder of the previous dichotomies only one color of point P will fit, so there'll be *one* new dichotomy for each old one.

Thus

$$C(p + 1, N) = C(p, N) + D \qquad (5.64)$$

where D is the number of the previous $C(p, N)$ dichotomies that could have had the dividing hyperplane drawn through P as well as the origin O. But this number is simply $C(p, N - 1)$, because constraining the hyperplanes to go through a particular point P makes the problem effectively $(N - 1)$-dimensional; as illustrated in Fig. 5.12, we can project the whole problem onto an $(N - 1)$-dimensional plane

---

[7]Nor is it well defined unless a distribution function is specified.

perpendicular to OP, since any displacement of a point along the OP direction cannot affect which side it is of any hyperplane containing OP.

We thereby obtain the **recursion relation**

$$C(p+1, N) = C(p, N) + C(p, N-1). \tag{5.65}$$

Iterating this equation for $p, p-1, p-2, \ldots, 1$ yields

$$C(p, N) = \binom{p-1}{0}C(1, N) + \binom{p-1}{1}C(1, N-1) + \ldots + \binom{p-1}{p-1}C(1, N-p+1). \tag{5.66}$$

For $p \leq N$ this is easy to handle, because $C(1, N) = 2$ for all N; one point can be colored red or black. For $p > N$ the second argument of C becomes 0 or negative in some terms, but these terms can be eliminated by taking $C(p, N) = 0$ for $N \leq 0$. It is easy to check that this choice is consistent with the recursion relation (5.65), and with $C(p, 1) = 2$ (in one dimension the only "hyperplane" is a point at the origin, allowing two dichotomies). Thus (5.66) makes sense for all values of p and N and can be written as

$$C(p, N) = 2 \sum_{i=0}^{N-1} \binom{p-1}{i} \tag{5.67}$$

if we use the standard convention that $\binom{n}{m} = 0$ for $m > n$. Equation (5.67) was used to plot Fig. 5.11, thus completing the demonstration.

It is actually easy to show from the symmetry $\binom{2n}{n-m} = \binom{2n}{n+m}$ of binomial coefficients that

$$C(2N, N) = 2^{p-1} \tag{5.68}$$

so the curve goes through $1/2$ at $p = 2N$. To show analytically that the transition sharpens up for increasing N, one can appeal to the large N Gaussian limit of the binomial coefficients, which leads to

$$C(p, N)/2^p \approx \frac{1}{2}\left[1 + \operatorname{erf}\left(\sqrt{\frac{p}{2}}\left(\frac{2N}{p} - 1\right)\right)\right] \tag{5.69}$$

for large N.

It is worth noting that $C(p, N) = 2^p$ if $p \leq N$ (this is shown on page 155). So *any* coloring of up to N points is linearly separable, provided only that the points are in general position. For N or fewer points general position is equivalent to linear independence, so the sufficient conditions for a solution are exactly the same in the threshold and continuous-valued networks. But this is not true, of course, for $p > N$.

SIX

Multi-Layer Networks

The limitations of a simple perceptron do not apply to feed-forward networks with intermediate or "hidden" layers between the input and output layer. In fact, as we will see later, a network with just one hidden layer can represent any Boolean function (including for example XOR). Although the greater power of multi-layer networks was realized long ago, it was only recently shown how to make them *learn* a particular function, using "back-propagation" or other methods. This absence of a learning rule—together with the demonstration by Minsky and Papert [1969] that only linearly separable functions could be represented by simple perceptrons—led to a waning of interest in layered networks until recently.

Throughout this chapter, like the previous one, we consider only *feed-forward* networks. More general networks are discussed in the next chapter.

6.1 Back-Propagation

The back-propagation algorithm is central to much current work on learning in neural networks. It was invented independently several times, by Bryson and Ho [1969], Werbos [1974], Parker [1985] and Rumelhart et al. [1986a, b]. A closely related approach was proposed by Le Cun [1985]. The algorithm gives a prescription for changing the weights w_{pq} in any feed-forward network to learn a training set of input-output pairs $\{\xi_k^\mu, \zeta_i^\mu\}$. The basis is simply gradient descent, as described in Sections 5.4 (linear) and 5.5 (nonlinear) for a simple perceptron.

We consider first a two-layer network such as that illustrated by Fig. 6.1. Our notational conventions are shown in the figure; output units are denoted by O_i, hidden units by V_j, and input terminals by ξ_k. There are connections w_{jk} from the

115

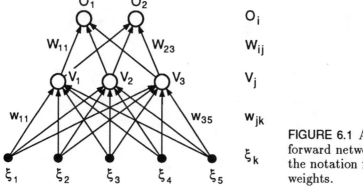

FIGURE 6.1 A two layer feed-forward network, showing the notation for units and weights.

inputs to the hidden units, and W_{ij} from the hidden units to the output units. Note that the index i always refers to an output unit, j to a hidden one, and k to an input terminal.

The inputs are always clamped to particular values. As in previous chapters, we label different patterns by a superscript μ, so input k is set to ξ_k^μ when pattern μ is being presented. The ξ_k^μ's can be binary (0/1, or ± 1) or continuous-valued. We use N for the number of *input* units and p, as before, for the number of input patterns ($\mu = 1, 2, \ldots, p$).

Given pattern μ, hidden unit j receives a net input

$$h_j^\mu = \sum_k w_{jk} \xi_k^\mu \tag{6.1}$$

and produces output

$$V_j^\mu = g(h_j^\mu) = g\left(\sum_k w_{jk} \xi_k^\mu\right). \tag{6.2}$$

Output unit i thus receives

$$h_i^\mu = \sum_j W_{ij} V_j^\mu = \sum_j W_{ij} g\left(\sum_k w_{jk} \xi_k^\mu\right) \tag{6.3}$$

and produces for the final output

$$O_i^\mu = g(h_i^\mu) = g\left(\sum_j W_{ij} V_j^\mu\right) = g\left(\sum_j W_{ij} g\left(\sum_k w_{jk} \xi_k^\mu\right)\right). \tag{6.4}$$

As in the previous chapter we have omitted the thresholds; they can be taken care of as usual by an extra input unit clamped to -1 and connected to all units in the network.

Our usual error measure or cost function

$$E[\mathbf{w}] = \frac{1}{2} \sum_{\mu i} [\zeta_i^\mu - O_i^\mu]^2 \tag{6.5}$$

now becomes

$$E[\mathbf{w}] = \frac{1}{2} \sum_{\mu i} \left[\zeta_i^\mu - g\left(\sum_j W_{ij} g\left(\sum_k w_{jk} \xi_k^\mu \right) \right) \right]^2. \tag{6.6}$$

This is clearly a continuous differentiable function of every weight, so we can use a gradient descent algorithm to learn appropriate weights. In one sense this is all there is to back-propagation, but there is great practical importance in the form of the resulting update rules.

For the hidden-to-output connections the gradient descent rule gives

$$\Delta W_{ij} = -\eta \frac{\partial E}{\partial W_{ij}} = \eta \sum_\mu [\zeta_i^\mu - O_i^\mu] g'(h_i^\mu) V_j^\mu$$

$$= \eta \sum_\mu \delta_i^\mu V_j^\mu \tag{6.7}$$

where we have defined

$$\delta_i^\mu = g'(h_i^\mu)[\zeta_i^\mu - O_i^\mu]. \tag{6.8}$$

The result is of course identical to that obtained earlier (equations (5.50) and (5.51)) for a single layer perceptron, with the output V_j^μ of the hidden units now playing the role of the perceptron input.

For the input-to-hidden connections Δw_{jk} we must differentiate with respect to the w_{jk}'s, which are more deeply embedded in (6.6). Using the chain rule, we obtain

$$\Delta w_{jk} = -\eta \frac{\partial E}{\partial w_{jk}} = -\eta \sum_\mu \frac{\partial E}{\partial V_j^\mu} \frac{\partial V_j^\mu}{\partial w_{jk}}$$

$$= \eta \sum_{\mu i} [\zeta_i^\mu - O_i^\mu] g'(h_i^\mu) W_{ij} g'(h_j^\mu) \xi_k^\mu$$

$$= \eta \sum_{\mu i} \delta_i^\mu W_{ij} g'(h_j^\mu) \xi_k^\mu$$

$$= \eta \sum_\mu \delta_j^\mu \xi_k^\mu \tag{6.9}$$

with

$$\delta_j^\mu = g'(h_j^\mu) \sum_i W_{ij} \delta_i^\mu. \tag{6.10}$$

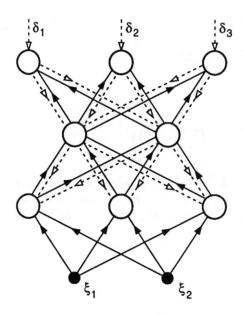

FIGURE 6.2 Back-propagation in a three-layer network. The solid lines show the forward propagation of signals and the dashed lines show the backward propagation of errors (δ's).

Note that (6.9) has the same form as (6.7), but with a different definition of the δ's. In general, with an arbitrary number of layers, the back-propagation update rule always has the form

$$\Delta w_{pq} = \eta \sum_{\text{patterns}} \delta_{\text{output}} \times V_{\text{input}} \tag{6.11}$$

where *output* and *input* refer to the two ends p and q of the connection concerned, and V stands for the appropriate input-end activation from a hidden unit or a real input. The meaning of δ depends on the layer concerned; for the last layer of connections it is given by (6.8), while for all other layers it is given by an equation like (6.10). It is easy to derive this generalized multi-layer result (6.11), simply by further application of the chain rule.

Equation (6.10) allows us to determine the δ for a given hidden unit V_j in terms of the δ's of the units O_i that it feeds. The coefficients are just the usual "forward" W_{ij}'s, but here they are propagating errors (δ's) backwards instead of signals forwards: hence the name **error back-propagation** or just **back-propagation**. We can therefore use the same network—or rather a bidirectional version of it—to compute both the output values and the δ's. Figure 6.2 illustrates this idea for a three-layer network.

Although we have written the update rules (6.7) and (6.9) as sums over all patterns μ, they are usually used incrementally: a pattern μ is presented at the input and then all weights are updated before the next pattern is considered. This clearly decreases the cost function (for small enough η) at each step, and lets successive steps adapt to the local gradient. If the patterns are chosen in random order it also makes the path through weight-space *stochastic*, allowing wider exploration of the cost surface. The alternative **batch mode**—taking (6.7) and (6.9) literally and only updating after all patterns have been presented—requires additional local storage for each connection. The relative effectiveness of the two approaches depends on the problem, but the incremental approach seems superior in most cases, especially for very regular or redundant training sets.

The fact that the appropriate cost function derivatives can be calculated by back-propagating errors is clearly attractive. But it also has two important consequences:

- The update rule (6.11) is *local*. To compute the weight change for a given connection we only need quantities available (after back-propagation of the δ's) at the two ends of that connection. This makes the back-propagation rule appropriate for parallel computation. It may even have some indirect relevance for neurobiology.[1]

- The computational complexity is less than might have been expected. If we have n connections in all, computation of the cost function (6.6) takes of order n operations, so calculating n derivatives directly would take order n^2 operations. In contrast the back-propagation scheme lets us calculate *all* the derivatives in order n operations.

It is normal to use a sigmoid function for the activation function $g(h)$. The function clearly *must* be differentiable, and we normally want it to saturate at both extremes. Either a $0/1$ or a ± 1 range can be used, with

$$g(h) \;=\; f_\beta(h) \;=\; \frac{1}{1 + \exp(-2\beta h)} \qquad (6.12)$$

and

$$g(h) = \tanh \beta h \qquad (6.13)$$

respectively for the activation function. The steepness parameter β is often set to 1, or 1/2 for (6.12). As we noted in Chapter 5, the derivatives of these functions are readily expressed in terms of the functions themselves as $g'(h) = 2\beta g(1 - g)$ for (6.12) and $g'(h) = \beta(1 - g^2)$ for (6.13). Thus one often sees (6.8), for example, written as

$$\delta_i^\mu = O_i^\mu (1 - O_i^\mu)(\zeta_i^\mu - O_i^\mu) \qquad (6.14)$$

for $0/1$ units with $\beta = 1/2$.

Because back-propagation is so important, we summarize the result in terms of a step-by-step procedure, taking one pattern μ at a time (i.e., incremental updates). We consider a network with M layers $m = 1, 2, \ldots, M$ and use V_i^m for the output

---

[1] Locality is necessary for biological implementation, but not sufficient. Bidirectional bifunctional connections are not biologically reasonable [Grossberg, 1987b], but can be avoided, allowing hypothetical neurophysiological implementations [Hecht-Nielsen, 1989]. Nevertheless back-propagation seems rather far-fetched as a biological learning mechanism [Crick, 1989].

of the ith unit in the mth layer. V_i^0 will be a synonym for ξ_i, the ith input. Note that superscript m's label layers, not patterns. We let w_{ij}^m mean the connection from V_j^{m-1} to V_i^m. Then the back-propagation procedure is:

1. Initialize the weights to small random values.

2. Choose a pattern ξ_k^μ and apply it to the input layer ($m = 0$) so that

$$V_k^0 = \xi_k^\mu \qquad \text{for all } k\,. \tag{6.15}$$

3. Propagate the signal forwards through the network using

$$V_i^m \;=\; g(h_i^m) \;=\; g\left(\sum_j w_{ij}^m V_j^{m-1}\right) \tag{6.16}$$

for each i and m until the final outputs V_i^M have all been calculated.

4. Compute the deltas for the output layer

$$\delta_i^M = g'(h_i^M)[\zeta_i^\mu - V_i^M] \tag{6.17}$$

by comparing the actual outputs V_i^M with the desired ones ζ_i^μ for the pattern μ being considered.

5. Compute the deltas for the preceding layers by propagating the errors backwards

$$\delta_i^{m-1} = g'(h_i^{m-1}) \sum_j w_{ji}^m \delta_j^m \tag{6.18}$$

for $m = M, M-1, \ldots, 2$ until a delta has been calculated for every unit.

6. Use

$$\Delta w_{ij}^m = \eta \delta_i^m V_j^{m-1} \tag{6.19}$$

to update all connections according to $w_{ij}^{\text{new}} = w_{ij}^{\text{old}} + \Delta w_{ij}$.

7. Go back to step 2 and repeat for the next pattern.

It is straightforward to generalize back-propagation to other kinds of networks where connections jump over one or more layers, such as the direct input-to-output connections in Fig. 6.5(b). This produces the same kind of error propagation scheme as long as the network is *feed-forward*, without any backward or lateral connections.

6.2 Variations on Back-Propagation

Back-propagation has been much studied in the past few years, and many extensions and modifications have been considered. The basic algorithm given above is exceedingly slow to converge in a multi-layer network, and many variations have been suggested to make it faster. Other goals have included avoidance of local minima, and improvement of generalization ability. Here we discuss mainly the issue of speed.

It is worth mentioning that speed comparisons between different techniques are not always clear-cut. Different authors have used different problems, different stopping criteria, different measures of computational speed, and different approaches to averaging. As one example of the difficulties, consider what to do with unsuccessful trials. Usually a problem is solved many times (e.g., from random starting weights, and with random update sequences), and in some cases a trial becomes stuck in a local minimum or on a very flat plateau. After some maximum time T the experimenter must give up on that trial, but how should it be reported in the average time $\langle t \rangle$ per trial? Some of the different solutions that have been used are:

- Count it like a successful trial, with time T.

- Discard it, reporting unsuccessful trials separately.

- Don't count it as a valid trial, but add the time T onto the time for the next trial, so the times averaged are the total times between successful outcomes.

- Average $1/t$ instead of t, taking $1/t = 0$ for unsuccessful trials.

Fahlman [1989] discusses these and other benchmarking issues at greater length.

There are many parameters one can consider varying within the general back-propagation framework, including the architecture (number of layers, number of units per layer), the size and nature of the training set, and the update rule. In this section we focus mainly on the update rule, keeping a fixed feed-forward architecture. Other issues and other architectures are discussed later.

Alternative Cost Functions

The quadratic cost function (6.5) is not the only possible choice. We can replace the $(\zeta_i^\mu - O_i^\mu)^2$ factor by any other differentiable function $F(\zeta_i^\mu, O_i^\mu)$ that is minimized when its arguments are equal, and derive a corresponding update rule. Direct differentiation shows that only the expression (6.8) for δ_i^μ in the *output* layer changes; all the other equations of back-propagation remain unchanged.

A particularly good choice for the cost function seems to be the entropic measure (5.52) that we discussed in Chapter 5 on page 109 [Solla et al., 1988]. This is for a ± 1 output range and, using $g(h) = \tanh h$, reduces (6.8) to

$$\delta_i^\mu = \zeta_i^\mu - O_i^\mu \tag{6.20}$$

—the g' factor disappears (for the output layer only). Since $g'(h)$ becomes very small when $|h|$ becomes large, leaving it out accelerates progress in large $|h|$ regions, where the cost surface is relatively flat. On the other hand it gives no acceleration—which could lead to overshoot and oscillation—when h is equivocating around 0 and the cost surface is more sharply curved.

In one study [Fahlman, 1989] it was found that the compromise

$$\delta_i^\mu = [g'(h_i^\mu) + 0.1](\zeta_i^\mu - O_i^\mu) \tag{6.21}$$

worked better than either (6.8) or (6.20). This form restores some of the effect of the g' term, but eliminates the flat spots by keeping δ_i^μ non-zero even when $|h_i^\mu|$ becomes large.

Another approach is to change the $\zeta_i^\mu - O_i^\mu$ term instead of (or as well as) the g' one, increasing δ_i^μ when $|\zeta_i^\mu - O_i^\mu|$ becomes large. For example

$$\delta_i^\mu = \operatorname{arctanh} \tfrac{1}{2}(\zeta_i^\mu - O_i^\mu) \tag{6.22}$$

(where $x = \operatorname{arctanh} y$ is the inverse function of $y = \tanh x$) goes to $\pm\infty$ when $O_i^\mu \to -\zeta_i^\mu$ (with a ± 1 range) [Fahlman, 1989]. This rule could be obtained by differentiating a suitable cost function, but in fact we may just as well start with the derivative.

Finally, we could add parameters to the cost function so as to adjust its steepness or roughness during training. It should be possible to smooth out the surface initially, to avoid local minima at the expense of loss of detail, and then gradually add back the detail once the network has reached the right general region of weight space. Simulated annealing employs stochastic units to achieve a similar goal. One example was suggested by Makram-Ebeid et al. [1989], who used

$$E = \sum_{\mu i} \left\{ \begin{array}{ll} \gamma(\zeta_i^\mu - O_i^\mu)^2 & \text{if } \operatorname{sgn} \zeta_i^\mu = \operatorname{sgn} O_i^\mu \\ (\zeta_i^\mu - O_i^\mu)^2 & \text{if } \operatorname{sgn} \zeta_i^\mu = -\operatorname{sgn} O_i^\mu \end{array} \right\} \tag{6.23}$$

with γ gradually increasing from 0 to 1. This has the effect of focusing initially just on getting the *sign* of ζ_i^μ right, and only later paying attention to the magnitude.

A different approach to training a layered network is based on optimizing the **internal representations** of the input patterns by the hidden layers. The idea is very simple: although the input patterns may not be linearly separable, the result of transforming them by one or more layers of processing may be. Taking the case of a single hidden layer for simplicity, if these transformed patters were known then both the input-to-hidden and the hidden-to-output weights could be found as for a simple perceptron. So this transforms the essential problem to one of determining an appropriate set of internal representations, which are therefore made explicit arguments of a cost function. This new cost function must be minimized with respect to the choice of internal representations as well the weights. Several different implementations of this idea have been explored [Grossman et al., 1989; Grossman, 1990; Krogh et al., 1990; Rohwer, 1990; Saad and Marom, 1990a, b]. In some limits these approaches become equivalent to ordinary back-propagation, but in other parts of their parameter spaces they can give faster learning.

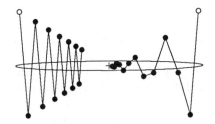

FIGURE 6.3 Gradient descent on the simple quadratic surface of Fig. 5.10. Both trajectories are for 12 steps with $\eta = 0.0476$, the best value in the absence of momentum. On the left there is no momentum ($\alpha = 0$), while $\alpha = 0.5$ on the right.

Momentum

We saw in Chapter 5 that gradient descent can be very slow if η is small, and can oscillate widely if η is too large; see Fig. 5.10 on page 106. The problem essentially comes from cost-surface valleys with steep sides but a shallow slope along the valley floor. There are a number of ways of dealing with this problem, including the replacement of gradient descent by more sophisticated minimization algorithms, as discussed below. However, a much simpler approach, the addition of a **momentum term** [Plaut et al., 1986], is often effective and is very commonly used.

The idea is to give each connection[2] w_{pq} some inertia or momentum, so that it tends to change in the direction of the *average* downhill "force" that it feels, instead of oscillating wildly with every little kick. Then the effective learning rate can be made larger without divergent oscillations occurring. This scheme is implemented by giving a contribution from the previous time step to each weight change:

$$\Delta w_{pq}(t+1) = -\eta \frac{\partial E}{\partial w_{pq}} + \alpha \Delta w_{pq}(t). \tag{6.24}$$

The **momentum parameter** α must be between 0 and 1; a value of 0.9 is often chosen.

If we are marching through a plateau region of the cost surface, then $\partial E / \partial w_{pq}$ will be about the same at each time-step and (6.24) will converge to

$$\Delta w_{pq} \approx -\frac{\eta}{1-\alpha} \frac{\partial E}{\partial w_{pq}} \tag{6.25}$$

with an effective learning rate of $\eta/(1-\alpha)$. On the other hand, in an oscillatory situation, Δw_{pq} responds only with coefficient η to instantaneous fluctuations of $\partial E/\partial w_{pq}$. The overall effect is to accelerate the long term trend by a factor of $1/(1-\alpha)$, without magnifying the oscillations. Figure 6.3 shows a simple example.

A momentum term is useful with either pattern-by-pattern or batch mode updating. It was first proposed for the pattern-by-pattern case, where it has the effect of partial averaging over the patterns. The averaging is not as complete as in batch mode however, and can thus leave some beneficial fluctuations in the trajectory.

---

[2] In this and subsequent discussions w_{pq} stands for *any* weight in the network, not just one between the input and the hidden layer.

Adaptive Parameters

It is not easy to choose appropriate values of the parameters η and α for a particular problem. Moreover the best values at the beginning of training may not be so good later on. Thus many authors have suggested adjusting the parameters automatically as learning progresses [e.g., Cater, 1987; Franzini, 1987; Vogl et al., 1988; Jacobs, 1988].

The usual approach is to check whether a particular weight update did actually decrease the cost function. If it didn't, then the process overshot, and η should be reduced. On the other hand if several steps in a row *have* decreased the cost, then perhaps we are being too conservative, and could try increasing η. It appears best to increase η by a constant, but decrease it geometrically to allow rapid decay when necessary. This gives the overall scheme

$$\Delta\eta = \begin{cases} +a & \text{if } \Delta E < 0 \text{ consistently;} \\ -b\eta & \text{if } \Delta E > 0 \,; \\ 0 & \text{otherwise} \end{cases} \qquad (6.26)$$

where ΔE is the cost function change, and a and b are appropriate constants. The meaning of *consistently* can be based on the last K steps, or on a weighted moving average of the observed ΔE's. When a bad step decreases η, it is also worthwhile "undoing" the step and setting $\alpha = 0$ until a good step is taken.

This kind of adaptive scheme can be made more effective by having several different learning rates. Cater [1987] suggested having parameters η^μ, one for each pattern μ, and Jacobs [1988] suggested an η_{pq} for each connection pq. Even without an adaptive rule it may be appropriate to have different η's to suit the architecture; the choice

$$\eta_{pq} \propto 1/(\text{fan-in of site } i) \qquad (6.27)$$

is often used [Plaut et al., 1986], though an even stronger dependence on fan-in may be better [Tesauro and Janssens, 1988].

Other Minimization Procedures ⋆

Gradient descent is one of the simplest optimization techniques, but not a very good one. There are many more powerful techniques available, though most are not well suited to a network implementation. They are nevertheless worth considering, perhaps for **off-line** training on an ordinary (or parallel) computer for later implementation as a physical network with predetermined weights.

To simplify notation let us use the vector $\mathbf{x}$ for the weight space we are searching; specifying $\mathbf{x}$ corresponds to specifying all the weights. Then expanding the cost function $E(\mathbf{x})$ about the current point $\mathbf{x}_0$ we obtain[3]

$$E(\mathbf{x}) = E_0 + (\mathbf{x} - \mathbf{x}_0) \cdot \nabla E(\mathbf{x}_0) + \tfrac{1}{2}(\mathbf{x} - \mathbf{x}_0) \cdot \mathbf{H} \cdot (\mathbf{x} - \mathbf{x}_0) + \cdots \qquad (6.28)$$

---

[3]The gradient ∇E means simply the vector with components $\partial E/\partial x_i$.

where **H** is the second derivative **Hessian matrix**

$$\mathbf{H}_{ij} = \frac{\partial^2 E}{\partial x_i \partial x_j} \tag{6.29}$$

evaluated at $\mathbf{x}_0$. Differentiating (6.28) gives a similar expansion

$$\nabla E(\mathbf{x}) = \nabla E(\mathbf{x}_0) + \mathbf{H} \cdot (\mathbf{x} - \mathbf{x}_0) + \cdots \tag{6.30}$$

for the gradient.

We want to find the minimum of $E(\mathbf{x})$, where of course $\nabla E(\mathbf{x}) = 0$. One approach is to set (6.30) to zero, ignoring the higher-order terms, to yield

$$\nabla E(\mathbf{x}_0) + \mathbf{H} \cdot (\mathbf{x} - \mathbf{x}_0) = 0 \tag{6.31}$$

or

$$\mathbf{x} = \mathbf{x}_0 - \mathbf{H}^{-1} \nabla E(\mathbf{x}_0) \tag{6.32}$$

as an estimate for the location of the minimum. We can use this equation iteratively, repeatedly using the previous estimate $\mathbf{x}$ as the new $\mathbf{x}_0$. This is **Newton's method**; in one dimension it becomes the familiar rule

$$x = x_0 - \frac{E'(x_0)}{E''(x_0)} \tag{6.33}$$

for finding a root of $E'(x) = 0$.

Newton's method is very expensive computationally, because for n weights we must invert an $n \times n$ matrix **H** at each iteration, taking order n^3 steps every time. It also requires computation of the second derivatives, which does not fit into the back-propagation framework. And it is numerically unstable unless we start close enough to the minimum, which is unlikely in a practical case. So in all it is not a practical technique in many dimensions. Nevertheless it serves as a useful reference point for other approaches.

Most practical minimization methods use only first derivative information, combined with **line searches** along selected directions. In a line search starting from $\mathbf{x}_0$ in direction **d** we stay on the line

$$\mathbf{x} = \mathbf{x}_0 + \lambda \mathbf{d} \tag{6.34}$$

and choose λ to minimize $E(\mathbf{x})$. There are many ways of performing this one-dimensional minimization, which often need not be done very accurately at first [Luenberger, 1986]. One simple approach is simply to repeat a fixed η step until E no longer decreases [Hush and Salas, 1988].

The simplest approach using line search is the **steepest descent method**, in which we choose $\mathbf{d} = -\nabla E(x_0)$. Thus we minimize along a line in the gradient direction, and then re-evaluate the gradient and repeat the process. The left-hand

FIGURE 6.4 Line minimization on the simple quadratic surface of Figs. 5.10 and 6.3. Steepest descent was used on the left, Polak-Ribiere conjugate gradient descent on the right.

side of Fig. 6.4 shows a simple example for the quadratic surface of Figs. 5.10 and 6.3. Although it seems to converge in fewer steps than any method yet discussed, note that each step involves a line minimization, which may itself take many function evaluations. But in fact steepest descent is usually considerably faster than ordinary back-propagation, though it sometimes fails to converge at all [Watrous, 1987; Kramer and Sangiovanni-Vincentelli, 1989].

Successive steps are necessarily perpendicular in steepest descent, because

$$0 = \frac{\partial}{\partial \lambda} E(\mathbf{x}_0 + \lambda \mathbf{d}^{\text{old}}) = \mathbf{d}^{\text{old}} \cdot \nabla E^{\text{new}} \qquad (6.35)$$

implies that the new gradient direction ∇E^{new} is perpendicular to the old direction $\mathbf{d}^{\text{old}}$ after the line minimization. Thus the approach to the minimum is a zigzag path, as we see in the figure. A better approach is to use as the new search direction a compromise between the gradient direction and the previous search direction:

$$\mathbf{d}^{\text{new}} = -\nabla E^{\text{new}} + \beta \mathbf{d}^{\text{old}} \qquad (6.36)$$

for some appropriate β. This is the basis of **conjugate gradient methods**, which choose β so that each new search direction spoils as little as possible the minimization achieved by the previous one. This means that the new search direction $\mathbf{d}^{\text{new}}$ should be such that it does not change (to first order) the component of the gradient along the previous direction, which was just made zero by (6.35). Thus we need (to first order in λ)

$$\mathbf{d}^{\text{old}} \cdot \nabla E(\mathbf{x}_0 + \lambda \mathbf{d}^{\text{new}}) = 0 \qquad (6.37)$$

or, using (6.30) and (6.35),

$$\mathbf{d}^{\text{old}} \cdot \mathbf{H} \cdot \mathbf{d}^{\text{new}} = 0. \qquad (6.38)$$

The vectors $\mathbf{d}^{\text{old}}$ and $\mathbf{d}^{\text{new}}$ are then said to be *conjugate*.

To complete the specification of a practical procedure we need to know how to choose β in (6.36) so as to satisfy (6.38). The **Polak-Ribiere rule**[4]

$$\beta = \frac{(\nabla E^{\text{new}} - \nabla E^{\text{old}}) \cdot \nabla E^{\text{new}}}{(\nabla E^{\text{old}})^2} \qquad (6.39)$$

---

[4]The term $\nabla E^{\text{old}} \cdot \nabla E^{\text{new}}$ is zero if the line minimization is done exactly, but helps when it is done only approximately.

achieves this goal, and in fact keeps the last n directions all mutually conjugate. The result is remarkable because no explicit knowledge of the Hessian **H** is needed. We omit the proof; see Press et al. [1986] for details.

In summary the conjugate gradient method (in the Polak-Ribiere variant) consists of a succession of line minimizations along directions given by (6.36) and (6.39). On a strictly quadratic surface in n dimensions it reaches the minimum in exactly n steps, as shown for $n = 2$ on the right-hand side of Fig. 6.4. It has been applied in a neural network context by Kramer and Sangiovanni-Vincentelli [1989] and by Makram-Ebeid et al. [1989], with advantages in speed and convergence over back-propagation or steepest descent. A parallel implementation is possible.

Another technique, also using successive line minimizations, is the **quasi-Newton** or **variable metric method**. This also reaches the bottom of a strictly quadratic valley in n steps, but by a different approach. The idea is to use the Newton rule (6.32) but with an *approximation* to the inverse Hessian $\mathbf{H}^{-1}$. This is achieved through an iterative scheme

$$\mathbf{x}^{\text{new}} = \mathbf{x}^{\text{old}} - \lambda \mathbf{G}^{\text{old}} \cdot \nabla E^{\text{old}} \tag{6.40}$$

$$\mathbf{G}^{\text{new}} = \mathbf{G}^{\text{old}} + \mathbf{F}(\mathbf{G}^{\text{old}}, \mathbf{x}^{\text{new}} - \mathbf{x}^{\text{old}}, \nabla E^{\text{new}} - \nabla E^{\text{old}}) \tag{6.41}$$

where **G** is the approximation to $\mathbf{H}^{-1}$ and the matrix **F** is a rather complicated function of its three arguments; see Press et al. [1986] or Luenberger [1986] for the details. Equation (6.40) is just the Newton rule with the approximate inverse Hessian **G** and an extra factor λ which we choose by line minimization. Equation (6.41) creates successive approximations (starting from a unit matrix) to the actual Hessian, without ever evaluating a second derivative.

The quasi-Newton techniques usually produce comparable results to the conjugate gradient techniques, with similar speed. They do however require more storage (for the matrix **G**), which may be a disadvantage in a parallel implementation. Watrous [1987] has studied their use in neural networks, finding an order of magnitude speed increase over back-propagation.

All these line search methods—steepest descent, conjugate gradient, and quasi-Newton—use only first derivative information. In a neural network problem the derivatives can therefore be calculated efficiently by back-propagation of the δ's, just as for the full back-propagation rule; we replace only the gradient descent rule $\Delta w_{ij} = -\eta \partial E / \partial w_{ij}$, not the gradient calculation.

Several authors have suggested using second derivatives too, approximating the full Newton rule (6.32) so that a matrix inversion is not required [Parker, 1987; Scalettar and Zee, 1988; Ricotti et al., 1988; Becker and Le Cun, 1989]. The simplest approach is to ignore the off-diagonal elements of the Hessian, yielding the **pseudo-Newton rule**

$$\Delta w_{ij} = -\frac{\partial E}{\partial w_{ij}} \Big/ \frac{\partial^2 E}{\partial w_{ij}^2} \tag{6.42}$$

which is equivalent to performing Newton's rule separately for each weight. In a two-layer network the diagonal second derivatives can actually be calculated by propagating two types of δ's back through the network, but for more than two layers or for off-diagonal elements the scheme breaks down and non-local quantities are required [Ricotti et al., 1988]. Alternatively the second derivatives could be calculated by finite-differencing—finding the change of $\partial E/\partial w_{ij}$ produced by a small change of w_{ij}.

As it stands, (6.42) is dangerous, because the denominator could be negative or close to zero, possibly leading to a giant step in the wrong direction. These problems can be largely cured by using

$$\Delta w_{ij} = -\frac{\partial E}{\partial w_{ij}} \bigg/ \left(\left| \frac{\partial^2 E}{\partial w_{ij}^2} \right| + \mu \right) \qquad (6.43)$$

with a small positive constant μ. Nevertheless this whole approach can fail to converge, and seems to give only a modest speed increase over simple back-propagation.

Owens and Filkin [1989] have suggested replacing the discrete gradient descent rule

$$\Delta w_{ij} = -\eta \frac{\partial E}{\partial w_{ij}} \qquad \text{for all } ij \qquad (6.44)$$

by the continuous differential equations

$$\frac{dw_{ij}}{dt} = -\eta \frac{\partial E}{\partial w_{ij}} \qquad \text{for all } ij \,. \qquad (6.45)$$

When we write out the right-hand side of (6.45) as an explicit function of the weights we obtain a set of coupled nonlinear ordinary differential equations (ODE's). The conventional gradient descent rule (6.44) can be viewed as a simple fixed-step "forward-Euler" integrator for these equations. But in fact there are much better ways of integrating such equations numerically, particularly if the equations are mathematically **stiff**, which is equivalent to there being shallow steep-sided valleys in the cost function.[5] Most neural network problems *do* seem to be stiff, so specialized stiff ODE solvers are appropriate, and are found to give very large speed increases over ordinary back-propagation.

A very different approach uses a **genetic algorithm** to search the weight space without use of any gradient information [Whitley and Hanson, 1989; Montana and Davis, 1989]. See Goldberg [1989] for an introduction. A complete set of weights is coded in a binary string (or chromosome), which has an associated "fitness" that depends on its effectiveness. For example the fitness could be given by $-E$ where E is the value of the cost function for that set of weights. Starting with a random population of such strings, successive generations are constructed using **genetic operators** such as mutation and crossover to construct new strings out of old ones, with some form of survival of the fittest; fitter strings are more likely to survive and to participate in mating (crossover) operations. The crossover operation combines

---

[5] Technically, stiffness depends on the ratio of the largest and smallest eigenvalues of the underlying Hessian, just as found in Chapter 5 for the maximum convergence rate of gradient descent.

part of one string with part of another, and can in principle bring together good building blocks—such as hidden units that compute particular logical functions—found by chance in different members of the population. The way in which the weights are coded into strings and the details of the genetic operators are both crucial in making this effective.

Genetic algorithms perform a *global* search and are thus not easily fooled by local minima. The fitness function does not need to be differentiable, so we can start with threshold units in Boolean problems, instead of having to use sigmoids that are later trained to saturation. On the other hand there is a high computational penalty for not using gradient information, particularly when it is so readily available by back-propagating errors. An initial genetic search followed by a gradient method might be an appropriate compromise. Or a gradient descent step can be included as one of the genetic operators [Montana and Davis, 1989]. There are also large costs in speed and storage for working with a whole population of networks, perhaps making genetic algorithms impractical for large network design.

There has not yet been sufficient comparative study to determine which of the techniques discussed here is best overall. Conventional back-propagation is slow compared to almost any of the improved techniques, but of course speed is not the only issue. We should also consider storage requirements, locality of quantities required (for a parallel or network implementation), reliability of convergence, and the problem of local minima. Probably there is no single best approach, and an optimal choice must depend on the problem and on the design criteria.

Local Minima

Gradient descent, and all of the other optimization techniques discussed above, can become stuck in **local minima** of the cost function. Actually local minima have not, in fact, been much of a problem in most cases studied empirically, though it is not really understood why this should be so. There certainly are local minima in some problems [McInerny et al., 1989], though apparent local minima sometimes turn out to be the bottoms of very shallow steep-sided valleys.

The size of the initial random weights is important. If they are too large the sigmoids will saturate from the beginning, and the system will become stuck in a local minimum (or very flat plateau) near the starting point. A sensible strategy is to choose the random weights so that the magnitude of the typical net input h_i to unit i is less than—but not too much less than—unity. This can be achieved by taking the weights w_{ij} to be of the order of $1/\sqrt{k_i}$ where k_i is the number of j's which feed forward to i (the fan-in of unit i).

A common type of local minimum is one in which two or more errors compensate each other. These minima are not very deep so just a little *noise* (random fluctuation) is needed to get out. As we mentioned earlier, a simple and very efficient approach is to use incremental updating (one pattern at a time), choosing the patterns in a random order from the training set. Then the average over patterns is avoided and the random order generates noise. If, on the other hand, the training set is cycled through in the same sequence all the time it is possible to get stuck because a series of weight changes cancel.

Another way of introducing noise is to let upward steps in E be allowed occasionally, as in the stochastic networks discussed earlier, with a temperature T controlling the probabilities. Annealing—a gradual lowering of T—is then performed. However, this approach tends to make learning very slow.

Alternatively it is possible to add noise explicitly by randomly changing the w's slightly [von Lehman et al., 1988] or by adding noise to the training set inputs ξ_k^μ, independently at each presentation [Sietsma and Dow, 1988]. In each case there seems to be an optimum amount of noise; a little helps, but too much hurts and slows down the learning process considerably.

6.3 Examples and Applications

Although our main focus in this book is on the theoretical aspects of neural computation, it is helpful at this stage to survey some examples. Back-propagation and its variants have been applied to a wide variety of problems, from which we select a few.

Our first three examples are "toy problems" which are often used for testing and benchmarking a network. Typically the training set contains *all* possible input patterns, so there is no question of generalization. In the later examples the training set is only a part of the problem domain, but the networks are able to generalize to previously unseen cases.

Most of the examples employ straightforward back-propagation learning by gradient descent in two-layer networks with full connectivity between layers. Some of them illustrate that one can solve some nontrivial real-world problems without special tricks or refinements of this standard, off-the-shelf network. A few examples, on the other hand, require a little more thinking about the problem and consequent modification of the algorithm for a satisfactory solution.

XOR

The XOR problem was described on page 96, where we saw that it could *not* be solved by a single layer network because it is not linearly separable. However, there are several solutions using one hidden layer [Rumelhart, McClelland, et al., 1986], two of which are shown in Fig. 6.5 for threshold units with 0/1 inputs and outputs. In solution (a) the two hidden units compute the logical OR (left) and AND (right) of the two inputs, and the output fires only when the OR unit is on and the AND unit is off. Solution (b) is not a conventional feed-forward architecture, but is interesting because it needs only two units; the hidden unit computes a logical AND to inhibit the output unit when both inputs are on.

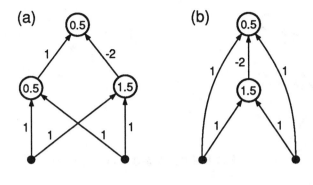

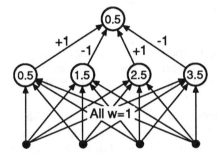

FIGURE 6.5 Two networks that can solve the XOR problem using 0/1 threshold units. Each unit is shown with its threshold.

FIGURE 6.6 A network to solve the $N = 4$ parity problem with 0/1 threshold units.

Solutions like (a) are readily found using back-propagation with continuous-valued units. However, the weight values found become much larger than those shown in the figure, so as to drive the sigmoid units to saturation. The training time from random starting weights is surprisingly long; hundreds of **epochs**—passes through the training set—are required to get good results.

Parity

The parity problem is essentially a generalization of the XOR problem to N inputs [Minsky and Papert, 1969]. The single output unit is required to be on if an odd number of inputs are on, and off otherwise. One hidden layer of N units suffices to solve the problem, as shown in Fig. 6.6. Hidden unit j is on when at least j input units are on, and either excites or inhibits the output unit depending on whether j is odd or even. Again back-propagation finds this solution using continuous-valued units, except that the weights become scaled up by a large factor.

The parity problem (or XOR, its $N = 2$ version) is often used for testing or evaluating network designs. It should be realized, however, that it is a very hard problem, because the output must change whenever any single input changes. This is untypical of most real-world classification problems, which usually have much more regularity and allow generalization within classes of similar input patterns [Fahlman, 1989].

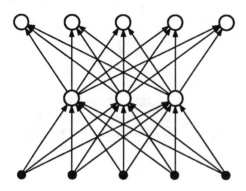

FIGURE 6.7 A 5–3–5 encoder network.

Encoder

The general encoding problem involves finding an efficient set of hidden unit patterns to encode a large number of input/output patterns. The number of hidden
units is intentionally made small to enforce an efficient encoding [Ackley, Hinton,
and Sejnowski, 1985].

The specific problem usually considered involves **auto-association**, using identical unary input and output patterns. We take a two-layer network of N inputs,
N output units, and M hidden units, with $M < N$. This is often referred to as
an N–M–N encoder; Fig. 6.7 shows the architecture of a 5–3–5 encoder. There are
exactly N members of the training set ($p = N$), each having *one* input and the
corresponding target on, and the rest off: $\xi_i^\mu = \zeta_i^\mu = \delta_{i\mu}$. A permutation of the
outputs could also be used—it makes no difference to the network.

One solution to the problem is to use binary coding on the hidden layer, so
that the activation pattern of the hidden units gives the binary representation of
μ, the pattern number. This can be achieved with connection strengths patterned
after the binary numbers. Clearly we need $M \geq \log_2 N$ for such a scheme. But
back-propagation tends to find alternative schemes, often using intermediate (non-
saturated) values for the hidden units, and can sometimes solve the problem even
when $M < \log_2 N$.

Encoder problems have been widely used for benchmarking networks. One advantage is that they can be scaled up to any desired size, and another is that their
difficulty can be varied by changing the ratio $M/\log_2 N$. They also have practical
applications for image compression, as discussed later on page 136.

NETtalk

In the preceding examples the training set normally includes *all* possible input
patterns, so no generalization issue arises. But from here onwards the training set
is only a *part* of the problem domain, and the networks are able to generalize to

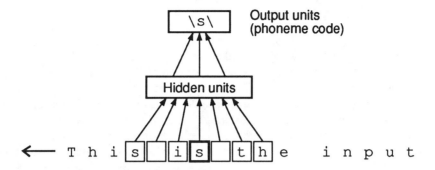

FIGURE 6.8 The NETtalk architecture.

previously unseen cases. In some cases the emergence of significant internal representations is also observed; some of the hidden units discover meaningful features in the data (meaningful to humans) and come to signify their presence or absence.

The NETtalk project aimed at training a network to pronounce English text [Sejnowski and Rosenberg, 1987]. As shown in Fig. 6.8, the input to the network consisted of 7 consecutive characters from some written text, presented in a moving window that gradually scanned the text. The desired output was a phoneme code which could be directed to a speech generator, giving the pronunciation of the letter at the center of the input window. The architecture employed 7×29 inputs encoding 7 characters (including punctuation), 80 hidden units, and 26 output units encoding phonemes.

The network was trained on 1024 words from a side-by-side English/phoneme source, obtaining intelligible speech after 10 training epochs and 95% accuracy after 50 epochs. It first learned gross features such as the division points between words and then gradually refined its discrimination, sounding rather like a child learning to talk. Some internal units were found to be representing meaningful properties of the input, such as the distinction between vowels and consonants. After training, the network was tested on a continuation of the side-by-side source, and achieved 78% accuracy on this generalization task, producing quite intelligible speech. Damaging the network by adding random noise to the connection strengths, or by removing some units, was found to degrade performance continuously (not catastrophically as expected for a digital computer), with rather rapid recovery upon retraining.

It is interesting to compare NETtalk with the commercially available DEC-talk, which is based on hand-coded linguistic rules. There is no doubt that DEC-talk performs better than NETtalk, but this should be seen in the context of the effort involved in creating each system. Whereas NETtalk simply learned from examples, the rules embodied in DEC-talk are the result of about a decade of analysis by many linguists. This exemplifies the utility of neural networks; they are easy to construct and can be used even when a problem is not fully understood. However, rule-based algorithms usually out-perform neural networks when enough understanding *is* available.

Protein Secondary Structure

The primary structure of a protein consists of a linear sequence of amino acid residues chosen from 20 possibilities, like text with an alphabet of 20 letters. The secondary structure involves the local configuration of this linear strand, most commonly as an α-helix or a β-sheet. This secondary structure is then folded into the tertiary structure. In an obvious parallel to NETtalk, Qian and Sejnowski [1988a] constructed a network which took as input a moving window of 13 amino acids and produced as output a prediction of *α-helix*, *β-sheet*, or *other* for the central part of the sequence. After training on extensive protein folding data it achieved an accuracy of 62% on previously unseen sequences, compared to about 53% for the best available alternative (the Robson method). This scheme was invented independently by Bohr et al. [1988] and developed further. A neural network approach is thus currently the method of choice in this problem. The accuracy obtained may in fact be very close to the best possible achievable from a *local* window; interactions from residues far along the primary sequence (but spatially nearby in the tertiary structure) may be needed to do any better.

Hyphenation Algorithms

Here the problem is to determine the points at which a word may be hyphenated. This depends on whole syllables, not isolated letter-pairs, as shown by the examples *prop·a·ga·tion*, *pro·pa·gan·da*, and *pro·pane*. Reasonably good deterministic algorithms (with exception dictionaries) exist for English [Liang, 1983], but there is a current lack of good methods for languages such as Danish and German. A neural network is likely to be the quickest way to produce a working solution, again with an architecture like that of NETtalk, and is currently under study [Brunak and Lautrup, 1989, 1990].

Sonar target recognition

In another interesting example, Gorman and Sejnowski [1988a, b] trained the same kind of standard two-layer perceptron to distinguish between the reflected sonar signals from two kinds of objects lying at the bottom of Chesapeake Bay: rocks and metal cylinders. The network inputs ξ_k were not taken directly from the raw signal x_t (where t is time), but were based on the frequency spectrum (Fourier transform) of that signal. This preprocessing was approximately linear

$$\xi_k = \sum_t a_{kt} x_t \tag{6.46}$$

(for discrete time and suitable coefficients a_{kt}), and so should not be necessary in principle; a network should be able to learn an appropriate linear transformation if one is needed. However, going to the frequency domain made it possible to use a much smaller number of input units than with raw time domain data.

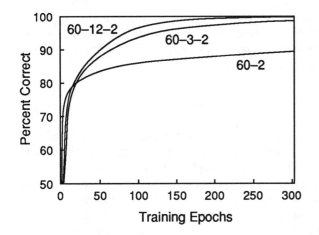

FIGURE 6.9 Learning curves for sonar target recognition (averaged over many trials and smoothed). The notation 60–12–2 means 60 inputs, 12 hidden units, and 2 output units.

The network had 60 input units and two output units, one for *rock* and one for *cylinder*. The number of hidden units was varied from none (one-layer only) to 24. Figure 6.9 shows the **learning curves**—the percentage of correct classifications as a function of the number of presentations of the training data set—for 0, 3, and 12 hidden units. Without any hidden units the network rapidly reached about 80% correct performance, but then improved only very slowly. With 12 hidden units the performance came close to 100% accuracy, on every trial. No improvement was visible in the results on increasing the number of hidden units from 12 to 24.

After training, the network was tested on new data not in the training set. About 85% correct classification was achieved with 12 hidden units, showing reasonable generalization ability. This was improved to about 90% when the training set was more carefully selected to contain examples of more of the possible signal patterns.

Navigation of a Car

Pomerleau has constructed a neural network controller for driving a car on a winding road [Pomerleau, 1989; Touretzky and Pomerleau, 1989]. The inputs to the network consist of a 30 × 32 pixel image from a video camera mounted on the roof of the car, and an 8 × 32 image from a range finder coding distances in a gray scale. These inputs are fed to a hidden layer of 29 units and from there to an output layer of 45 units arranged in a line. The output unit in the center represents "drive straight ahead" while gradually sharper left and right turns are represented by the units to the left and right.

The network was trained on 1200 simulated road images using back-propagation. After training on each image about 40 times the network performed well, and the car could drive with a speed of about 5 km/hr (3 mph) on a road through a wooded area near the Carnegie-Mellon campus. The speed is limited by the time taken by the small Sun-3 computer in the car to do a forward pass through the large network, and could obviously be increased by using some dedicated parallel hardware. Nevertheless the speed was about twice as fast as that obtained using any of the other (non-network) algorithms which were tried.

Image Compression

In high-definition television, the channel capacity is too small to allow transmission of all the information (intensity and color) characterizing every pixel on the screen. Practical transmission schemes must therefore exploit the redundancy that naturally exists in most images, encoding the picture in a much smaller number of bits than the total required to describe it exactly. The encoded or "compressed" image must then be decoded at the receiver into a full-sized picture. Ideally the encoding and decoding should be done in a way that optimizes the quality of the decoded picture. A great many *ad hoc* schemes have been tried on this problem.

We can formulate this as a supervised learning problem by making the targets equal to the inputs (auto-association), and taking the compressed signal from the hidden layer. Then we just have a scaled-up version of the encoder task discussed earlier (page 132 and Fig. 6.7), sometimes called **self-supervised backpropagation** in this context. The input-to-hidden connections perform the encoding and the hidden-to-output connections do the decoding.

This approach was tried by Cottrell et al. [1987]. The network they studied most took input from 8 × 8-pixel regions of the image (64 input units, each specified to 8-bit precision) and had 16 hidden units. It was trained by standard back-propagation on randomly selected patches of a particular image for typically 150,000 training steps. Then it was tested on the entire image patch by patch, using a complete set of nonoverlapping patches. Cottrell et al. obtained near state-of-the-art results from this simple procedure. As one might expect, the performance was somewhat worse when the network was tested on very different pictures, but it was still respectable. The fact that one can do so well with the simplest kind of architecture imaginable suggests that more complex networks could do even better.

An interesting aspect of this problem is that nonlinearity in the hidden units is theoretically of no help [Bourland and Kamp, 1988], and indeed Cottrell et al. found that nonlinearity conferred no advantage in their simulations. Using linear units allows a detailed theoretical analysis [Baldi and Hornik, 1989], which shows that the network projects the input onto the subspace spanned by the first M **principal components** of the input, where M is the number of hidden units. Principal components are discussed in detail in Section 8.3. A projection onto principal components discards as little information as possible, by retaining those components of the input vector which vary the most.

Backgammon

In the game of backgammon each player in turn rolls two dice and then typically has a choice of around 20 possible moves consistent with the dice. Look-ahead is difficult because of the number of possible rolls and the number of possible moves for each roll. A network was trained to score, on a scale of -100 (terrible move) to $+100$ (best move), triples {*current position, dice values, possible move*} using a set of 3000 examples hand-scored by an expert player [Tesauro and Sejnowski, 1988]. The input to the network consisted of the triple itself, suitably coded, plus some precomputed features such as **pip count** and **degree of trapping**, with 459 units in all. There were *two* hidden layers of 24 units each, and a single continuous-valued output unit for the score. Noise was added to the training data by including some positions with a randomly chosen score. During development notable errors made by the network were corrected by adding hand-crafted examples to the training set.

Playing against Sun Microsystems' *gammontool* program, the final network won about 59% of the time. Note that generalization is almost always necessary; after the first few moves of a game it would be rare to encounter an example from the training set. The network exhibited a great deal of "common sense," almost invariably choosing the best move in situations that were transparent or intuitively clear to a human player. The success rate dropped to 41% without the precomputed features, showing their importance. Without the training set noise (but with the precomputed features) it was 45%, indicating that noise actually helps the system.

A later version of "Neurogammon" won the gold medal at the computer olympiad in London, 1989. It defeated *all* other programs (five commercial and two non-commercial), but lost the game to a human expert by a score of 2–7 [Tesauro, 1990].

Signal Prediction and Forecasting

The problem of signal prediction can be treated in a purely empirical fashion, but the theoretical foundations of the problem are also of fundamental conceptual importance. Since our emphasis in this book is on theory, we digress for a few paragraphs to summarize the important ideas before taking up the operational problem and its neural network implementation.

In many situations in science and technology we want to predict the future evolution of a system from past measurements on it. This is in fact the central procedure of classical physics: we make a mathematical model of a system by writing equations of motion and try to integrate them forward in time to predict the future state. Everyone who has taken an elementary physics course is familiar with this paradigm in systems with a small number of degrees of freedom. Mathematically, we would say that we describe the state of the system by a point $\mathbf{x}$ in a multi-dimensional space Γ (with one dimension for each degree of freedom), and the dynamics can be characterized as the motion of $\mathbf{x}$ in Γ.

This procedure runs into trouble, however, in nonlinear systems with many degrees of freedom, like a turbulent fluid, the weather, or the economy. It is not practical to try to solve the equations of fluid dynamics explicitly except for rather simple special situations; we just cannot keep track of motion in such a high-dimensional space. But all is not lost. Studies of the dynamics of apparently chaotic systems with many degrees of freedom reveal that dissipation (e.g., viscosity) can reduce the number of effectively relevant degrees of freedom to a small number. That is, the motion of the system, which in principle occurs in a very high-dimensional space Γ, becomes confined after some time to a subspace Γ_A of low dimensionality called an **attractor**. The attractor Γ_A is often a rather strange **fractal** object, with a dimensionality d that is not even an integer [Mandelbrot, 1982].

If we can somehow identify the coordinates which characterize the attractor Γ_A, then our problem becomes like one in simple physics, with only a few degrees of freedom. This sounds at first difficult or impossible: when there are so many degrees of freedom in the original description, how can we possibly know—without already having solved the problem—which combinations are the few relevant variables? But in fact this is not a difficulty at all. It is not at all crucial how we choose the new variables as long as there are enough of them, and a set of previous values of the quantity to be predicted is in fact a completely adequate choice.

The physical argument [Packard et al., 1980] for expecting a result like this is that most measurements will probe some combination of the relevant variables (since the motion of the system lies on the attractor), and measurements at different times will in general probe different combinations. Thus a sequence of m such measurements should contain enough information to predict the motion on the attractor, provided m is sufficiently large compared to the attractor dimensionality d. Packard et al. tested this idea in numerical experiments and found support for it.

There is also a rigorous theorem, proved by Takens [1981], which says that there exists a smooth function of at most $2d + 1$ past measurements that correctly predicts the future value of the variable in question. The prediction is just as good as the one we would have made as if we had been able to solve the complete system with its millions of degrees of freedom. Furthermore, if the measured values are known with infinite precision, then the result is insensitive to both the time interval between the past measurements and to how far ahead we want to predict the future measurement. But in practice noise and imprecision in the measurements limit how freely these quantities can be chosen.

Thus we know that we can in principle reduce the prediction problem for a complex dynamical system whose motion lies on a low-dimensional attractor to something like an elementary physics problem. What Takens' theorem does not give us is the explicit form of the function which accomplishes the desired extrapolation. It is here that neural networks come into the picture. The idea is to train a feed-forward network using the set of variables

$$x(t), \ x(t - \Delta), \ x(t - 2\Delta), \ \dots, x(t - (m - 1)\Delta) \tag{6.47}$$

as the input pattern, and the (known) value $x(t+T)$ as target, for many past values of t. In this way one is approximating the true extrapolation mapping by a function like (6.4), parameterized by the weights and thresholds of the network. Once the network has been trained, it can be used for prediction a time T into the future.

Lapedes and Farber [1988] tested this idea using a signal $x(t)$ produced by the numerical solution of the Mackey-Glass differential-delay equation [Mackey and Glass, 1977]

$$\frac{dx}{dt} = -0.1x(t) + \frac{0.2x(t - \tau)}{1 + x^{10}(t - \tau)} . \tag{6.48}$$

Because of the delay τ, this is an infinite-dimensional system—one has to know the initial value of the function on a continuous interval. τ controls how chaotic the motion is. The larger τ, the larger the dimensionality of the attractor. For $\tau = 17$, one finds $d \approx 2.1$ and for $\tau = 30$ one gets $d \approx 3.5$.

The Lapedes-Farber network had *two* hidden layers, because of arguments discussed later about how many layers it takes to approximate an arbitrary function efficiently. Furthermore, the single output unit was taken to be linear ($g(x) = x$) because the target takes on a continuous range of real values; there is no need for the saturation given by the usual sigmoidal nonlinearity.

Training this network by ordinary back-propagation is very slow because of the numerical accuracy desired and the presence of two hidden layers. Lapedes and Farber were able to speed up the convergence by using a conjugate gradient technique. Still, they had to use a Cray supercomputer to achieve a prediction accuracy comparable to that achieved by another new method due to Farmer and Sidorowich [1987, 1988]. In the Farmer and Sidorowich approach one keeps a file of thousands of previous input vectors (6.47), stored in a way that makes it easy to find those vectors close to any chosen point in the m-dimensional space. When a new input vector $x(t)$ is presented as the basis for a prediction of $x(t + T)$, the program looks up the closest vectors in the file and notes where *they* evolved to after a time T. Then it does a linear interpolation from those examples, creating a linear fit to the $t \rightarrow t + T$ mapping in the neighborhood of $x(t)$, and uses that to predict $x(t + T)$.

Both methods predict deterministic chaotic time series much better than traditional methods. The best results can often be achieved with a hybrid of the local-map and neural network schemes; this is discussed at the end of Chapter 9, since it involves unsupervised as well as supervised learning.

Recognizing Hand-Written ZIP Codes

A back-propagation network has been designed to recognize handwritten ZIP codes (numerical postal codes) from the U.S. mail [Le Cun et al., 1989]. The network employs many interesting ideas, some of which we will return to in the next section. Almost 10,000 digits recorded from the mail were used in training and testing the system. These digits was located on the envelopes and segmented into digits by another system, which in itself had to solve a very difficult task.

The network input was a 16×16 array that received a pixel image of a particular handwritten digit, scaled to a standard size. As sketched in Fig. 6.10, this fed forward through *three* hidden layers to the 10 output units, each of which signified one of the digits 0–9.

10 output units

30 units

12 feature
detectors
(4 by 4)

12 feature
detectors
(8 by 8)

16 by 16 input

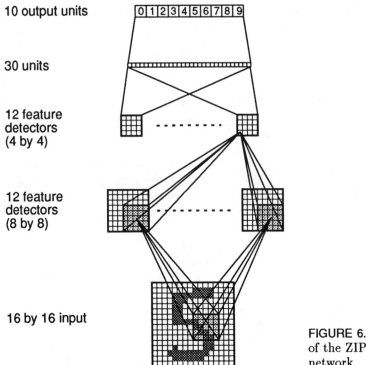

FIGURE 6.10 Architecture
of the ZIP-code reading
network.

The first two hidden layers consisted of trainable feature detectors. The first hidden layer had 12 groups of units with 64 units per group. Each unit in a group had connections to a 5 × 5 square in the input array, with the location of the square shifting by two input pixels between neighbors in the hidden layer. All 64 units in a group had the same 25 weight values, so they all detected the same feature in different places on the retina. This **weight sharing**, and the 5 × 5 receptive fields, reduced the number of free parameters for the first hidden layer from almost 200,000 for fully connected layers to only (25 + 64) × 12 = 1068 (including independent thresholds for every unit).

The second hidden layer was a very similar set of trainable feature detectors consisting of 12 groups of 16 units, again using 5 × 5 receptive fields, a 50% scale reduction, and weight sharing. Inputs to different units were taken from different combinations of 8 of the 12 groups in the first hidden layer, making 8 × 25 shared weights per second-layer group. The third hidden layer consisted of 30 units fully connected to all units in the previous layer, and the 10 output units were in turn fully connected to the all third-layer units. In all there were 1256 units and 9760 independent parameters.

The network was trained by back-propagation, accelerated with the pseudo-Newton rule (6.43). It was trained on 7300 digits and tested on 2000, giving an error rate of about 1% on the training set and 5% on the test set. The error on the test set could be reduced to about 1% by rejecting marginal cases with insufficient difference between the strongest and next strongest outputs. This procedure led to a rejection rate of about 12%.

To obtain good generalization it is important to limit the number of free parameters of the network, as discussed later in this chapter. The weight sharing already reduces this number greatly, but the group has also been experimenting with what they call "optimal brain damage" to remove further weights from the network [Le Cun, Denker, and Solla, 1990]. Using information theoretic ideas a method was developed to make an optimal selection of unnecessary weights. These were then removed and the network was retrained. The resulting network had only around 1/4 as many free parameters as that described above, and worked even better: the rejection rate necessary to achieve 99% correct classification of the test set was reduced to about 9% [Le Cun, Boser, et al., 1990].

Speech Recognition

Speech recognition is one of the most studied tasks in artificial intelligence and neural networks. Speech is difficult to recognize for several reasons:

- Speech is continuous; often there is no pause between words.
- The speed of speech is varying.
- The meaning and pronunciation of a word is highly dependent on context.
- Pronunciation, speed, and syntax are speaker dependent.

Current methods generally perform poorly on speaker-independent continuous speech recognition. Many groups have attempted to train various networks to perform this task. Because of the temporal structure of speech it is natural to consider using *recurrent* networks, as we describe in Section 7.3. Feed-forward networks have also been applied to some speech recognition problems, such as distinguishing among a set of words. The scheme is simple: input some representation of a spoken word and train the perceptron to recognize it. But in practice this only works on tasks limited to a small vocabulary and separate words. Lippmann [1989] gives a good review of the status of speech recognition by neural networks.

6.4 Performance of Multi-Layer Feed-Forward Networks

There are many theoretical questions concerning what multi-layer feed-forward networks can and cannot do, and what they can and cannot *learn* to do. How many layers are needed for a given task? How many units per layer? To what extent does the representation matter? When can we expect a network to generalize? What do we really mean by generalization? How do answers to these questions depend on the details of learning rate, momentum terms, decay terms, etc.?

Some of these questions, and similar problems, can be answered definitively, often with solid theorems. Answers to others are not yet in, but are under intensive study. The following discussion draws strongly on papers by Lapedes and Farber [1987, 1988] and Denker et al. [1987]. See also Lippmann [1987].

The Necessary Number of Hidden Units

First consider layered networks of continuous-valued units, with activation function $g(u) = 1/(1 + e^{-u})$ for hidden units and $g(u) = u$ for output units. Overall such a network implements a set of functions $y_i = F_i\{x_k\}$ from input variables x_k to output variables y_i, where $\{x_k\}$ means $x_1, x_2, \ldots, x_N$. Explicitly, a network with no hidden layers computes

$$y_i = \sum_i w_{ik} x_k - \theta_i \tag{6.49}$$

and one with one hidden layer computes

$$y_i = \sum_j W_{ij} g \left(\sum_k w_{jk} x_k - \phi_j \right) - \theta_i \tag{6.50}$$

and so on. We have included explicit thresholds θ_i and ϕ_j for convenience.

Suppose we want to approximate a particular set of functions $F_i\{x_k\}$ to a given accuracy; how many hidden layers and how many units per layer do we need? The answer is *at most two* hidden layers, with arbitrary accuracy being obtainable given enough units per layer [Cybenko, 1988]. It has also been proved that only *one* hidden layer is enough to approximate any *continuous* function [Cybenko, 1989; Hornik et al., 1989]. The utility of these results depends, of course, on how many hidden units are necessary, and this is not known in general. In many cases it may grow exponentially with the number of input units, as occurs in the case of the general *Boolean* functions discussed later.

Here we give a non-rigorous proof due to Lapedes and Farber [1988] that two hidden layers are enough. The essential points are that

1. any "reasonable" function $F_i\{x_k\}$ can be represented by a linear combination of localized **bumps** that are each non-zero only in a small region of the domain $\{x_k\}$; and

2. such bumps can be constructed with two hidden layers.

With one variable x the function $g(x) - g(x - c)$ obviously gives a peak at $x = c/2$ and is zero far from there. A sharp peak anywhere can be produced with $g(ax + b) - g(ax + c)$. In two or more dimensions we can add together one such function for each dimension, producing a highest peak wherever desired, but also some secondary peaks and valleys. All but the highest peak can be suppressed, however, by another application of $g(u)$ with a suitable threshold. In two dimensions, for example, the function

$$g(A[g(ax + b) - g(ax + c) + g(ay + d) - g(ay + e)] - B) \tag{6.51}$$

can be used to produce a localized bump at any desired (x, y). This is of course exactly the type of function that we can calculate with two hidden layers. Following the same scheme in N dimensions, we need $2N$ units in the first hidden layer and one in the second hidden layer for each bump. The output layer then sums the bumps to produce the desired function(s), in a manner similar to Fourier analysis or Green's function representation.

The bump approach may not be the best one for any particular problem, but it is only intended as an existence proof. More than two hidden layers may well permit a solution with fewer units in all, or may speed up learning. In fact the construction says nothing about learning (or generalization), and it is possible that some functions are representable but not learnable with two hidden layers, perhaps because of local minima.

It is possible to construct units that themselves have a localized bump-like response, each becoming activated only for inputs in some small region of the input space. Not surprisingly, only *one* hidden layer of such units is needed to represent any reasonable function [Hartman et al., 1990]. These *radial basis function* units are discussed further in Section 9.7.

Now let us consider Boolean functions, using $x_k = \pm 1$, $(k = 1, 2, \ldots, N)$ for inputs, $g(u) = \text{sgn}(u)$ on the single output unit, and $g(u) = \tanh(u)$ for the hidden units. How many layers and units/layer do we need to represent any Boolean function? We have already seen (page 94) that only linearly separable functions can be represented with *no* hidden layers. But just *one* hidden layer suffices to represent *any* Boolean function! A one sentence proof for computer scientists consists of the observation that such a network in the deterministic threshold limit contains a programmable logic array (PLA) as a special case, a PLA contains a read-only memory (ROM) as a special case, and a ROM with N-bit addressing can obviously implement any Boolean function on N bits.

For the rest of us, the proof is by construction. Use 2^N units in the hidden layer, $j = 0, 1, \ldots, 2^N - 1$, and set $w_{jk} = +b$ if the kth digit in the binary representation of j is a 1 or $w_{jk} = -b$ if it is a zero. Use a threshold of $N(b - 1)$ at each hidden unit. Then one of the hidden units (the one coded in binary by the input) receives a net input of $h_j = +b$, while all others have $h_j \le -b$. Making b large enough thus turns one hidden unit on ($\sim +1$) and the rest off (~ -1) for each input pattern. The hidden units are said to act as **match filters** or **grandmother cells**.[6] The second stage consists of a link $w_{ij} = +c$ to the output unit i from grandmother cells representing inputs for which the answer is to be $+1$, and links $w_{ij} = -c$ from the rest. The threshold on the output unit is set to $\sum_j w_{ij}$, giving $h_i = \pm 2c$ for the two answers.

Of course this solution is not likely to be practical in the real world. The number of hidden units, and all fan-ins and fan-outs, grow exponentially with the number of input bits N. It is therefore useful to define *efficient* network solutions

---

[6]The term comes from discussion as to whether your brain might contain cells that fire *only* when you encounter your maternal grandmother, or whether such higher level concepts are more distributed.

as those which use a total number of units that grows only polynomially in N. This immediately raises the question of what Boolean functions can be represented by such networks. Denker et al. [1987] call such functions NERFs, for Network Efficiently Representable Functions, and compare them to low order polynomials in curve fitting. Little is yet known as to what this class includes and excludes. It is actually rather hard to find useful Boolean functions that are definitely not NERFs.

Input Representation

Another issue is the importance of input representations. Do they really matter, or can we expect a network to be able to learn a task from any representation? The answer is that they certainly do matter. Consider first two predicates concerned with integers n:

1. n is odd.

2. n has an odd number of prime factors.

If n is presented to a network in binary then predicate 1 is easy; you just look at the lowest order bit. Indeed a simple perceptron is guaranteed to learn such a task from a small number of examples. On the other hand 2 is hard; factoring a number is a very hard problem in computer science and there is no reason to expect a network to learn it from a small training set. But suppose we change the representation? In base 3 rather than base 2, predicate 1 becomes hard too; every bit counts. If however we represented numbers by specifying their prime factors, then predicate 2 would become easy.

Representation is thus crucial. Indeed, we can prove a silly theorem: *learning will always succeed, given the right preprocessor*. The proof makes clear why it is silly: let the preprocessor compute the answer and attach it as additional bits to the raw input; then a simple perceptron (or any generalization thereof) is guaranteed to learn to copy the answer bits and ignore the rest.

It is also important to ask whether the information apparently represented in the input is actually available to the network. With fully connected layers, for example, the network has no inherent sense of the order of the inputs; a permutation of the input bits is a symmetry of the network. A problem that might appear easy for us, given a particular ordering of the input bits, might be much harder for such a network. An example studied by Denker et al. [1987] is the *two-or-more-clumps* predicate, in which the result is to be $+1$ if there are two or more consecutive sequences of $+1$'s in the (ordered) input. Permutation makes the predicate quite obscure, as seen in table 6.1 for the permutation $(0123456789) \rightarrow (3120459786)$. Actually a fully connected network learns this 10-input predicate fairly quickly, but it gets rapidly harder as the number of inputs nodes is made larger [Solla, 1989].

If ordering or another geometrical or topological property of the input is important, then the network should be told about it in some way. For example some input bits may be known to be related or "near" to each other, in the sense that their

TABLE 6.1 The *two-or-more-clumps* predicate

Original input	Result	Permuted input
---+++----	-1	+---++----
---++-+---	+1	+---+-+---
-+++++++++	-1	+++-++++++
+++--++--+	+1	-+++-+++--
----------	-1	----------

correlation is likely to affect the output, whereas others may be "far" from each other, each having a more-or-less independent influence on the output. This occurs for instance when the input bits are representing pixel illuminations in a spatial array. In such cases one may limit the spatial range of the input-to-hidden connections. Solla et al. [1988] found for the *two-or-more-clumps* predicate that this can improve both learning and generalization, and this kind of limited receptive field also proved useful in the ZIP-code reading network described earlier. Another idea is to add a penalty term to the cost function,

$$E = E_0 + \lambda \sum_{jk} |w_{jk}|^2 K(j,k) \qquad (6.52)$$

where $K(j,k)$ is a suitable kernel that increases as the distance between j (in the hidden layer) and k (in the input layer) increases, thus penalizing "distant" connections more strongly.

There may be *internal* symmetries of the architecture too. Permutation of the units within a layer is an obvious example. One can also invert any given unit, changing the sign of all its input and output connections and its threshold. These are discrete symmetries, but there are often continuous symmetries as well, at least approximately. In any situation where the input connections and the threshold scale with some parameter b, the precise value of b is immaterial as long as it is large enough to give saturation. Moreover, b can be different on different units, so we have a multi-dimensional continuous symmetry. The various symmetries give the cost-function landscape $E[\mathbf{w}]$ periodicities, multiple minima, (almost) flat valleys, and (almost) flat plateaus. The last are the most troublesome, because the system can get stuck on such a plateau during training and take an immense time to find its way down.

Generalization

We gave some examples of generalization in the previous section. Some seemed quite impressive; the networks generalized in very "sensible" ways. But it is important to be clear just what it is we are expecting a network to do when we look for generalization. Given the various symmetries in the network and in the problem,

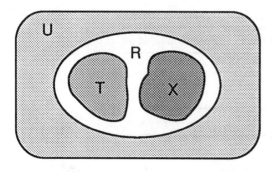

FIGURE 6.11 Rule Extraction
(after Denker et al. [1987]).

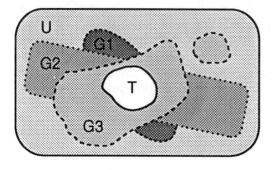

FIGURE 6.12 Generalization (after Denker et al. [1987]).

and perhaps the lack of relevant geometric information, it should not surprise us if a network finds a non-human solution that does not appear to generalize in the way we wish. It may in fact generalize in another way, perhaps equally valid.

Suppose first that we want a network to extract a *rule* from some examples. Figure 6.11 from Denker et al. [1987] defines the situation. There is a universe U of possible input-output pairs, some of which are consistent with a rule R. We select *some* examples of these in a training or memorization set T and try to teach them to the network. Then we test its generalization performance on a *disjoint* subset X of R, an extraction set. Performance on X measures generalization; performance on T measures only memorization. The sets T and X should both be representative of R, probably randomly chosen.

The network only knows about T, not X or R. So it is perfectly valid for it to make any generalization that is consistent with T, as shown in Fig. 6.12, also from Denker et al. [1987]. Here $G1$, $G2$, and $G3$ are all valid generalizations of the memorization set T. It is worth asking how many possible such generalizations there are. If we have N input bits and one output bit there are 2^N input patterns, each of which might have either output value, so there are 2^{2^N} possible rules. If we train the network with p distinct examples we fix p out of 2^N rows of the truth table and thus leave 2^{2^N-p} rules consistent with T. All of these are valid generalizations. The numbers $N = 30$, $p = 1000$ are realistic and give over 2^{10^9} generalizations!

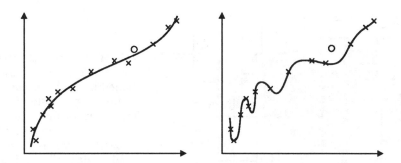

FIGURE 6.13 (a) A good fit to noisy data. (b) Overfitting of the same data: the fit is perfect on the "training set" (x's), but is likely to be poor on a "test set" represented by the circle.

With these numbers one must wonder why networks *ever* find the generalization we prefer. The answer of course must lie in the very special nature of the rules that we normally consider. The information needed to specify an arbitrary rule on N bits is clearly 2^N bits, but "reasonable" rules are specified by no more than N^k bits, for some small k. Similarly the network itself cannot possibly represent most arbitrary rules unless it has an exponentially large number of hidden units. Again we should consider NERFs (page 144), not general Boolean functions.

Another important lesson about generalization can be learned from statistics and curve-fitting; too many free parameters results in **overfitting**. As illustrated in Fig. 6.13, a curve fitted with too many parameters follows all the small details or noise but is very poor for interpolation and extrapolation. The same is true for neural networks: too many weights in a network give poor generalization.

6.5 A Theoretical Framework for Generalization ⋆

The preceding discussion suggests the possibility of quantitative estimates of what networks can and cannot do as a function of their architecture and the nature and size of their training set. There are actually several ways to quantify generalization; we will consider measures of

1. the average number of alternative generalizations of a training set;

2. the probability that the trained network generates the right output for a randomly chosen input, on average; and

3. the same as (2), but in the worst case.

In case (2) we average over possible networks that are consistent with the training set T, but possibly different on other inputs. In case (3) we take the worst network (among those consistent with T), so as to obtain a bound on the probability of

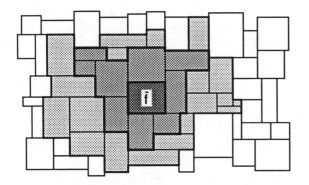

FIGURE 6.14 Weight space. The whole space is partitioned into regions corresponding to different input-output functions. The shaded areas show how the allowed volume is successively reduced by training on examples of f.

error. In each case we can, at least in principle, use the generalization measure to estimate how large a training set is needed for "good" generalization, defining *good* by a requirement that the probable error is smaller than some convenient number.

All the material in this section is also applicable to other deterministic networks, not just feed-forward ones.

The Average Generalization Ability

We discuss first a theoretical framework constructed by Schwartz et al. [1990]. The principal result is surprising; one can calculate the average probability of correct generalization for any training set size p if one knows a certain function that can (in principle) be calculated before training begins. However, the averaging is over all possible networks consistent with the training set, and does not necessarily represent the typical situation encountered in a specific training scheme.

Suppose that we have a class of networks with a certain fixed architecture, specified by the number of layers, numbers of units in each layer, and so on. Each particular case has some specific set of connection weights and thresholds, which we denote collectively by $\mathbf{w}$. We think of a specific $\mathbf{w}$ as a point in an abstract space of all possible $\mathbf{w}$'s, which we call **weight space**. Our averages over possible networks will actually be averages over weight space, with some *a priori* density factor $\rho(\mathbf{w})$; we might for example have the same density for any weights that lie between -10 and $+10$, and zero density otherwise. The total available "volume" V_0 of weight space is just

$$V_0 = \int d\mathbf{w}\, \rho(\mathbf{w}). \tag{6.53}$$

Every vector $\mathbf{w}$ in weight space represents a network implementing a function $f_\mathbf{w}(\boldsymbol{\xi})$ that gives the network output for each input vector $\boldsymbol{\xi}$. Note that different $\mathbf{w}$'s might produce the same function, but different functions cannot have the same $\mathbf{w}$. Thus the whole of weight space is partitioned into a set of disjoint regions, one for each function $f(\mathbf{w})$ that the class of networks can implement, as sketched in Fig. 6.14. This diagram assumes that $\mathbf{w}$ can vary within some region and still implement the same input-output mapping $f(\boldsymbol{\xi})$, as expected for binary output units, but the idea can be extended to continuous-valued outputs: instead of saying whether a set of weights implements a given function we need a measure of the accuracy. We can also have several output units—this makes $f_\mathbf{w}(\boldsymbol{\xi})$ a vector function, but causes no other complication.

The volume of the region of weight space that implements a particular function f is

$$V_0(f) = \int d\mathbf{w}\, \rho(\mathbf{w})\Theta_f(\mathbf{w}) \tag{6.54}$$

where the **indicator function** $\Theta_f(\mathbf{w})$ restricts the nonzero contribution to regions of $\mathbf{w}$ where $f_\mathbf{w} = f$:

$$\Theta_f(\mathbf{w}) = \begin{cases} 1 & \text{if } f_\mathbf{w}(\boldsymbol{\xi}) = f(\boldsymbol{\xi}) \text{ for all } \boldsymbol{\xi}; \\ 0 & \text{otherwise.} \end{cases} \tag{6.55}$$

Note that V_0 and $V_0(f)$ mean different things. Indeed their ratio

$$R_0(f) = \frac{V_0(f)}{V_0} \tag{6.56}$$

is the fraction of weight space that implements a given function f, or the probability of getting that function if we choose random weights with density $\rho(\mathbf{w})$.

Summing over all functions, we can define an information-theoretic entropy

$$S_0 = -\sum_f R_0(f) \log_2 R_0(f) \tag{6.57}$$

which measures the **functional diversity** of the architecture; the larger S_0 the more information is required to specify a particular function. If there were K possible functions with equal volume $V_0(f)$ we would have $R_0(f) = 1/K$ for those functions (and $R_0(f) = 0$ for any other), giving $S_0 = \log_2 K$ or $2^{S_0} = K$. Even when the volumes are not equal, 2^{S_0} is a good measure of the effective number of functions, weighted according to their occurrence probability. We can estimate an upper bound for S_0 by counting how many bits of information are required to specify a set of weights and thresholds.

Under supervised learning, examples of input-output pairs $(\boldsymbol{\xi}^\mu, \zeta^\mu)$ satisfying a particular function $\bar{f}$ are presented to the network; $\zeta^\mu = \bar{f}(\boldsymbol{\xi}^\mu)$. Assuming that learning is successful, the weight vector $\mathbf{w}$ eventually lies within the region of weight space that is compatible with the presented examples. If p examples are learned, then the volume of this remaining region is

$$V_p = \int d\mathbf{w}\, \rho(\mathbf{w}) \prod_{\mu=1}^{p} I(f_\mathbf{w}, \boldsymbol{\xi}^\mu) \tag{6.58}$$

where

$$I(f_\mathbf{w}, \boldsymbol{\xi}^\mu) = \begin{cases} 1 & \text{if } f_\mathbf{w}(\boldsymbol{\xi}^\mu) = \bar{f}(\boldsymbol{\xi}^\mu); \\ 0 & \text{otherwise.} \end{cases} \tag{6.59}$$

V_p includes the region belonging to the desired function $\bar{f}$, plus regions corresponding to other functions that agree with $\bar{f}$ on the training set. As we increase p we expect fewer such alternate functions, and we can think of learning as a continual reduction of the volume of allowed weight space, $V_0 \geq V_1 \geq V_2 \geq \ldots \geq V_p$; see Fig. 6.14.

The fraction of the remaining weight space belonging to a particular function f is modified after learning p examples from $R_0(f)$ to

$$R_p(f) = \frac{V_p(f)}{V_p} \tag{6.60}$$

where $V_p(f)$ is the volume of weight space consistent with both f and the training examples:

$$V_p(f) = \int d\mathbf{w}\, \rho(\mathbf{w}) \Theta_f(\mathbf{w}) \prod_{\mu=1}^{p} I(f_\mathbf{w}, \boldsymbol{\xi}^\mu) \tag{6.61}$$

$$= V_0(f) \prod_{\mu=1}^{p} I(f, \boldsymbol{\xi}^\mu). \tag{6.62}$$

Note that we took the $I(f, \boldsymbol{\xi}^\mu)$ factors outside of the integral—which then reduced to $V_0(f)$—because the $\Theta_f(\mathbf{w})$ factor makes $f_\mathbf{w}$ equal to f in all non-vanishing cases. The result (6.62) shows that $V_p(f)$ is either equal to $V_0(f)$ or is 0, according to whether or not f agrees with the training examples.

The corresponding entropy

$$S_p = -\sum_f R_p(f) \log_2 R_p(f) \tag{6.63}$$

is a measure of how many implementable functions are compatible with the training set. As training proceeds it decreases steadily, and would go to zero if we ever reached the stage where only the desired function $\bar{f}$ was possible. Since S_p is actually a measure of the information required to specify a particular function, the difference $S_{p-1} - S_p$ tells us how much information (in bits) is gained by training on the pth example. This cannot be more than the information required to specify the output ζ^μ, which is one bit for a single binary output, so we would expect $S_p = S_0 - p$ if training were perfectly efficient. We can use this idea, along with an initial estimate of S_0, in a couple of ways: to bound the number of training examples needed to learn a function $\bar{f}$, or to estimate the actually efficiency of training [Denker et al., 1987].

In this discussion we have discriminated sharply between weights which are consistent and inconsistent with particular examples, using the $I(f, \boldsymbol{\xi}^\mu)$ factors in (6.61) and (6.62). For continuous-valued outputs we would have to relax this assumption, replacing the sharp $I(f, \boldsymbol{\xi}^\mu)$ by a smooth function $\exp(-\beta\varepsilon_\mu)$ of the error ε_μ in the μth example. This function falls off gradually from 1 if there is no error to 0 for large error, with a rate governed by the parameter β. Introducing such exponential factors makes (6.61) and (6.62) look exactly like partition function calculations, and indeed they can be treated by statistical mechanics methods [Tishby et al., 1989]. However, we restrict ourselves in this book to the sharp yes/no case.

So far we have said nothing about the training sequence $\xi^1, \xi^2, \ldots, \xi^p$. Let us now assume that each each input pattern ξ^μ is chosen randomly from some distribution $P(\xi)$ of possible inputs, without any dependence on previous choices or on success rate. Then each factor $I(f, \xi^\mu)$ in (6.62) is independent of the others, and we can average over possible training sequences to obtain

$$\langle V_p(f) \rangle \;=\; V_0(f) \left\langle \prod_{\mu=1}^{p} I(f, \xi^\mu) \right\rangle \;=\; V_0(f) g(f)^p \,. \tag{6.64}$$

Here the averages are over ξ^1 to ξ^p, with the appropriate weights $P(\xi^\mu)$, and

$$g(f) \;=\; \langle I(f, \xi) \rangle \;=\; \mathrm{Prob}\big(f(\xi) = \bar{f}(\xi)\big) \tag{6.65}$$

is the probability that a particular function f agrees with $\bar{f}$ for an input ξ randomly chosen from $P(\xi)$.

The quantity $g(f)$ is usually called the **generalization ability** of f; it tells us how well f conforms to $\bar{f}$. Note that it is between 0 and 1, and is independent of the training examples. For example, suppose that f and $\bar{f}$ are Boolean functions of N inputs, and that all 2^N input patterns are equally likely. The functions can be each defined by the 2^N bits in their truth tables. Then $g(f)$ is the fraction of those bits where the two truth tables agree.

Now we consider the probability $P_p(f)$ that a particular function f can be implemented after training on p examples of $\bar{f}$. Our basic approach is to take such probabilities as proportional to weight-space volume, so $P_p(f)$ is equal to the average fraction of the remaining weight space that f occupies:

$$P_p(f) \;=\; \left\langle \frac{V_p(f)}{V_p} \right\rangle \;\approx\; \frac{\langle V_p(f) \rangle}{\langle V_p \rangle} \,. \tag{6.66}$$

The approximation is based on the assumption that V_p will not vary much with the particular training sequence, so $V_p \approx \langle V_p \rangle$ for each probable sequence; we say that V_p is **self-averaging**. This assumption is expected to be good as long as p is small compared to the total number of possible inputs.

We can use (6.66) to compute something more useful; the distribution of generalization ability $g(f)$ across all possible f's:

$$\rho_p(g) \;\equiv\; \sum_f P_p(f) \delta(g - g(f)) \;\propto\; \sum_f \langle V_p(f) \rangle \delta(g - g(f))$$

$$=\; g^p \sum_f V_0(f) \delta(g - g(f))$$

$$\propto\; g^p \rho_0(g) \,. \tag{6.67}$$

Here we used (6.64) and (6.66), and omitted g-independent factors like $\langle V_p \rangle$ because it is easier to normalize at the end, giving

$$\rho_p(g) = \frac{g^p \rho_0(g)}{\int_0^1 (g')^p \rho_0(g') \, dg'} .\tag{6.68}$$

This remarkable result shows that we can calculate the distribution $\rho_p(g)$ of generalization ability after p training examples if we know it before learning begins. The initial distribution

$$\rho_0(g) = V_0^{-1} \sum_f V_0(f) \delta(g - g(f))\tag{6.69}$$

depends only on the architecture and the *a priori* constraints incorporated in $\rho(\mathbf{w})$. Just knowing $\rho_0(g)$ we can find essentially all we want to know about the learning process.

Particularly useful is the *average* generalization ability

$$G(p) = \int_0^1 g\rho_p(g) \, dg = \frac{\int_0^1 g^{p+1} \rho_0(g) \, dg}{\int_0^1 g^p \rho_0(g) \, dg} .\tag{6.70}$$

This gives us the entire **learning curve**—the average expected success rate as a function of p; see Fig. 6.9 for an example. We can use $G(p)$ to predict how many examples will be necessary to train the network to a given average performance.

The form of (6.68) shows that the distribution $\rho_p(g)$ tends to get concentrated at higher and higher values of g as more and more examples are learned. Thus the shrinking of the allowed volume of weight (or function) space under learning is such as to leave remaining regions where the generalization ability is large.

The asymptotic (large p) behavior of $\rho_p(g)$, and hence of $G(p)$, is determined by the form of the initial distribution $\rho_0(g)$ near $g = 1$. There are two cases:

- If there is a finite gap ε between $g = 1$ and the next highest g for which $\rho_0(g)$ is nonzero, then $G(p)$ approaches 1 exponentially:

$$1 - G(p) \propto e^{-p/\varepsilon} .\tag{6.71}$$

- If, on the other hand, there is no such gap in $\rho_0(g)$, then $G(p)$ approaches 1 algebraically:

$$1 - G(p) \propto 1/p .\tag{6.72}$$

It is easy to verify both these results with simple examples, taking for example $\rho_0(g) = \frac{1}{2}\delta(g-1) + \frac{1}{2}\delta(g-1+\varepsilon)$ for the first case, and $\rho_0(g) = 1$ for the second.

The difficulty with applying this theoretical framework is that it requires knowledge of the *a priori* distribution $\rho_0(g)$, which we can calculate analytically only for some very simple problems. One approach is to compute it by exhaustive enumeration for small networks. Schwartz et al. did this for the *two-or-more-clumps* problem (page 144) with 9 to 11 input units. They used the resulting $\rho_0(g)$ in (6.70) and compared the result with that from direct calculation of $G(p)$. Although there were quantitative differences due to an accumulation of numerical errors in the estimated $\rho_0(g)$, the qualitative shape of the learning curves was the same, including the exponential approach (6.71) to saturation.

Bounding the Probability of Poor Generalization

The preceding calculation focuses on the *average* generalization ability, averaging over the region of weight space consistent with the training set. But it is not really clear that this average is appropriate, because the learning rule might favor some regions over others. After all, a particular training run consists of a *path* through weight space as we gradually adjust the weights, not a random selection of a new set of weights constrained by the training set. The initial density $\rho(\mathbf{w})$ can perhaps incorporate some of this effect, but probably not all. It is therefore worthwhile considering another approach that tells us about the generalization ability in the *worst* case, rather than the average case.

The following theory applies not only to neural networks, but to any approximation scheme. We draw heavily on a short pedagogical paper by Abu-Mostafa [1989] that explains the basics. The more detailed treatment can be found in Vapnik and Chervonenkis [1971], Vapnik [1982], and Blumer et al. [1986].

We consider only Boolean functions, appropriate for a network with a single binary output. Again we focus on a particular function $\bar{f}(\boldsymbol{\xi})$ of interest, with value ± 1 (say) for each $\boldsymbol{\xi}$. The generalization ability $g(f)$ of any other function $f(\boldsymbol{\xi})$ is defined by (6.65); $g(f)$ is the probability that $f(\boldsymbol{\xi}) = \bar{f}(\boldsymbol{\xi})$ on a randomly chosen example $\boldsymbol{\xi}$ drawn from some distribution $P(\boldsymbol{\xi})$. Note that $g(f)$ measures how well f approximates $\bar{f}$ and is *independent* of any particular set of examples. We would like to know $g(f)$ for the functions that our network implements, because it tells us how well we are doing—we could use a stopping criterion based on $g(f)$ exceeding some value like 0.95.

Let us assume that we have a training set of p input-output pairs $(\boldsymbol{\xi}^\mu, \zeta^\mu)$ with $\mu = 1, 2, \ldots, p$, where $\zeta^\mu = \bar{f}(\boldsymbol{\xi}^\mu)$. We evaluate a particular network (or other approximation scheme) by how well it performs on this training set; let $g_p(f_\mathbf{w})$ be the fraction of the training set correctly classified by the function $f_\mathbf{w}$ implemented by the network. Our training rule typically adjusts the weights to maximize $g_p(f_\mathbf{w})$, usually achieving $g_p(f_\mathbf{w}) = 1$ if this is possible.

Now compare $g(f)$ and $g_p(f)$. These both measure how well f approximates $\bar{f}$, but $g(f)$ is the average over *all* inputs whereas $g_p(f)$ is the average over a specific *sample* of p inputs, the training set. We might think of using $g_p(f)$ as an estimate of $g(f)$, and we would certainly expect $g_p(f) \to g(f)$ as $p \to \infty$. But unfortunately the sample is *biased* for the functions $f_\mathbf{w}$ that our learning rule produces, since those functions have been chosen with reference to the sample. So in fact we expect $g_p(f_\mathbf{w}) > g(f_\mathbf{w})$ for the functions chosen by the learning rule, and cannot immediately use $g_p(f_\mathbf{w})$ as an unbiased estimate of $g(f_\mathbf{w})$. Indeed it might be possible to find a function without errors on the training set, $g_p(f_\mathbf{w}) = 1$, but far from right on most other inputs, $g(f) \approx 0$, as in the curve fitting example in Fig. 6.13.

However, the sample value $g_p(f)$ is *not* biased if we consider an arbitrary function f among all those that the network can implement, instead of a specific one $f_\mathbf{w}$ associated with the training set. This allows the application of some powerful statistical ideas to tell us how bad the estimate could be in the worst case. And the worst-case result applies to *any* implementable function f, so it also bounds the error for our specific function $f_\mathbf{w}$. The key result is a bound

$$\text{Prob}\big(\max_f |g_p(f) - g(f)| > \varepsilon\big) \leq 4m(2p)e^{-\varepsilon^2 p/8} \tag{6.73}$$

proved by Vapnik and Chervonenkis [1971]. The left-hand side is the probability that the worst case estimation error exceeds some small number ε, for any possible function f implementable by the network. So if we could make the right-hand side small, less than 0.01 say, then would know with 99% probability that $g_p(f)$ and $g(f)$ are within ε of each other for *any* implementable function f. In particular, we would know (with 99% certainty) that

$$g(f_\mathbf{w}) > 1 - \varepsilon \tag{6.74}$$

if we obtained a perfect result $g_p(f_\mathbf{w}) = 1$ on the training set.

The right-hand side of (6.73) involves the **growth function** $m(p)$, which is defined to be the maximum number of *different* binary functions that could be implemented by the network on *any* set of p examples ξ^μ. There are a total of 2^p different binary functions on p points (consider the different red/black colorings of the points), so $m(p) \leq 2^p$. But it is not necessarily the case that a particular architecture can implement all 2^p cases as we vary the weights, no matter how the p points are chosen. For example a simple perceptron can only implement linearly separable functions, and for large enough p there are less than 2^p of those. In some cases the total number of possible functions is limited by the architecture, giving an upper bound for $m(p)$. Consider for instance the effect of limiting each connection weight to one of k values; clearly $m(p) \leq k^W$ where W is the total number of weights.

If $m(p)$ grows less rapidly than exponential in p, then the right-hand of (6.73) can be made arbitrarily small by choosing p large enough. This lets us make the generalization error as small as desired. On the other hand, the bound (6.73) is not very useful if $m(p)$ grows exponentially for ever.

Vapnik and Chervonenkis proved that the growth function $m(p)$ always looks as shown in Fig. 6.15. It is equal to 2^p up to some point, $p = d_{VC}$, where the growth starts to slow down. d_{VC} is called the Vapnik-Chervonenkis dimension, or just the VC dimension. It can be infinite, in which case $m(p) = 2^p$ for all p and the network will never generalize. But if d_{VC} is finite it can be shown that $m(p)$ obeys the inequality

$$m(p) \leq p^{d_{VC}} + 1 \tag{6.75}$$

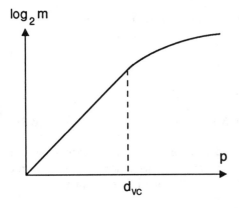

log$_2$ m

FIGURE 6.15 The growth function always starts out equal to 2^p and then bends over at d_{VC} (unless d_{VC} is infinite).

and so does *not* continue growing exponentially. Then (6.73) becomes useful in bounding the generalization error.

Let us consider some examples. A straightforward case is the simple perceptron with N inputs, for which

$$m(p) = C(p, N) = 2\sum_{i=0}^{N-1} \binom{p-1}{i} \tag{6.76}$$

from (5.67); the number $m(p)$ of implementable Boolean functions on p points is just the same as the number $C(p, N)$ of dichotomies on p points, because every set of weights corresponds to a dichotomy.

The binomial formula leads to the identity

$$2^n = (1+1)^n = \sum_{m=0}^{n} \binom{n}{m} 1^m 1^{n-m} = \sum_{m=0}^{n} \binom{n}{m} \tag{6.77}$$

so for $p \le N$ we obtain, recalling $\binom{n}{m} = 0$ for $m > n$,

$$C(p, N) = 2\sum_{i=0}^{N-1} \binom{p-1}{i} = 2\sum_{i=0}^{p-1} \binom{p-1}{i} = 2 \cdot 2^{p-1} = 2^p . \tag{6.78}$$

But for $p > N$, $C(p, N)$ is less than 2^p, because some terms in the sum (6.77) are missing. Thus the VC dimension is N for the simple perceptron.

By inserting $C(p, N)$ into (6.73) one can calculate numerically the number of examples needed (in the worst case) to obtain good generalization with a simple perceptron. We can get an upper bound on this number using (6.73) and (6.75):

$$\text{Prob}\left(\max_{f} |g_p(f) - g(f)| > \varepsilon\right) \le 4[(2p)^N + 1]e^{-\varepsilon^2 p/8} \approx 4e^{N\log(2p)-\varepsilon^2 p/8}. \tag{6.79}$$

For this bound to be smaller than some small number δ we must have

$$N < \frac{\varepsilon^2 p - 8\log(4/\delta)}{8\log(2p)}. \tag{6.80}$$

In the limit of large N and p this gives approximately

$$p > 8N\log(N)/\varepsilon^2. \tag{6.81}$$

Thus the number of training examples needed to train an N-input perceptron and obtain good generalization in the worst case scales up as $N\log N$. This doesn't seem too bad—considering the fact that there are 2^N possible examples—but the prefactor $8/\varepsilon^2$ dampens any enthusiasm for small ε.

As a second example consider a general feed-forward network with M threshold nodes and W weights (including thresholds). Baum and Haussler [1989] calculated an upper bound for the VC dimension of any such network, obtaining

$$d_{VC} \le 2W\log_2(eM) \tag{6.82}$$

where e is the base of the natural logarithm, $e = \exp(1)$. From this they derived an upper bound on the size p of the training set needed to get a good generalization probability. If the error on the training set is less than $\varepsilon/2$ they showed that at most of the order of $\frac{W}{\varepsilon}\log\frac{M}{\varepsilon}$ examples are needed to obtain a generalization error less than ε.

For a network with N inputs and one fully connected hidden layer of H units they also derived a lower bound

$$d_{VC} \ge 2\lfloor H/2 \rfloor N \simeq W \qquad \text{(for large } H) \tag{6.83}$$

where $\lfloor x \rfloor$ means the largest integer not greater than x. They used this to show that one needs (loosely speaking) of the order of W/ε training examples to expect a generalization error less than ε.

6.6 Optimal Network Architectures

We have seen that the network architecture is very important, and each application requires its own architecture. To obtain good generalization ability one has to build into the network as much knowledge about the problem as possible (e.g., the topology of the input space) and limit the number of connections appropriately. It is therefore desirable to find algorithms that not only optimize the weights for a given architecture, but also optimize the architecture itself. This means in particular optimizing the number of layers and the number of units per layer.

Of course there are various different criteria for *optimal*, including generalization ability, learning time, number of units, and so on. In fact, given various hardware restrictions, there may be quite a complicated cost function for the architecture itself. We focus mainly on using as few units as possible; this should not only reduce computational costs and perhaps training time, but should also improve generalization.

It is of course possible to mount a search in the space of possible architectures. We have to train each architecture separately by (say) back-propagation, and then evaluate it with an appropriate cost function that incorporates both performance and number of units. Such a search can be carried out by a **genetic algorithm**, so that good building blocks found in one trial architecture are likely to survive and be combined with good building blocks from other trials [Harp et al., 1990; Miller et al., 1989]. However, this kind of search seems unlikely to be practical for applications requiring large networks, where training just one architecture often requires massive CPU power.

More promising are approaches in which we construct or modify an architecture to suit a particular task, proceeding incrementally. There are two such ways to reach as few units as possible: start with too many and take some away; or start with too few and add some more. We consider examples of each approach.

Pruning and Weight Decay

We have already briefly described one way of optimizing the architecture in the ZIP-code reading network on page 139. There the network was trimmed by removing unimportant connections. It is also possible to prune unimportant *units* [Sietsma and Dow, 1988]. In either case it is necessary to retrain the network after the "brain damage", though this retraining is usually rather fast.

Another approach is to have the network itself remove non-useful connections during training. This can be achieved by giving each connection w_{ij} a tendency to decay to zero, so that connections disappear unless reinforced [Hinton, 1986; Scalettar and Zee, 1988; Kramer and Sangiovanni-Vincentelli, 1989]. The simplest method is to use

$$w_{ij}^{\text{new}} = (1 - \varepsilon)w_{ij}^{\text{old}} \tag{6.84}$$

after each update of w_{ij}, for some small parameter ε. This is equivalent to adding a penalty term w_{ij}^2 to the original cost function E_0

$$E = E_0 + \gamma \sum_{(ij)} w_{ij}^2 \tag{6.85}$$

and performing gradient descent $\Delta w_{ij} = -\eta \partial E / \partial w_{ij}$ on the resulting total E. The ε parameter is then just $\gamma\eta/2$.

While (6.85) clearly penalizes use of more w_{ij}'s than necessary, it overly discourages use of large weights; one large weight costs much more than many small ones. This can be cured by using a different penalty term, such as

$$E = E_0 + \gamma \sum_{(ij)} \frac{w_{ij}^2}{1 + w_{ij}^2} \tag{6.86}$$

which is equivalent to making ε in (6.84) dependent on w_{ij}

$$\varepsilon_{ij} = \frac{\gamma\eta/2}{\left(1 + w_{ij}^2\right)^2} \tag{6.87}$$

so that small w_{ij}'s decay more rapidly than large ones.

These decay rules perform well in removing unnecessary *connections*, but often we want to remove whole *units*. Then we can start with an excess of hidden units and later discard those not needed. It is easy to encourage this by making the weight decay rates larger for units that have small outputs, or that already have small incoming weights [Hanson and Pratt, 1989; Chauvin, 1989]. For example we could replace (6.87) by

$$\varepsilon_i = \frac{\gamma\eta/2}{\left(1 + \sum_j w_{ij}^2\right)^2} \tag{6.88}$$

and use this same ε_i for all connections feeding unit i.

Network Construction Algorithms

Rather than starting with too large a network and performing some pruning, it is more appealing to start with a small network and gradually grow one of the appropriate size. There have been several attempts in this direction; we outline three of them and discuss one (the tiling algorithm) in more detail.

Figure 6.16 shows the way that three different algorithms construct networks. In each case the aim is to construct a network that correctly evaluates a Boolean function from N binary inputs to a single binary output, given by a training set of p input-output pairs. We assume that the training set has no internal conflicts (different outputs for the same input). Extensions to multiple outputs are possible, but not discussed here. Threshold units are used in all layers. Each algorithm tries to construct a network using as few units as possible within a particular construction scheme.

Marchand et al. [1990] propose an algorithm than constructs a solution using a single hidden layer, as shown in Fig. 6.16(a). Hidden units are added one-by-one, each separating out one or more of the p patterns, which are then removed from consideration for the following hidden units. Specifically, each hidden unit is chosen so that it has the same output (say $+1$) for *all* remaining patterns with one target (say $+1$), and the opposite output (-1) for *at least one* of the remaining patterns with the opposite target (-1); this can always be done. The latter one or more patterns are then removed from consideration for the following hidden units. This process terminates when all remaining patterns have the same target.

The resulting patterns, or **internal representations**, on the hidden layer each have a unique target ± 1. Moreover, the hetero-association problem from these internal representations to their targets can be shown to be linearly separable. It can therefore be performed with just one more layer, using the perceptron learning rule.

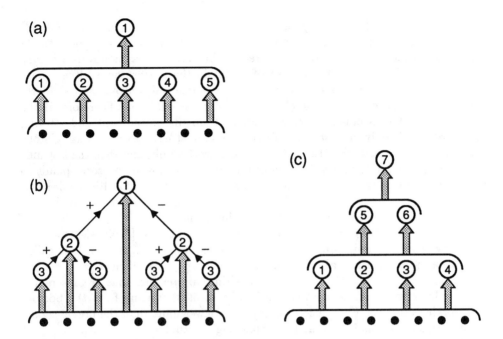

FIGURE 6.16 Network construction algorithms. The black dots are inputs, while the numbered circles are threshold units, *numbered in order of their creation*. Shaded arrows represent connections from all the units or inputs at the arrow's tail. (a) Marchand et al. [1990]. (b) Frean [1990]. (c) Mézard and Nadal [1989].

Figure 6.16(b) shows the kind of architecture generated by the **upstart algorithm** of Frean [1990]. It is specifically for off/on 0/1 units. First we do the best we can with a single output unit 1, directly connected to the input. Then we note all the cases in which the output is wrong, and create two more units 2 (if needed), one to correct the *wrongly on* cases and one to correct the *wrongly off* ones. The subsidiary units 2 are connected to the output units with large positive or negative weights, so that they override the previous output when activated. The subsidiary units 2 are directly connected to the input, and are trained to do the best they can on their own problem of correcting the wrongly on or wrongly off patterns, without upsetting the correct patterns. If necessary we create further units 3 to correct *their* mistakes, and so on. Each additional unit created reduces the number of incorrectly classified patterns by at least one, so the process must eventually cease.

The upstart algorithm generates an unusual hierarchical architecture, but in fact this can be converted into an equivalent two-layer network. All the units of the hierarchical arrangement are placed in the hidden layer, removing the connections between them. Then a new output unit is created and fully connected to the hidden layer; appropriate connections can be found to regenerate the desired targets.

Sirat and Nadal [1990] independently proposed exactly the same way of dividing up the input space. In their approach the "daughter" units are not actually connected to their "parent" unit to correct it, but are used as nodes in a binary decision tree. Note that if the output of a unit in the upstart network is 1 only the daughter unit that corrects *wrongly on* can change this output, and similarly if the output is 0 only the *wrongly-off* daughter can affect it. Therefore, if the units are updated in opposite order—starting from the output unit and working backwards—it is only necessary to update *one* unit at each level. Updating stops when there is no daughter of the right type to correct a unit, and then the last unit that *was* updated determines the class of the input. This procedure corresponds to traversing a binary tree, called a **neural tree** by Sirat and Nadal. Binary decision trees are frequently used for classification, and this one is characterized by having a simple one-layer perceptron at each branching point.

Another interesting algorithm by Fahlman and Lebiere [1990] also builds a hierarchy of hidden units similar to the one in the upstart algorithm. It is called the **cascade-correlation algorithm**, and seems to be very efficient.

Mézard and Nadal [1989] proposed a **tiling algorithm** that creates multi-layer architectures such as that shown in Fig. 6.16(c), starting from the bottom and working upwards. Each successive layer has fewer units than the previous one, so the process eventually terminates with a single output unit.

For any such architecture to be successful it is clear that the patterns (internal representations) on every layer must be **faithful representations** of the input patterns. That is, if two input patterns have different targets at the output layer, then their internal representations must be different on every hidden layer. The idea of the tiling algorithm is to start each layer with a **master unit** that does as well as possible on the target task, and then add further **ancillary units** until the representation on that layer is faithful. The next layer is constructed in just the same way, using the output of the previous layer as its input. Eventually a master unit itself classifies all patterns correctly, and is therefore the desired output unit.

The master unit in each layer is trained so as to produce the correct target output (± 1) on as many of its input patterns as possible. This can be done by a variant of the usual perceptron learning rule (5.19) called the **pocket algorithm** [Gallant, 1986]. The unmodified perceptron learning rule wanders through weight space when the problem is *not* linearly separable, spending most time in regions giving the fewest errors, but not staying there. So the pocket algorithm modification consists simply in storing (or "putting in your pocket") the set of weights which has had the longest unmodified run of successes so far. The algorithm is stopped after some chosen time t, which is the only free parameter in the tiling algorithm.

The ancillary units in each layer are also trained using the pocket algorithm, but only on subsets of the patterns. Whenever the representation on the latest layer is *not* faithful, then that layer has at least one activation pattern without a unique target; the subset of input patterns that produces the ambiguous pattern includes both targets. So we train a new ancillary unit on this subset, trying to separate it as far as possible. Then we look again for ambiguous patterns, and repeat as often

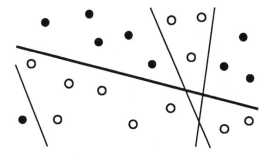

FIGURE 6.17 Tiling the input space. The master unit (heavy line) does the best possible separation of the points. The ancillary units (thinner lines) make sure that the wrongly classified points are separated into classes with the same target.

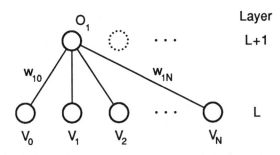

FIGURE 6.18 Notation for convergence proof of the tiling algorithm. V_1 and O_1 are master units, and V_0 is a bias unit.

as necessary until the representation is faithful. Fig. 6.17 shows how this divides up the input space.

To prove that the whole process converges we construct a set of weights that lets the master unit in layer $L + 1$ classify at least one more pattern correctly than the master unit in layer L. Of course the pocket algorithm will probably not choose our specific weights, but it will certainly do no worse in the number of misclassifications. Thus each master unit correctly classifies more patterns than the last, and so the network construction process eventually terminates.

We use the notation shown in Fig. 6.18; units (or inputs) V_j in layer L are connected to the master unit O_1 in layer $L + 1$ by weights w_{1j}. It is essential to include an explicit bias or threshold, which we do by fixing $V_0 = 1$. We assume that the layer L master unit V_1 classifies q input patterns correctly ($V_1^\mu = \zeta^\mu$), with $q < p$. Now consider a pattern ν that is *not* classified correctly by V_1, so that $V_1^\nu = -\zeta^\nu$. Choose the weights

$$w_{1j} = \begin{cases} 1 & \text{if } j = 1; \\ \varepsilon \zeta^\nu V_j^\nu & \text{otherwise} \end{cases} \tag{6.89}$$

with

$$\frac{1}{N} < \varepsilon < \frac{1}{N - 2} \tag{6.90}$$

where N is the number of units in layer L besides the bias unit V_0. This makes O_1

classify pattern ν correctly

$$O_1^\nu = \text{sgn}\left(\sum_j w_{1j} V_j^\nu\right) = \text{sgn}\left(-\zeta^\nu + \varepsilon\zeta^\nu N\right) = \zeta^\nu \tag{6.91}$$

since $N\varepsilon > 1$. But it also leaves intact the q correct classifications by V_1, for which we find

$$O_1^\mu = \text{sgn}\left(\sum_j w_{1j} V_j^\mu\right) = \text{sgn}\left(\zeta^\mu + \varepsilon\zeta^\nu \sum_{j\neq 1} V_j^\mu V_j^\nu\right). \tag{6.92}$$

Because $(N-2)\varepsilon < 1$, the second term in the sign function could only change the correct sign of the first term ζ^μ if $|X| = N$ where

$$X = \sum_{j\neq 1} V_j^\mu V_j^\nu . \tag{6.93}$$

But X cannot be $-N$ because $V_0^\mu = V_0^\nu = 1$, and it cannot be $+N$ because V_j^μ and V_j^ν are not identical if the level L representation is faithful. So the q originally correct patterns are not upset, and the new unit O_i classifies at least $q+1$ patterns correctly, as claimed.

It is not yet clear which of the three methods described is best for a given problem. All of them have given encouraging results on simple test problems, both in terms of generalization and in terms of finding efficient architectures. In one comparative test the upstart algorithm used fewer units than the tiling algorithm [Frean, 1990], but more wide-ranging studies are needed. It is also likely that further construction algorithms will be proposed in the future.

SEVEN

Recurrent Networks

The preceding chapter was concerned strictly with supervised learning in *feed-forward* networks. We now turn to supervised learning in more general networks, with connections allowed both ways between a pair of units, and even from a unit to itself. These are usually called **recurrent networks**. They do not necessarily settle down to a stable state even with constant input. Symmetric connections ($w_{ij} = w_{ji}$) ensure a stable state of course (as seen in the Hopfield networks), and the Boltzmann machines discussed first are limited to the symmetric case. In the second section we consider networks without the symmetry constraint, but only treat those that *do* reach a stable state. Then we examine networks that can learn to recognize or reproduce time sequences; some of these produce cyclic output rather than a steady state. We conclude with a discussion of reinforcement learning in recurrent and non-recurrent networks.

7.1 Boltzmann Machines

Hinton and Sejnowski [Hinton and Sejnowski, 1983, 1986; Ackley, Hinton and Sejnowski, 1985] introduced a general learning rule applicable to any *stochastic* network with *symmetric* connections, $w_{ij} = w_{ji}$. They called this type of network a **Boltzmann machine** because the probability of the states of the system is given by the Boltzmann distribution of statistical mechanics. Boltzmann machines may be seen as an extension of Hopfield networks to include hidden units. Just as in feed-forward networks with hidden units, the problem is to find the right connections to the hidden units without knowing from the training patterns what the hidden units should represent.

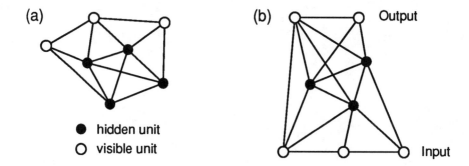

FIGURE 7.1 (a) A Boltzmann machine has the units divided into visible and hidden ones. (b) The visible units can be divided into input and output units.

In its original form, the Boltzmann learning algorithm is very slow because of the need for extensive averaging over stochastic variables. There have been therefore few applications compared to back-propagation. But a deterministic "mean field" version of the Boltzmann machine speeds up the learning considerably and may lead to more applications.

Note that stochastic networks are sometimes called Boltzmann machines even if the weights are chosen *a priori* and there is no learning involved. We avoid this usage, and discuss stochastic networks along with their deterministic counterparts in most chapters of this book.

Stochastic Units

The units S_i are divided into **visible** and **hidden units** as shown in Fig. 7.1(a). The visible units can, but need not, be further divided into separate input and output units as in Fig. 7.1(b). The hidden units have no connection to the outside world. The connections between units may be complete—every pair—or structured in some convenient way. For example, one might have everything except direct input-output connections. But whatever the choice, all connections must be symmetric, $w_{ij} = w_{ji}$.

The units are stochastic, taking output value $S_i = +1$ with probability $g(h_i)$ and value $S_i = -1$ with probability $1 - g(h_i)$, where

$$h_i = \sum_j w_{ij} S_j \tag{7.1}$$

and

$$g(h) = \frac{1}{1 + \exp(-2\beta h)} \tag{7.2}$$

just as in (2.48). Here $\beta = 1/T$ as usual, and we omit thresholds for convenience. Because of the symmetric connections (cf. Chapter 2) there is an energy function

$$H\{S_i\} = -\frac{1}{2} \sum_{ij} w_{ij} S_i S_j \qquad (7.3)$$

which has minima whenever there is a stable state characterized by $S_i = \text{sgn}(h_i)$. The Boltzmann-Gibbs distribution from statistical mechanics,

$$P\{S_i\} = e^{-\beta H\{S_i\}}/Z \qquad (7.4)$$

gives, at least in principle, the probability of finding the system in a particular state $\{S_i\}$ after equilibrium is reached. The denominator Z, called the partition function, is the appropriate normalization factor

$$Z = \sum_{\{S_i\}} e^{-\beta H\{S_i\}} . \qquad (7.5)$$

A more detailed discussion of the Boltzmann-Gibbs distribution is provided in the Appendix. We can use (7.4) to compute the average $\langle X \rangle$ of any quantity $X\{S_i\}$ that depends just on the state $\{S_i\}$ of the system:

$$\langle X \rangle = \sum_{\{S_i\}} P\{S_i\} X\{S_i\} . \qquad (7.6)$$

In Boltzmann learning we attempt to adjust the connections w_{ij} to give the states of the *visible* units a particular desired probability distribution. We might, for example, want only a few of the possible states to have appreciable probability of occurring, and want those few to have equal probability. At low temperature we can use such a scheme for **pattern completion**, in which missing bits of a partial pattern are filled in by the system. Or we might divide the visible units into *input* and *output* units and do ordinary association. Our information as to the "correct" output for each input pattern might itself be probabilistic (or *fuzzy*), as for example in medical diagnosis of diseases from patterns of symptoms [Hopfield, 1987]. Asking for an approximation to a probability distribution is a general goal covering these and other cases.

The task is similar to that given an auto-associative (Hopfield) network, but the architecture of the Boltzmann machine differs in having hidden units. Without hidden units we can do no more than specify all $\langle S_i \rangle$'s and all $\langle S_i S_j \rangle$'s; higher-order correlations cannot be chosen independently. Thus, for example, a three-unit network without hidden units could not learn to produce the four patterns +-+, -++, ---, ++- with probability $\frac{1}{4}$ each, and the remaining four patterns not at all, because the first- and second-order correlations here are $\langle S_i \rangle = \langle S_i S_j \rangle = 0$ for all i and j, exactly as in the set of all 8 states. With hidden units this task (which is actually our old friend the XOR problem in disguise) *can* be learned. But note that the $w_{ij} \propto \xi_i \xi_j$ rule that we used in the Hopfield network gives us no information about how to treat links involving hidden units; we need a new rule.

Let us label the *states* of the visible units by an index α and those of the hidden units by β.[1] With N visible units and K hidden units α and β run from 1 to 2^N and from 1 to 2^K respectively. A state of the whole system is uniquely specified by an α and a β, with 2^{N+K} possibilities. The probability $P_{\alpha\beta}$ of finding each of these states is given by (7.4). The probability P_α of finding the visible units in state α irrespective of β is then obtained from

$$
\begin{aligned}
P_\alpha &= \sum_\beta P_{\alpha\beta} \\
&= \sum_\beta e^{-\beta H_{\alpha\beta}} / Z
\end{aligned}
\tag{7.7}
$$

where

$$
Z = \sum_{\alpha\beta} e^{-\beta H_{\alpha\beta}}
\tag{7.8}
$$

and

$$
H_{\alpha\beta} = -\frac{1}{2} \sum_{ij} w_{ij} S_i^{\alpha\beta} S_j^{\alpha\beta}
\tag{7.9}
$$

is the energy of the system in state $\alpha\beta$, in which $S_i^{\alpha\beta}$ is the value (± 1) of S_i.

Equation (7.7) gives the actual probability P_α of finding the visible units in state α in the freely running system, and is determined by the w_{ij}'s. We, on the other hand, have a set of **desired probabilities** R_α for these states. A suitable measure of the difference between the distributions P_α and R_α, properly weighted by the occurrence probabilities R_α, is the relative entropy

$$
E = \sum_\alpha R_\alpha \log \frac{R_\alpha}{P_\alpha} .
\tag{7.10}
$$

This may be derived from information-theoretic arguments as for (5.52), though we will discuss an alternative statistical mechanics interpretation below. E is always positive or zero, and can only be zero if $P_\alpha = R_\alpha$ for all α.[2] We therefore minimize E, using gradient descent:

$$
\Delta w_{ij} = -\eta \frac{\partial E}{\partial w_{ij}} = \eta \sum_\alpha \frac{R_\alpha}{P_\alpha} \frac{\partial P_\alpha}{\partial w_{ij}} .
\tag{7.11}
$$

---

[1] We use β in this sense as well as to mean $1/T$, but the latter never occurs as an index or subscript.

[2] It is easy to show from the integral definition of the logarithm that $\log X \geq 1 - 1/X$. Therefore $E \geq \sum_\alpha R_\alpha (1 - P_\alpha/R_\alpha) = \sum_\alpha (R_\alpha - P_\alpha) = 0$.

Using (7.7)–(7.9) we find

$$\frac{\partial P_\alpha}{\partial w_{ij}} = \frac{\beta \sum_\beta e^{-\beta H_{\alpha\beta}} S_i^{\alpha\beta} S_j^{\alpha\beta}}{Z} - \frac{(\sum_\beta e^{-\beta H_{\alpha\beta}}) \beta \sum_{\lambda\mu} e^{-\beta H_{\lambda\mu}} S_i^{\lambda\mu} S_j^{\lambda\mu}}{Z^2}$$

$$= \beta \left[\sum_\beta S_i^{\alpha\beta} S_j^{\alpha\beta} P_{\alpha\beta} - P_\alpha \langle S_i S_j \rangle \right]. \tag{7.12}$$

Thus

$$\Delta w_{ij} = \eta\beta \left[\sum_\alpha \frac{R_\alpha}{P_\alpha} \sum_\beta S_i^{\alpha\beta} S_j^{\alpha\beta} P_{\alpha\beta} - \sum_\alpha R_\alpha \langle S_i S_j \rangle \right]$$

$$= \eta\beta \left[\sum_{\alpha\beta} R_\alpha P_{\beta|\alpha} S_i^{\alpha\beta} S_j^{\alpha\beta} - \langle S_i S_j \rangle \right]$$

$$= \eta\beta \left[\overline{\langle S_i S_j \rangle}_{\text{clamped}} - \langle S_i S_j \rangle_{\text{free}} \right] \tag{7.13}$$

where we have defined the conditional probability $P_{\beta|\alpha}$ of hidden state β given visible state α by

$$P_{\alpha\beta} = P_{\beta|\alpha} P_\alpha \tag{7.14}$$

and have identified

$$\overline{\langle S_i S_j \rangle}_{\text{clamped}} = \sum_{\alpha\beta} R_\alpha P_{\beta|\alpha} S_i^{\alpha\beta} S_j^{\alpha\beta} \tag{7.15}$$

as the value of $\langle S_i S_j \rangle$ when the visible units are clamped in state α, averaged over α's according to their probabilities R_α.

Equation (7.13) is the central rule of Boltzmann learning. The first term is essentially a Hebb term, with the visible units clamped, while the second term corresponds to Hebbian **unlearning** with the system free running. The process converges when the free unit-unit correlations $\langle S_i S_j \rangle$ are equal to the clamped ones. It is crucial that Hebb-like learning in a network employ both these terms, or have some other way of telling the hidden units whether or not the visible units are clamped. Otherwise an attempt to adjust w_{ij} to increase the correlation $\langle S_i S_j \rangle$ between two units will "learn" internally generated fluctuations as well as externally applied signals; the network will become absorbed in its own fantasies.

To operate a Boltzmann machine we need to be in *equilibrium* at some temperature $T > 0$. The state $\{S_i\}$ of the whole system then fluctuates, and we measure the correlations $\langle S_i S_j \rangle$ by taking a time average of $S_i S_j$. To obtain all the information needed for the update rule (7.13) we need to do this once with the visible units clamped in each of their states α for which $R_\alpha > 0$, and once with them unclamped. In each case we must bring the system to equilibrium anew before taking an average.

Some specialized electronic [Alspector and Allen, 1987] and optoelectronic [Farhat, 1987; Ticknor and Barrett, 1987] hardware has been developed for the

Boltzmann machine. But in the absence of such hardware we must study the system by **Monte Carlo simulation**,[3] selecting units at random and updating them according to the probability (7.2). This is equivalent (as in (4.4)) to flipping their state $(S_i \rightarrow -S_i)$ with probability

$$\text{Prob}(S_i \rightarrow -S_i) = \frac{1}{1 + \exp(\beta \Delta H_i)} \tag{7.16}$$

where ΔH_i is the energy change produced by such a flip. Unfortunately this procedure takes a *very* long time to come to equilibrium at low T because the system tends to get trapped in local minima of the (free) energy. But we *do* need fairly low T's, to represent a wide range of probabilities. The solution is to use a **simulated annealing** procedure, with a gradual lowering of the temperature from a high initial value to the desired value T, as discussed on page 75. This achieves equilibrium at the working temperature T much more rapidly than starting from there initially.

The Boltzmann machine is clearly very computation intensive, even with simulated annealing to help us reach equilibrium. In effect there are four nested loops:

1. At the outermost level we must adjust the weights many times for convergence, using (7.13) for each update.

2. For each of these we must calculate $\langle S_i S_j \rangle$ in an unclamped state, and with the visible units clamped in each desired pattern.

3. For each of these averages we must come to equilibrium using an **annealing schedule** $T(t)$ of gradually decreasing temperatures.

4. At each of these temperatures we must sample many units and update them according to (7.16).

Note that the weight updates depend on the difference between two averages, each of which has fluctuations. Using a poorly equilibrated system or a short averaging time can reduce the time per update, but produces poor Δw_{ij}'s and ultimately requires more updating cycles.

Choice of the annealing schedule and the final working temperature are part of an active research field [see e.g., Salamon et al., 1988]. In practice various *ad hoc* schedules are used. Ideally one can speed up simulated annealing greatly, and use a T_0/t schedule, by considering occasional **multiple flips** of many units at a time, using the same rule (7.16) to decide whether to accept such moves [Szu, 1986]. This is known as **fast simulated annealing** or a **Cauchy machine**. However it is not clear how to generate such moves in practice and a neural network implementation has not yet been demonstrated.

Boltzmann machines have been applied to a number of problems, including constraint satisfaction problems in vision [Hinton and Sejnowski, 1983], the encoder problem [Ackley et al., 1985; Parks, 1987], learning symmetries in two dimensions [Sejnowski et al., 1986], statistical pattern recognition [Kohonen et al.,

---

[3]A good introduction to Monte Carlo methods is Binder and Heerman [1988]. Mazaika [1987] discusses selection and update rules for the Boltzmann machine.

1988], speech recognition [see Lippmann, 1989], and combinatorial optimization problems [Gutzmann, 1987]. Although extremely slow they are found to be very effective. In a detailed comparison on a statistical decision task, Kohonen et al. [1988] found that a Boltzmann machine achieved considerably better accuracy than a back-propagation network, and came close to the theoretical Bayes limit.

Variations on Boltzmann Machines

Several minor variations on the Boltzmann rule are often used in practice [Ackley et al., 1985; Derthick, 1984]:

- Weight decay terms can be added, as described in Chapter 6 on page 157. This seems to improve performance considerably. Hinton [1989] notes that it also makes the weights automatically become symmetric ($w_{ij} = w_{ji}$) even if we start from an asymmetric state, because the updates (7.13) are always symmetric.

- A commonly used modification, convenient to implement, is the **incremental rule**, in which we *add* a small constant ε to w_{ij} each time S_i and S_j are on together during the clamped phase, and *subtract* ε each time they are on together during the free-running phase. This can be expressed by our standard rule (7.13) if we replace S_i by $n_i = \frac{1}{2}(1 + S_i)$, which is 0 or 1. Note that this rule uses **co-occurrences** of n_i and n_j, not their full *correlation*. The use of a constant step-size ε, instead of one dependent on the local gradient of E, helps the algorithm to make progress along shallow valleys while avoiding oscillations across such valleys (cf. Fig. 5.10). It is usually appropriate to decrease ε gradually during the training process.

- If (as is usual) some of the possible states α have zero desired probability R_α, it may be worthwhile increasing this to a small nonzero value. This avoids the need for infinite weights, which are the only way that the network can make $P_\alpha \equiv 0$ if $T > 0$. One way of producing all states α at least occasionally is to add noise to the desired patterns before using them for clamping.

Thus far we have discussed learning a set of patterns on the visible units, with associated probabilities. We can also consider a network having distinguishable input, output, and hidden units, with states labelled by γ, α, and β respectively. Then we want the network to learn associations $\gamma \to \alpha$; for each γ we want to be able to make the network's distribution $P_{\alpha|\gamma}$ as close as possible to a desired distribution $R_{\alpha|\gamma}$. If the possible inputs γ occur with probabilities p_γ a suitable error measure is

$$E = \sum_\gamma p_\gamma \sum_\alpha R_{\alpha|\gamma} \log \frac{R_{\alpha|\gamma}}{P_{\alpha|\gamma}}. \tag{7.17}$$

This leads to the learning rule

$$\Delta w_{ij} = \eta \beta \left[\overline{\langle S_i S_j \rangle}_{\text{I,O clamped}} - \overline{\langle S_i S_j \rangle}_{\text{I clamped}} \right] \tag{7.18}$$

in which both the inputs and outputs are clamped in the Hebb term, while only the inputs are clamped in the unlearning term, with averages over the inputs taken in both cases [Hopfield, 1987].

Smolensky [1986] introduced a related network called **harmonium**, based on his **harmony theory**. Harmonium is effectively a two-layer version of a Boltzmann machine, with connections only between the layers, though "connections" involving more than two units are also allowed. The harmony function is essentially equivalent to minus the free energy F discussed below.

Hopfield et al. [1983] proposed a learning procedure very similar to the Boltzmann learning rule (7.13) for networks *without hidden units*. In this case the clamped term just reduces to the Hebb recipe and does not require any simulation; we simply average $\xi_i^\mu \xi_j^\mu$ over patterns μ. Instead of the unlearning term in (7.13), Hopfield et al. proposed starting from random configurations, letting the network relax at zero temperature, and averaging $S_i S_j$ over the resulting local energy minima. This weighting is different from that in the Boltzmann rule; different states contribute proportionally to the sizes of their basins of attraction rather than according to their thermal weights $\exp(-\beta H)/Z$. Although this is not as elegant as the Boltzmann learning and does not guarantee $\Delta w_{ij} = 0$ when the equilibrium correlations are the same as the average clamped ones, it is much faster to implement. Recent work by van Hemmen et al. [1990] shows that this addition of unlearning to the Hopfield model produces a large increase of memory capacity and the ability to deal automatically with biased patterns.

Statistical Mechanics Reformulation ⋆

Boltzmann learning may be reinterpreted from a statistical mechanics viewpoint (see the Appendix). First consider the probability P_α of finding the visible units in state α given by (7.7). The numerator $\sum_\beta e^{-\beta H_{\alpha\beta}}$ is a partition function $Z_{\text{clamped}}^\alpha$ for the hidden units when the visible units are clamped in state α. Using the general correspondence between free energy F and partition function Z,

$$Z = e^{-\beta F} \tag{7.19}$$

we can write (7.7) as

$$P_\alpha = \frac{Z_{\text{clamped}}^\alpha}{Z} = \frac{e^{-\beta F_{\text{clamped}}^\alpha}}{e^{-\beta F}} = e^{-\beta(F_{\text{clamped}}^\alpha - F)} . \tag{7.20}$$

Since the cost function (7.10) is just a constant E_0 minus $\sum_\alpha R_\alpha \log P_\alpha$ it becomes

$$E = E_0 + \beta \left[\overline{F_{\text{clamped}}^\alpha} - F \right] . \tag{7.21}$$

We now see that we are attempting to minimize the difference between the average clamped free energy and the full free energy (which is necessarily lower).

The learning rule (7.13) now follows immediately from (7.21) using the identity

$$\langle S_i S_j \rangle = -\frac{\partial F}{\partial w_{ij}} \tag{7.22}$$

which also holds for the case of clamped visible units. To prove (7.22), we just perform the differentiation using (7.19):

$$-\frac{\partial F}{\partial w_{ij}} = T\frac{\partial \log Z}{\partial w_{ij}} = \frac{T}{Z}\frac{\partial Z}{\partial w_{ij}} = \sum_{\alpha\beta} S_i^{\alpha\beta} S_j^{\alpha\beta} \frac{e^{-\beta H_{\alpha\beta}}}{Z}. \tag{7.23}$$

The fraction in the last sum is the probability $P_{\alpha\beta}$, so (7.22) follows.

Deterministic Boltzmann Machines

A new mean field method suggested by Peterson and Anderson [1987] appears to circumvent the problem of excessive computer time for Boltzmann machines. Instead of calculating $\langle S_i S_j \rangle$ with Monte Carlo simulation and simulated annealing, they suggest using **mean field annealing** (see also Soukoulis et al. [1983] and Bilbro et al. [1989]). Specifically they take

$$\langle S_i S_j \rangle \approx m_i m_j \tag{7.24}$$

where $m_i \equiv \langle S_i \rangle$ is given by (2.45):

$$m_i = \tanh\left(\beta \sum_j w_{ij} m_j\right). \tag{7.25}$$

This is the natural mean field approximation, equivalent to (2.50). There is one such equation for each free unit i, while clamped units have m_i set to ± 1. Solving these simultaneous nonlinear equations is still not easy, but can be accomplished by iteration,

$$m_i^{\text{new}} = \tanh\left(\beta \sum_j w_{ij} m_j^{\text{old}}\right) \tag{7.26}$$

combined if necessary with a gradual lowering of T (annealing). Peterson and Anderson [1987] found this procedure to be 10–30 times faster than the Monte Carlo approach on some test problems. It actually gave somewhat better results as well.

Statistical mechanics again makes an alternative formulation possible. The free energy of the system is given by $F = H - TS$ where S is the entropy (see Appendix). In the mean field approximation the units are treated as independent, so the entropy is

$$S = -\sum_i \left(p_i^+ \log p_i^+ + p_i^- \log p_i^-\right) \tag{7.27}$$

where $p_i^\pm$ is the probability that S_i is ± 1. Since $p_i^+ + p_i^- = 1$ and $m_i = p_i^+ - p_i^-$ it is easy to see that

$$p_i^\pm = \frac{1 \pm m_i}{2}. \tag{7.28}$$

Thus the mean field free energy is

$$
\begin{aligned}
F_{\mathrm{MF}} = & -\frac{1}{2} \sum_{ij} w_{ij} m_i m_j \\
& + T \sum_i \left(\frac{1 + m_i}{2} \log \frac{1 + m_i}{2} + \frac{1 - m_i}{2} \log \frac{1 - m_i}{2} \right)
\end{aligned} \tag{7.29}
$$

using $S_i \to m_i$ in the energy (7.3). A more formal derivation gives the same result [Peterson and Anderson, 1987]. A similar expression for F_{MF}^α is found for the clamped states, the only difference being that the visible m_i's are fixed at ± 1. Minimizing F_{MF} with respect to m_i by setting $\partial F_{\mathrm{MF}} / \partial m_i = 0$ gives (7.25) again.

We may also regard this deterministic version of the Boltzmann machine independently of the original stochastic version, simply looking at the problem of minimizing the cost function

$$E_{\mathrm{MF}} = E_0 + \beta \left[\overline{F_{\mathrm{MF}}^\alpha} - F_{\mathrm{MF}} \right] \tag{7.30}$$

with respect to the w_{ij}'s [Hinton, 1989]. Equation (7.30) is just the mean field approximation to (7.21). Note that a double minimization is still involved; for each choice of the w_{ij}'s we must minimize (7.29) with respect to the m_i's by iteration of equations (7.26). Using gradient descent for the w_{ij} minimization gives directly

$$\Delta w_{ij} = -\eta \frac{\partial E_{\mathrm{MF}}}{\partial w_{ij}} = \eta \beta \left[\overline{m_i^\alpha m_j^\alpha} - m_i m_j \right] \tag{7.31}$$

which is the standard result (7.13) within the approximation (7.24).

7.2 Recurrent Back-Propagation

Pineda [1987, 1988, 1989], Almeida [1987, 1988], and Rohwer and Forrest [1987] have independently pointed out that back-propagation can be extended to arbitrary networks as long as they converge to stable states. At first sight it will appear that an $N \times N$ matrix inversion is required for each learning step, but Pineda and Almeida each showed that a modified version of the network itself can calculate what is required much more rapidly. The algorithm is usually called **recurrent back-propagation**.

Consider N continuous-valued units V_i with connections w_{ij} and activation function $g(h)$. Some of these units may be designated as input units and have input

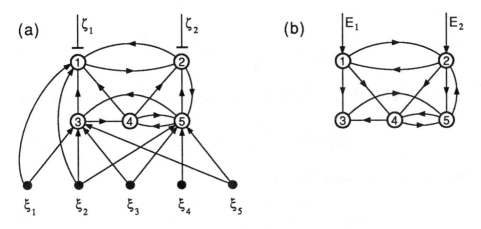

FIGURE 7.2 Recurrent back-propagation can handle arbitrary connections between the units. (a) Units 1 and 2 are output units with targets ζ_1 and ζ_2. Units 1, 3, and 5 are input units. Note that unit 1 is both an input and an output unit, while unit 4 is neither. (b) The corresponding error-propagation network, with reversed connections and error inputs E_1 and E_2.

values ξ_i^μ specified in pattern μ. We define $\xi_i^\mu = 0$ for the rest. Similarly some may be output units, with desired training values ζ_i^μ. Figure 7.2(a) provides an example. Henceforth we drop the pattern index μ for convenience; appropriate sums over μ are implicit.

Various dynamical evolution rules may be imposed on the network, such as

$$\tau\frac{dV_i}{dt} = -V_i + g\left(\sum_j w_{ij}V_j + \xi_i\right) \tag{7.32}$$

which is based on (3.31). It is easily seen that this dynamical rule leads to the right fixed points, where $dV_i/dt = 0$, given by

$$V_i = g\left(\sum_j w_{ij}V_j + \xi_i\right). \tag{7.33}$$

We *assume* that at least one such fixed point exists and is a stable attractor. Alternatives such as limit cycles and chaotic trajectories will be considered in the next section.

A natural error measure for the fixed point is the quadratic one

$$E = \frac{1}{2}\sum_k E_k^2 \tag{7.34}$$

where

$$E_k = \begin{cases} \zeta_k - V_k & \text{if } k \text{ is an output unit;} \\ 0 & \text{otherwise.} \end{cases} \tag{7.35}$$

Gradient descent gives

$$\Delta w_{pq} = -\eta \frac{\partial E}{\partial w_{pq}} = \eta \sum_k E_k \frac{\partial V_k}{\partial w_{pq}} \tag{7.36}$$

and to evaluate $\partial V_k / \partial w_{pq}$ we differentiate the fixed point equation (7.33) to obtain

$$\frac{\partial V_i}{\partial w_{pq}} = g'(h_i)\left[\delta_{ip} V_q + \sum_j w_{ij} \frac{\partial V_j}{\partial w_{pq}}\right]. \tag{7.37}$$

Collecting terms this may be written as

$$\sum_j \mathsf{L}_{ij} \frac{\partial V_j}{\partial w_{pq}} = \delta_{ip} g'(h_i) V_q \tag{7.38}$$

where

$$\mathsf{L}_{ij} = \delta_{ij} - g'(h_i) w_{ij} \tag{7.39}$$

and

$$h_i = \sum_j w_{ij} V_j + \xi_i \tag{7.40}$$

is the net input to unit i when the network is *at* the attractor. Inverting the linear equations (7.38) gives

$$\frac{\partial V_k}{\partial w_{pq}} = (\mathsf{L}^{-1})_{kp} g'(h_p) V_q \tag{7.41}$$

and thus from (7.36)

$$\Delta w_{pq} = \eta \sum_k E_k (\mathsf{L}^{-1})_{kp} g'(h_p) V_q . \tag{7.42}$$

This may be written in the familiar delta-rule form

$$\Delta w_{pq} = \eta \delta_p V_q \tag{7.43}$$

if we define

$$\delta_p = g'(h_p) \sum_k E_k (\mathsf{L}^{-1})_{kp} . \tag{7.44}$$

Equation (7.43) is our new learning rule. As it stands it requires a matrix inversion for the δ's, which could be done numerically [Rohwer and Forrest, 1987]. If however we write

$$\delta_p = g'(h_p) Y_p \tag{7.45}$$

so that

$$Y_p = \sum_k E_k (\mathsf{L}^{-1})_{kp} \tag{7.46}$$

then we can undo the inversion and obtain linear equations for Y_p

$$\sum_p \mathsf{L}_{pi} Y_p = E_i \tag{7.47}$$

or, using (7.39),

$$Y_i - \sum_p g'(h_p) w_{pi} Y_p = E_i . \tag{7.48}$$

This equation is of the same form as the original fixed point equation (7.33), and can be "solved" in the same way, by the evolution of a new **error-propagation network** with a dynamical equation analogous to (7.32):

$$\tau \frac{dY_i}{dt} = -Y_i + \sum_p g'(h_p) w_{pi} Y_p + E_i . \tag{7.49}$$

The topology of the required error-propagation network is the same as that of the original network, with the coupling w_{ij} from j to i replaced by $g'(h_i) w_{ij}$ from i to j, a simple linear transfer function $g(x) = x$, and an input term E_i (the error at unit i in the original network) instead of ξ_i. Figure 7.2(b) shows the error-propagation network corresponding to Fig. 7.2(a). In electrical network theory the two networks would be called **network transpositions** of one another.

The whole procedure is thus:

1. relax the original network with (7.32) to find the V_i's;

2. compare with the targets to find the E_i's from (7.35);

3. relax the error-propagation network with (7.49) to find the Y_i's; and

4. update the weights using (7.43) and (7.45).

This allows us to extend back-propagation to general networks without requiring any non-local operations (such as matrix inversion) that go outside the original connectivity of the network. Back-propagation itself is a special case of the general procedure; without recurrent connections the relaxation steps can be replaced by the corresponding fixed-point equations (7.33) and (7.48).

One can also have the two networks running together. The original network continuously supplies the error-propagation network with the error signals E_i, and the error-propagation network in turn adjusts the weights in the original network. A similar idea was put forward earlier by Lapedes and Farber [1986a, 1986b] in their **master-slave network**, where the master network calculates the weights for the slave. However, they had one master unit for each connection in the slave network (N^2 master units for N slave units), and made the master network *calculate* appropriate weights without using the slave for feedback. The problem solved by the master—find weights for the slave to minimize the total sum-of-squares error— was essentially an optimization problem, and was treated in much the same way as those described in Chapter 4.

It is worth noting that (7.48) is a *stable* attractor of (7.49) if (7.33) is a stable attractor of (7.32) (as we assumed). This may be shown by linearizing the dynamical equations about their respective fixed points, just as we did in (3.36). Writing $V_i = V_i^* + \varepsilon_i$ in (7.32) and $Y_i = Y_i^* + \eta_i$ in (7.48), where V_i^* and Y_i^* are respectively the solutions of (7.33) and (7.48), we obtain to linear order

$$\tau \frac{d\varepsilon_i}{dt} = -\varepsilon_i + g'(h_i) \sum_j w_{ij}\varepsilon_j = -\sum_j \mathsf{L}_{ij}\varepsilon_j \qquad (7.50)$$

and

$$\tau \frac{d\eta_i}{dt} = -\eta_i + \sum_p g'(h_p)w_{pi}\eta_p = -\sum_p \mathsf{L}_{ip}^T \eta_p . \qquad (7.51)$$

These linear equations are congruent except that L is replaced by its transpose L^T. But L and L^T have the same eigenvalues, so the local stability of the two equations is the same. Thus both attractors are stable (eigenvalues of L all positive) if one is. A further analysis of stability and convergence properties has been provided by Simard et al. [1989].

The scaling properties of the algorithm are interesting even on a sequential digital computer where no parallelism or hardware network is exploited [Pineda, 1989]. If we consider a fully connected network of N units, then L is an $N \times N$ matrix, and using (7.44) with direct matrix inversion would take a time proportional to N^3 using standard methods. In comparison, the time required to integrate (7.49) numerically scales as only N^2 provided the fixed points are stable. This translates into a great saving for large N. Of course both methods are much better than direct numerical differentiation of the error measure (7.34) with respect to all N^2 parameters, which would scale as N^4 (a relaxation of (7.32), taking order N^2 steps, would be needed for each parameter).

Using $g(x) = \tanh(x)$ and choosing the entropic error measure (5.52) instead of the quadratic one (7.34) simplifies the derivation a little and leads to slightly different equations.

Almeida [1987] has shown that recurrent networks give a large improvement in performance over normal feed-forward networks for a number of problems such as pattern completion. Qian and Sejnowski [1988b] have used recurrent back-propagation to train a network to calculate stereo disparity in random-dot stereograms. The network configures itself to use the Marr and Poggio [1976] algorithm in some cases, and finds a new algorithm in others. Barhen et al. [1989] have applied a modified version of recurrent back-propagation, including terminal attractors, to inverse kinematics problems for the control of robot manipulators.

7.3 Learning Time Sequences

In Section 3.5 we discussed ways of making a network generate a temporal sequence of states, usually in a limit cycle. This naturally required a recurrent network with asymmetric connections; neither a feed-forward network nor a network with symmetric connections will do, because they necessarily go to a stationary state. The desired sequences were embedded into the network by design; the w_{ij}'s were calculated, not learned.

We now turn to the problem of *learning* sequences. There are actually three distinct tasks:

- **Sequence Recognition.** Here one wants to produce a particular output pattern when (or perhaps just after) a specific input sequence is seen. There is no need to reproduce the input sequence. This is appropriate, for example, for speech recognition problems, where the output might indicate the word just spoken.

- **Sequence Reproduction.** In this case the network must be able to generate the rest of a sequence itself when it sees part of it. This is the generalization of auto-association or pattern completion to dynamic patterns. It would be appropriate for learning a set of songs, or for predicting the future course of a time series from examples.

- **Temporal Association.** In this general case a particular output *sequence* must be produced in response to a specific input sequence. The input and output sequences might be quite different, so this is the generalization of hetero-association to dynamic patterns. It includes as special cases pure sequence generation and the previous two cases.

We discuss in this section several approaches that can learn to do one or more of these tasks. The first task—sequence recognition—does not necessarily require a recurrent network, but belongs naturally with the other cases, which clearly do need recurrency.

Tapped Delay Lines

The simplest way to perform sequence recognition (but not sequence reproduction or general temporal association) is to turn the temporal sequence into a spatial pattern on the input layer of a network. Then conventional back-propagation methods can be used to learn and recognize sequences.

We already studied examples of this approach in Chapter 6. In NET-talk (page 132) a text string was moved along in front of a 7-character window, which thus received at a particular time characters $n, n-1, \ldots, n-6$ from the string. In our signal processing example (page 137) the values $x(t), x(t-\Delta), \ldots, x(t-(m-1)\Delta)$ from a signal $x(t)$ were presented *simultaneously* at the input of a network. In a practical network these values could be obtained by feeding the signal into a **delay line** that was tapped at various intervals, as sketched in Fig. 7.3. Or, in a synchronous network, a **shift register** could be used, in effect keeping several "old" values in a buffer [Kohonen, 1989]. The resulting architectures are sometimes called **time-delay neural networks**.

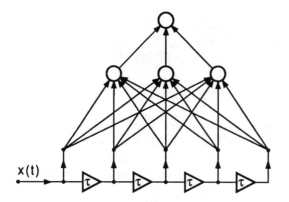

FIGURE 7.3 A time-delay neural network. Only one input $x(t)$ is shown. $x(t)$, $x(t - \tau)$, ..., $x(t - 4\tau)$ are fully connected to a hidden layer.

The tapped delay line scheme has been widely applied to the problem of speech recognition [e.g., McClelland and Elman, 1986; Elman and Zipser, 1988; Tank and Hopfield, 1987b; Waibel et al., 1989; Waibel, 1989; Lippmann, 1989]. Here the signal at a particular time is a set of spectral coefficients (strength in each of several frequency bands), so with a set of different time delays one has input patterns in a two-dimensional time-frequency plane. Different phonemes show different patterns in this plane and can be learned by an ordinary feed-forward network. Similar approaches can be applied to other units of speech or written language, such as syllables, letters, and words.

There are several drawbacks to this general approach to sequence recognition [Mozer, 1989]. The length of the delay line or shift register must be chosen in advance to accommodate the longest possible sequence; we cannot work with arbitrary-length sequences. The large number of units in the input layer leads to slow computation (unless fully parallel) and requires a lot of training examples for successful learning and generalization. And, perhaps most importantly, the input signal must be properly registered in time with the clock controlling the shift register or delay line, and must arrive at exactly the correct rate.

Tank and Hopfield [1987a] suggested a way of compensating for the last of these problems, thus making the scheme more robust. Given a raw signal $\mathbf{x}(t)$ the usual delay line technique would be to use $\mathbf{x}(t)$, $\mathbf{x}(t - \tau_1)$, $\mathbf{x}(t - \tau_2)$, ..., $\mathbf{x}(t - \tau_m)$ for the network inputs at time t. Tank and Hopfield suggest replacing the fixed delays by filters that broaden the signal in time as well as delaying it, with greater broadening for longer delays. They used

$$\mathbf{y}(t; \tau_i) = \int_{-\infty}^{t} G(t - t'; \tau_i)\mathbf{x}(t') \, dt' \tag{7.52}$$

for the network inputs at time t, with

$$G(t; \tau_i) = \left(\frac{t}{\tau_i}\right)^{\alpha} e^{\alpha(1 - t/\tau_i)} \tag{7.53}$$

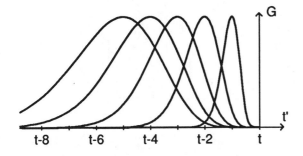

FIGURE 7.4 The function $G(t - t'; \tau_i)$ given by (7.53) for $\tau_i = 1, 2, 3, 4, 5$ with $\alpha = 10$. The network inputs at time t are averages of the past signal weighted by these functions.

for each τ_i. This is normalized to have maximum value 1 when $t = \tau_i$. The parameter α controls the broadening; for larger α the function G has a narrower peak. Figure 7.4 shows an example at $\alpha = 10$. This scheme can successfully compensate for phase and speed variations if they are not too large. Note too that it works in continuous time; there is no need for precise synchronization by a central clock. Schemes like this using a variety of delays are biologically feasible, and may be used in the brain, in auditory cortex for example [Hopfield and Tank, 1989]. Shift registers might also occur biologically [Anderson and Van Essen, 1987].

Context Units

A popular way to recognize (and sometimes reproduce) sequences has been to use **partially recurrent networks**. The connections are mainly feed-forward, but include a carefully chosen set of feedback connections. The recurrency lets the network remember cues from the recent past, but does not appreciably complicate the training. In most cases the feedback connections are fixed, not trainable, so back-propagation may easily be used for training. The updating is synchronous, with one update for all units at each time step. Such networks are sometimes referred to as **sequential networks**.

Figure 7.5 shows several architectures that have been used. In each case there is one special set of **context units** C_i, either a whole layer or a part thereof, that receives feedback signals. The forward propagation is assumed to occur quickly, or without reference to time, while the feedback signal is clocked. Thus at time t the context units have some signals coming from the network state at time $t - 1$, which sets a context for processing at time t. The context units remember some aspects of the past, and so the state of the whole network at a particular time depends on an aggregate of previous states as well as on the current input. The network can therefore recognize sequences on the basis of its state at the end of the sequence. In some cases it can generate sequences too.

Elman [1990] suggested the architecture shown in Fig. 7.5(a). Similar ideas were proposed by Kohonen; see Kohonen [1989]. The input layer is divided into two parts: the true input units and the context units. The context units simply hold a

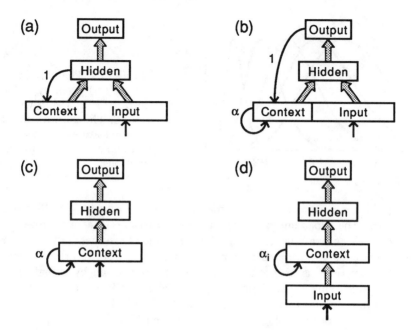

FIGURE 7.5 Architectures with context units. Single arrows represent connections only from the ith unit in the source layer to the ith unit in the destination layer, whereas shaded arrows represent fully connected layers.

copy of the activations of the hidden units from the previous time step. The modifiable connections are all feed-forward, and can be trained by conventional back-propagation methods; this is done at each time step in the case of input sequences. Note that we treat the context units like inputs in applying back-propagation, and do *not* introduce terms like $\partial C_k / \partial w_{ij}$ into the derivation. The network is able to recognize sequences, and even to produce short continuations of known sequences. Cleeremans et al. [1989] have shown that it can learn to mimic an existing finite state automaton, with different states of the hidden units representing the internal states of the automaton.

Figure 7.5(b) shows the Jordan [1986, 1989] architecture. It differs from that of Fig. 7.5(a) by having the context units fed from the output layer (instead of the hidden layer), and also from themselves. Similar behavior can probably be obtained whether the feedback is from a hidden or from an output layer, though particular problems might be better suited to one rather than the other. One could even use multiple sets of context units, one set from each layer [Bengio et al., 1990]. More importantly, the self-connections give the context units C_i themselves some individual memory, or inertia. Their updating rule is

$$C_i(t+1) = \alpha C_i(t) + O_i(t) \tag{7.54}$$

where the O_i are the output units and α is the strength of the self-connections. We require $\alpha < 1$. If the outputs O_i were fixed, then the C_i would clearly decay exponentially towards $O_i/(1 - \alpha)$, thus gradually forgetting their previous values. Such units are sometimes called **decay units, integrating units,** or **capacitive units.** Iterating (7.54) we obtain

$$
\begin{aligned}
C_i(t+1) &= O_i(t) + \alpha O_i(t-1) + \alpha^2 O_i(t-2) + \cdots \\
&= \sum_{t'=0}^{t} \alpha^{t-t'} O_i(t')
\end{aligned}
\tag{7.55}
$$

or, in the continuum limit,

$$
C_i(t) = \int_0^t e^{-\gamma(t-t')} O_i(t') \, dt'
\tag{7.56}
$$

where $\gamma = |\log \alpha|$. Thus in general context units accumulate a weighted moving average or **trace** of the past values they see, in this case of O_i. By making α closer to 1 the memory can be made to extend further back into the past, at the expense of loss of sensitivity to detail. In general the value of α should be chosen so that the decay rate matches the characteristic time scale of the input sequence [Stornetta et al., 1988]. In a problem with a wide range of time scales it may be useful to have several context units for each output unit, with a different decay rate (α value) for each [Anderson et al., 1989].

Note in passing that the result (7.56) is exactly of the form (3.57) with the exponential decay (3.60), so context units could be used to implement the moving averages needed in Section 3.5.

With *fixed* input, the network of Fig. 7.5(b) can be trained to generate a set of output sequences, with different input patterns triggering different output sequences [Jordan, 1986, 1989]. With an input *sequence*, the network can be trained to recognize and distinguish different input sequences. Anderson et al. [1989] have applied this to categorizing a class of English syllables, and find that a network trained by one group of speakers can successfully generalize to another group.

Figure 7.5(c) shows a simpler architecture, due to Stornetta et al. [1988], that can also perform sequence recognition tasks. Note that the only feedback is now from the context units to themselves, giving them decay properties as in (7.54). But their input is now the network input itself, which *only* reaches the rest of the network via the context units. In effect the input signal is *preprocessed* by the context units by the formation of a weighted moving average of past values; technically it is equivalent to processing by an *infinite impulse response* digital filter with transfer function $1/(1 - \alpha z^{-1})$. This preprocessing serves to include past features of the input signal into the present context values, thus letting the network recognize and distinguish different sequences. However, this architecture is less well suited than Fig. 7.5(a) or (b) to generating or reproducing sequences.

Mozer [1989] suggested the architecture shown in Fig. 7.5(d). This looks at first similar to case (c), but differs in two important ways. First, there is full connectivity with modifiable connections w_{ij} between input and context units instead of only one-to-one connections. This gives the network greater flexibility to use the context units in a way appropriate to the problem. Second, the self-connections to the context units are no longer fixed, but are trained like other connections. The context units still act fundamentally as integrators, but the decay rate can be found during training to match the time scales of the input.

Now however we need a new learning rule; conventional back-propagation does not allow for the modifiable recurrent connections. Recurrent back-propagation will not suffice either, since it assumes constant inputs and approach to an attractor, whereas here we are interested in input and/or output sequences. Mozer [1989] derived a new learning rule by differentiating an appropriate cost function and using gradient descent. The result is *almost* a special case of the "real-time recurrent learning" rule of Williams and Zipser [1989a] which we will treat shortly, so we omit the details here. The "almost" comes about because Mozer uses the update rule

$$C_i(t+1) = \alpha_i C_i(t) + g\left(\sum_j w_{ij}\xi_j\right) \qquad (7.57)$$

for the context units (where ξ_j is an input pattern), whereas the Williams and Zipser approach for fully recurrent networks would apply strictly only to

$$C_i(t+1) = g\left(\alpha_i C_i(t) + \sum_j w_{ij}\xi_j\right). \qquad (7.58)$$

The modification to the learning algorithm is minor however. Equation (7.57) seems to work somewhat better than (7.58), which has itself been studied by Bachrach [1988].

Like case (c), the architecture of case (d) is better suited to recognizing sequences than to generating or reproducing them. It works well for distinguishing letter sequences for example. When applying it to a problem involving reproducing a sequence after a delay, Mozer [1989] added an additional set of context units to the input layer to echo the output layer, in the style of case (a) or (b).

Other architectures, not illustrated in Fig. 7.5, are of course possible, and some have been tried. Shimohara et al. [1988] categorize the possibilities and examine several.

Back-Propagation Through Time

The networks just discussed were partially recurrent. Let us now turn to fully recurrent networks, in which any unit V_i may be connected to any other. We retain for now synchronous dynamics and discrete time, so the appropriate update rule is

$$V_i(t+1) = g(h_i(t)) = g\left(\sum_j w_{ij}V_j(t) + \xi_i(t)\right). \qquad (7.59)$$

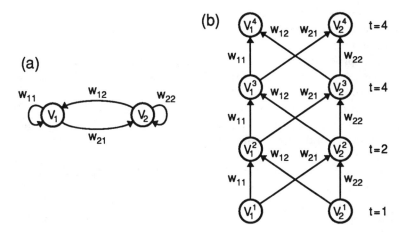

FIGURE 7.6 Back-propagation through time. (a) A recurrent network. (b) A feed-forward network that behaves identically for 4 time steps.

Here $\xi_i(t)$ is the input, if any, at node i at time t. If there is no such input at that node at that time we put $\xi_i(t) = 0$.

The general temporal association task is to produce particular output sequences $\zeta_i(t)$ in response to specific input sequences $\xi_i(t)$. Individual units may be input units, output units, neither, or both; a target $\zeta_i(t)$ is usually only defined on certain units at certain times.

If we are interested in sequences of a small maximum length T, we can use a trick to turn an arbitrary recurrent network into an equivalent feed-forward network. The idea, called **back-propagation through time** or **unfolding of time**, was originally suggested by Minsky and Papert [1969] and combined with back-propagation by Rumelhart, Hinton, and Williams [1986b]. For sequences spanning the time steps $t = 1, 2, \ldots, T$ we simply duplicate all units T times, so that a separate unit V_i^t holds the state $V_i(t)$ of the equivalent recurrent network at time t. Figure 7.6 shows the idea for a general two-unit recurrent network and $T = 4$ time steps. Note that the connection w_{ij} from V_j^t to V_i^{t+1} is independent of t; the *same* weight values must be used in each layer. It is easy to see that the two networks will behave identically for T time steps.

The resulting unfolded network is strictly feed-forward and can be trained by a slightly modified form of back-propagation. There is no longer any need to clock the network; as in other feed-forward networks we can just wait for every unit to set its output according to its net input. It is simple to allow for input and output *sequences*; the input or output specified for unit i at time t is applied to unit V_i^t. Note in the case of outputs that error signals may be produced in any layer, not just the last, and are propagated backwards from the layer in which they originate. The only real complication is the constraint that all "copies" of each connection w_{ij} must remain identical, whereas back-propagation would normally produce different increments Δw_{ij} for each particular copy. The usual solution is simply to add together the individual increments and then change all copies of w_{ij} by the total amount.

The main problem with this back-propagation through time approach is the need for large computer resources. Storage needs, computer time for simulation, and number of training examples needed are all enlarged by the duplication of units. For long sequences, or for sequences of unknown length, the approach becomes impractical. Note however that the duplication only applies to the training phase; once trained in the unfolded version, the recurrent network may be used for the temporal association task.

Back-propagation through time has not been widely applied, and has been largely superceded by the other approaches described in this chapter. Rumelhart, Hinton, and Williams [1986b] showed that it worked well for the task of learning to be a shift register and for a sequence completion task. Nowlan [1988] obtained good results on a constraint satisfaction problem.

Real-Time Recurrent Learning

Williams and Zipser [1989a, b] showed how to construct a learning rule for general recurrent networks without duplicating the units. See also Robinson and Fallside [1988] and Rohwer [1990]. One version of this rule can be run *on-line*, learning while sequences are being presented rather than after they are complete. It can thus deal with sequences of arbitrary length—there are no requirements to allocate memory proportional to the maximum sequence length.

We assume the same dynamics as for back-propagation through time,

$$V_i(t) \;=\; g(h_i(t-1)) \;=\; g\!\left(\sum_j w_{ij} V_j(t-1) + \xi_i(t-1)\right) \qquad (7.60)$$

with target outputs $\zeta_k(t)$ defined for some k's and t's as before. An appropriate error measure on unit k at time t is then

$$E_k(t) = \begin{cases} \zeta_k(t) - V_k(t) & \text{if } \zeta_k \text{ is defined at time } t\,; \\ 0 & \text{otherwise} \end{cases} \qquad (7.61)$$

as in (7.35). Then the total cost function is

$$E = \sum_{t=0}^{T} E(t) \qquad (7.62)$$

if the domain of interest is $t = 0, 1, \ldots, T$, where

$$E(t) = \frac{1}{2}\sum_k [E_k(t)]^2 \,. \qquad (7.63)$$

The gradient of E separates in time, so to do gradient descent we can define

$$\Delta w_{pq}(t) = -\eta \frac{\partial E(t)}{\partial w_{pq}} = \eta \sum_k E_k(t) \frac{\partial V_k(t)}{\partial w_{pq}} \qquad (7.64)$$

and take the full change in w_{pq} as the time-sum of $\Delta w_{pq}(t)$. The last derivative in (7.64) can now be found by differentiating the dynamical rule (7.60), just as in (7.37),

$$\frac{\partial V_i(t)}{\partial w_{pq}} = g'(h_i(t-1)) \left[\delta_{ip} V_q(t-1) + \sum_j w_{ij} \frac{\partial V_j(t-1)}{\partial w_{pq}} \right]. \qquad (7.65)$$

This relates the derivatives $\partial V_i / \partial w_{pq}$ at time t to those at time $t-1$. If we were interested only in a stable attractor we could set these equal and then solve the resulting linear equations, exactly as in recurrent back-propagation. Here instead we can iterate it forward from the reasonable initial condition

$$\frac{\partial V_i(0)}{\partial w_{pq}} = 0 \qquad (7.66)$$

at $t = 0$. We can do this at each time step along the way, after updating the units themselves according to (7.60). The time and memory requirements are nevertheless very large; for N fully recurrent units there are N^3 derivatives to maintain, and updating each takes a time proportional to N for each derivative, or N^4 in all at each time step.

Given all the derivatives, one finds $\Delta w_{pq}(t)$ from (7.64) at each t. After the full time interval has been traversed the total weight changes are found by summing all these $\Delta w_{pq}(t)$. The whole procedure has to be repeated until the system remembers the correct temporal associations. One can probably use tricks known from back-propagation to speed up the learning, such as momentum terms.

Williams and Zipser [1989a] find that updating the weights after each time step instead of waiting until the sequence is ended (at $t = T$) works well if η is sufficiently small. They call the resulting algorithm **real-time recurrent learning** (RTRL), because the weight changes occur in real time during the presentation of the input and output sequences. This avoids any array storage requirements of size depending on the maximum time T, and is especially simple to implement.

A modification often found to be useful is **teacher forcing**. Here one corrects units that are wrong whenever a target value is available, always replacing $V_k(t)$ by $\zeta_k(t)$ if the latter is known. This must be done of course *after* computing the error $E_k(t)$ and the set of derivatives. It is easy to show that the appropriate modification of the gradient descent calculation is to set $\partial V_i(t)/\partial w_{pq}$ to zero for all pq in the iterative use of (7.65) whenever unit i is forced to its target value, again *after* the weight changes have been computed. The teacher forcing procedure keeps the network closer to the desired track, and usually seems to speed up learning considerably. It can be used in other dynamic networks too, such as that of Fig. 7.5(b). There is however one potential disadvantage; if we train a network to have an attractor under some constraint conditions such as teacher forcing, there is no guarantee that the attractor will remain an attractor when the constraints are removed. In fact an attractor can turn into a repeller upon the removal of constraints. The terminal attractors described in Section 3.3 can be used to eliminate this problem.

Several examples investigated by Williams and Zipser [1989b] demonstrate the power and generality of the RTRL method. It works well for sequence recognition tasks, though these could also be performed by simpler networks. It can learn to be a flip-flop, so as to output a signal only after a symbol A and then—after an arbitrary interval—another symbol B has been seen. Clearly a tapped delay line approach could never do this. Most impressively, it can learn to be a finite state machine. By observing only the *actions* (not the internal states) of a Turing machine performing the parentheses-balancing task, a recurrent network of only 12 units learned the task. Finally the RTRL method with teacher forcing (but not without) can be used to teach a network to be a square wave or sine wave oscillator.

Time-Dependent Recurrent Back-Propagation

Pearlmutter [1989a, b] has developed a related algorithm for training a general *continuous-time* recurrent network (see also Werbos [1988] and Sato [1990]). The units evolve according to

$$\tau_i \frac{dV_i}{dt} = -V_i + g\left(\sum_j w_{ij} V_j\right) + \xi_i(t) \tag{7.67}$$

which is similar to the recurrent back-propagation dynamics (7.32). The inputs $\xi_i(t)$ are continuous functions of time, as are the desired outputs $\zeta_k(t)$. An appropriate error function for an overall time domain 0–T is

$$E = \frac{1}{2} \int_0^T \sum_{k \in O} [V_k(t) - \zeta_k(t)]^2 \, dt \tag{7.68}$$

where the sum is only over those "output" units with specified values of $\zeta_k(t)$. As usual we must differentiate E with respect to the weights w_{ij}, which here enter E through $V_k(t)$. We thus need the **functional derivatives**

$$E_k(t) = \frac{\delta E}{\delta V_k(t)} = [V_k(t) - \zeta_k(t)]. \tag{7.69}$$

It is now possible to obtain the derivative $\partial E/\partial w_{ij}$ needed for gradient descent in the form

$$\frac{\partial E}{\partial w_{ij}} = \frac{1}{\tau_i} \int_0^T Y_i g'(h_i) V_j \, dt \tag{7.70}$$

where $h_i(t) = \sum_j w_{ij} V_j(t)$ as usual and $Y_i(t)$ is the solution of the dynamical equation

$$\frac{dY_i}{dt} = -\frac{1}{\tau_i} Y_i + \sum_j \frac{1}{\tau_j} w_{ji} g'(h_j) Y_j + E_i(t) \qquad (7.71)$$

with the boundary condition $Y_i(T) = 0$ for all i at the end point. We omit the derivation of this result, which may be obtained using the calculus of variations and Lagrange multipliers, or from a finite difference approximation [Pearlmutter, 1989a]. It is possible to obtain $\partial E / \partial \tau_i$ too, and thus optimize with respect to the time constants τ_i.

To use these equations we first integrate (7.67) forwards from $t = 0$ to $t = T$ and store the resulting $V_i(t)$'s. Then we integrate (7.71) backwards, from $t = T$ to $t = 0$, to obtain the $Y_i(t)$'s. Finally (or by accumulation during the backward pass) we evaluate the integrals (7.70) to find the appropriate gradient descent increments $\Delta w_{ij} = -\eta \partial E / \partial w_{ij}$. In practice all the integration must be done by finite difference methods, and the stored functions are represented by arrays.

The technique is related in various ways to many of the others discussed in this chapter. It can be used to derive the Williams and Zipser RTRL approach [Pearlmutter, 1989a]. It may also be seen as a continuous-time extension of back-propagation through time, or as an extension of recurrent back-propagation to dynamic sequences. Notice the similarity of the backwards equation (7.71) to the transposed equation (7.49) in recurrent back-propagation, but recall that there we only cared about finding the stable attractor to which the equation led. The name **time-dependent recurrent back-propagation** is sometimes used for the present approach.

It is clear that the time and storage requirements of this technique are large, but they are generally not as extreme as for the Williams and Zipser RTRL approach. For N fully recurrent units and K time steps between 0 and T the time per forward-backward pass scales as $N^2 K$, compared to $N^4 K$ for RTRL. However, the appropriate K would often be much larger in the current case where we are discretizing continuous time. Memory requirements here scale as $aNK + bN^2$ compared to N^3 for RTRL.

Pearlmutter [1989a] applied the forwards-backwards technique to training a network with two output units $x \equiv V_1$, $y \equiv V_2$ to follow various trajectories in the (x,y) plane, including circles and figure eights. After training they also tried perturbation experiments, kicking the system (7.67) off its regular trajectory with random perturbations. They found that it returned to its proper trajectory, which was therefore a stable limit cycle.

Overall, the Pearlmutter approach is probably best unless on-line learning is needed. It can be discretized fairly coarsely when the input and output are discrete time series rather than continuous trajectories. The RTRL approach of Williams and Zipser is needed instead when on-line learning is desired. But for many temporal sequence association problems a partially recurrent architecture with context units may suffice, and is much less costly to implement.

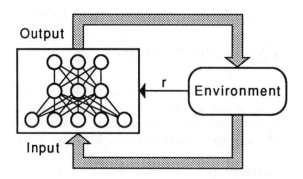

FIGURE 7.7 A neural network interacting with its environment. Shaded arrows represent multi-component signals, whereas the reinforcement signal r is a simple scalar.

7.4 Reinforcement Learning

Thus far we have assumed in all of our supervised learning discussion that correct "target" output values are known for each input pattern. But in some situations there is less detailed information available. In the extreme case there is only a single bit of information, saying whether the output is *right* or *wrong*. In this section we look at some simple examples of the **reinforcement learning** procedures applicable to this extreme case. Most of the networks involved can either be recurrent or non-recurrent; we discuss them in this chapter for convenience.

Reinforcement learning is a form of supervised learning because the network does, after all, get *some* feedback from its environment. But that feedback—the single yes/no **reinforcement signal**—is only evaluative, not instructive. Reinforcement learning is sometimes called **learning with a critic** as opposed to "learning with a teacher."

If the reinforcement signal says that a particular output is wrong it gives no hint as to what the right answer should be; in terms of a cost function there is no gradient information. It is therefore important in a reinforcement learning network for there to be some source of *randomness* in the network, so that the space of possible outputs can be explored until a correct value is found. This is usually done by using stochastic units, such as we have described in Sections 2.4 and 5.6.

In reinforcement learning problems it is common to think explicitly of a network functioning in an **environment**, as sketched in Fig. 7.7. The environment supplies the inputs to the network, receives its output, and then provides the reinforcement signal. There are several different reinforcement learning problems, depending on the nature of the environment:

- **Class I.** In the simplest case, the reinforcement signal is always the same for a given input-output pair. Thus there is a definite input-output mapping that the network must learn, or one of several if there are multiple correct outputs for a given input. Moreover the input patterns are chosen randomly or in some fixed schedule by the environment, without reference to previous outputs. This is the situation already familiar to us from other supervised learning examples.

■   **Class II.** A common extension is to a **stochastic environment**. Here a particular input-output pair determines only the *probability* of positive reinforcement. This probability is however fixed (stationary) for each input-output pair, and again the input sequence does not depend on past history. Problems like this arise frequently in modelling animal learning, economic systems, and some simple games. They are highly non-trivial even in the "non-associative" case of a single fixed input (or equivalently no input at all); how can we best determine the output pattern with the highest probability of positive reinforcement from only a finite set of trials? The case of a single two-valued output in this non-associative situation is known as the **two-armed bandit problem**, in which we imagine being able to pull either handle of a two-armed slot machine without prior knowledge of the payoff probabilities. There has been a great deal of theoretical work on such problems, known as **stochastic learning automata**; see e.g., Narendra and Thathachar [1989] for an introduction.

■   **Class III.** In the most general case the environment may itself be governed by a complicated dynamical process. Both reinforcement signals and input patterns may depend arbitrarily on the past history of the network output. The classic application is to game theory, where the "environment" is actually another player or players. Consider for example a neural network to play chess; it would only receive a reinforcement signal (win or lose) after a long sequence of moves. In such situations the **credit assignment problem** becomes severe; how are we to assign credit or blame individually to each move in such a sequence for an eventual victory or loss? This is a *temporal* credit assignment problem, as opposed to the more familiar *structural* credit assignment problem of attributing network error to different units or weights.

Associative Reward-Penalty

We discuss first a reinforcement learning procedure applicable to class I and II problems. Due originally to Barto and Anandan [1985], it is usually known as the **associative reward-penalty** algorithm, or A_{RP}. Its essential ingredients are *stochastic* output units and a particular learning rule; the rest of the network architecture is arbitrary and can be chosen to suit the problem at hand.

Let us consider a set of output units $S_i = \pm 1$ governed by the standard stochastic dynamical rule (2.48)

$$\text{Prob}(S_i = \pm 1) = g(\pm h_i) = f_\beta(\pm h_i) = \frac{1}{1 + \exp(\mp 2\beta h_i)} \qquad (7.72)$$

with $h_i = \sum_j w_{ij} V_j$ as usual. The V_j's might be the activations of some hidden units, or might be the network inputs ξ_j themselves. Our problem now is to determine an appropriate error estimate δ_i^μ for each of the output units when their inputs are set to a particular pattern V_j^μ.

Instead of known target patterns, we are given only the reinforcement signal r. We take $r = +1$ to mean positive reinforcement (reward) and $r = -1$ to mean negative reinforcement (penalty). Now we can construct our own target patterns ζ_i^μ by using

$$\zeta_i^\mu = \begin{cases} S_i^\mu & \text{if } r^\mu = +1 \text{ (reward)}; \\ -S_i^\mu & \text{if } r^\mu = -1 \text{ (penalty)} \end{cases} \tag{7.73}$$

[Widrow et al., 1973]. This suggests to the network that it should be more likely to do what it just did if that was rewarded, and more likely to do the opposite if not. Note that in the case of a *single* binary output and a class I problem (deterministic r), ζ_i^μ is the correct output because $-S_i^\mu$ must be correct if S_i^μ is wrong. In that situation reinforcement learning is the same as ordinary supervised learning with known targets, and the present approach reduces to the ordinary delta rule.

To construct our learning rule we compare the target ζ_i^μ with the *average* output value $\langle S_i^\mu \rangle$ to compute the appropriate δ_i^μ, just as in (5.58):

$$\delta_i^\mu = \zeta_i^\mu - \langle S_i^\mu \rangle . \tag{7.74}$$

We could find $\langle S_i^\mu \rangle$ by averaging for a while, or simply by calculation from (7.72) as in Chapter 2:

$$\langle S_i^\mu \rangle = (+1)g(h_i^\mu) + (-1)[1 - g(h_i^\mu)] = \tanh \beta h_i^\mu . \tag{7.75}$$

Then our update rule for the output weights is the usual delta rule

$$\Delta w_{ij} = \eta(r^\mu)\delta_i^\mu V_j^\mu \tag{7.76}$$

in which we let the learning rate η depend on whether the output was right or wrong. Typically $\eta(+1)$ is taken as 10–100 times larger than $\eta(-1)$. We could have put a $g'(h_i^\mu)$ factor into (7.76) too, as appropriate for minimizing $\sum_i (\delta_i^\mu)^2$, but choose to leave it out as for the entropic cost function (cf. (5.53)).

Overall our learning rule reads

$$\Delta w_{ij} = \begin{cases} \eta^+[S_i^\mu - \langle S_i^\mu \rangle]V_j^\mu & \text{if } r^\mu = +1 \text{ (reward)}; \\ \eta^-[-S_i^\mu - \langle S_i^\mu \rangle]V_j^\mu & \text{if } r^\mu = -1 \text{ (penalty)} \end{cases} \tag{7.77}$$

with $\eta^\pm = \eta(\pm 1)$. This is the standard form of the A$_{\mathrm{RP}}$ rule. It only tells us how to adjust the weights going to the *output* units, but we can generate appropriate weight adjustments for other connections in a multi-layer network simply by back-propagating the output errors δ_i^μ given by (7.74). Similarly in a recurrent network we can use recurrent back-propagation (Section 7.2) or one of the time-dependent techniques described in Section 7.3.

In practice (7.77) does work, for both deterministic and stochastic reinforcement, but is very slow compared to supervised learning. The speed is dependent on the size of the output space and the fraction of that space which is "correct" for a given input pattern; if there is only one correct answer for each input the search can be extremely time-consuming. Nevertheless it is faster (for large enough problems) than a trial-and-error search in cases where the network's generalization ability can help [Ackley and Littman, 1990].

When learning converges on a correct solution the $\langle S_i^\mu \rangle$'s approach ± 1, making the units effectively deterministic. This is true even for class II problems; the outputs converge to the state that provides the largest *average* reinforcement. But the stochastic units are essential earlier on during learning, to generate exploration of the output space.

Some variations on the standard rule (7.77) are worth noting:

- For 0/1 units instead of ± 1 units the $-S_i^\mu$ in the penalty case is replaced by $1 - S_i^\mu$, and $\langle S_i^\mu \rangle$ is often written as p_i^μ, the probability of getting a 1.

- If the reward r is a *continuous-valued* scalar evaluation signal, in the range $0 \leq r \leq 1$ (with 0 meaning terrible and 1 meaning excellent), we can generalize (7.77) to [Barto and Jordan, 1987]

$$\Delta w_{ij} = \eta (r^\mu) \big(r^\mu [S_i^\mu - \langle S_i^\mu \rangle] + (1 - r^\mu)[-S_i^\mu - \langle S_i^\mu \rangle] \big) V_j^\mu. \tag{7.78}$$

- Some authors prefer to use $\langle S_i^\mu \rangle - S_i^\mu$ instead of $-S_i^\mu - \langle S_i^\mu \rangle$ for the penalty case, giving simply

$$\Delta w_{ij} = \eta r^\mu [S_i^\mu - \langle S_i^\mu \rangle] V_j^\mu \tag{7.79}$$

if we leave out the dependence of η on r. This always gives Δw_{ij} the same sign as does (7.77), but the magnitude is different. On receipt of $r = -1$, in effect (7.79) tells the network not to do what it is doing, while (7.77) tells it to do the opposite of what it is doing. In practice (7.77) seems to work better, probably because it provides larger weight changes when $\langle S_i^\mu \rangle - S_i^\mu$ is near the wrong ± 1 extreme. But (7.79) is more amenable to theoretical analysis, as we will soon see.

- It seems to improve learning speed to present a given pattern μ several times before moving on to the next one [Barto and Jordan, 1987; Ackley and Littman, 1990]. It is important to do this in **batch mode**, accumulating all the weight changes and then adding the accumulated sum to the weights at the end. This procedure allows a better estimate of the local gradient of the cost function discussed below.

A reinforcement learning approach might prove useful even when targets *are* available. Barto and Jordan [1987] suggest using the A_{RP} approach for training the hidden layers of a multi-layer network, while using the ordinary delta-rule for the output units. This inevitably proves to be slower than using back-propagation, but might have an advantage for a hardware implementation. Instead of needing circuits for back-propagation of individual errors, it is only necessary to feed a global signal r (based on the mean square error at the output) to each unit. A similar strategy was used by Grossman [1990] in his algorithm for finding effective internal representations in a layered net. In general this leads to a picture of a network of

self-interested or **hedonistic units** achieving a global purpose through individual maximization of r [Klopf, 1982].

Theory of Associative Reward-Penalty ⋆

It would be comforting to prove that these procedures actually do converge to a solution. Unfortunately that goal has not been achieved except in some very special cases. Barto and Anandan [1985] proved convergence for the case of a *single* binary output unit and a set of *linearly independent patterns* V_j^μ, provided the learning rates η^+ and η^- decay to zero in the right way. Having only one binary output is only of interest with a stochastic environment (class II)—it is then the two-armed bandit problem. The proof cannot be extended to multiple output units simply by thinking of them independently, because there is only one reinforcement signal r which depends on all their actions; they must cooperate to be rewarded.

Our usual approach to discussing convergence has been to show the existence of a cost function which is steadily decreased by the learning rule. Then the process must eventually stop, at least at a local minimum, assuming that the cost is bounded below. We can do the same sort of thing here, but in a less satisfactory way because the process is stochastic.

A suitable cost function is minus the average (expected) value of the reinforcement r. Minimizing the cost then corresponds to maximizing the average reinforcement, which is clearly a sensible goal. For simplicity let us take a single layer network, replacing the general V_j's in the above analysis by a set of inputs ξ_j. Let us also choose a deterministic environment (class I), so that the reinforcement signal r is a well-defined function $r(\mathbf{S}, \boldsymbol{\xi})$ of the output pattern $\mathbf{S}$ and the input pattern $\boldsymbol{\xi}$; we use vectors to represent the whole set of inputs or outputs. Then, for a given pattern ξ_j^μ, the average reinforcement $\langle r \rangle^\mu$ is

$$\langle r \rangle^\mu = \sum_{\mathbf{S}} P(\mathbf{S}|\mathbf{w}, \boldsymbol{\xi}^\mu) r(\mathbf{S}, \boldsymbol{\xi}^\mu) \tag{7.80}$$

where the sum is over all possible output configurations $\mathbf{S}$, each of which has a probability $P(\mathbf{S}|\mathbf{w}, \boldsymbol{\xi}^\mu)$ that depends on the input and all the weights $\mathbf{w}$. In fact, since output unit i chooses $S_i = \pm 1$ independently of what all the other other units are doing, this probability factorizes:

$$P(\mathbf{S}|\mathbf{w}, \boldsymbol{\xi}^\mu) = \prod_k \left\{ \begin{matrix} g(h_k^\mu) & \text{if } S_k = +1 \\ 1 - g(h_k^\mu) & \text{if } S_k = -1 \end{matrix} \right\} \tag{7.81}$$

where $h_k^\mu = \sum_j w_{kj} \xi_j^\mu$ as usual.

Now we want to differentiate (7.80) with respect to some particular weight w_{ij}, in order to show that our learning rule goes in the gradient direction. So we need to differentiate the product in (7.81) with respect to w_{ij}:

$$\frac{\partial P(\mathbf{S}|\mathbf{w},\boldsymbol{\xi}^\mu)}{\partial w_{ij}} = \left(\prod_{k\neq i}\left\{\begin{matrix} g(h_k^\mu) & \text{if } S_k = +1 \\ 1 - g(h_k^\mu) & \text{if } S_k = -1 \end{matrix}\right\}\right)\left\{\begin{matrix} g'(h_i^\mu)V_j^\mu & \text{if } S_i = +1 \\ -g'(h_i^\mu)V_j^\mu & \text{if } S_i = -1 \end{matrix}\right\}$$

$$= P(\mathbf{S}|\mathbf{w},\boldsymbol{\xi}^\mu)\left\{\begin{matrix} g'(h_i^\mu)/g(h_i^\mu) & \text{if } S_i = +1 \\ -g'(h_i^\mu)/[1 - g(h_i^\mu)] & \text{if } S_i = -1 \end{matrix}\right\}V_j^\mu. \qquad (7.82)$$

A little algebra is now required. Using the logistic function $f_\beta(h)$ for $g(h)$ implies that $g'(h) = 2\beta g(h)[1 - g(h)]$, and (7.75) allows us to write $g(h) = \frac{1}{2}(\langle S \rangle + 1)$. These lead easily to

$$\left\{\begin{matrix} g'(h_i^\mu)/g(h_i^\mu) & \text{if } S_i = +1 \\ -g'(h_i^\mu)/[1 - g(h_i^\mu)] & \text{if } S_i = -1 \end{matrix}\right\} = \beta[S_i^\mu - \langle S_i^\mu\rangle]. \qquad (7.83)$$

Putting together (7.80), (7.82), and (7.83) we obtain

$$\frac{\partial \langle r \rangle^\mu}{\partial w_{ij}} = \beta \sum_{\mathbf{S}} P(\mathbf{S}|\mathbf{w},\boldsymbol{\xi}^\mu)r(\mathbf{S},\boldsymbol{\xi}^\mu)[S_i^\mu - \langle S_i^\mu\rangle]V_j^\mu$$

$$= \beta\langle r(\mathbf{S},\boldsymbol{\xi}^\mu)[S_i^\mu - \langle S_i^\mu\rangle]\rangle V_j^\mu \qquad (7.84)$$

where the averages $\langle\cdots\rangle$ are over all output configurations, weighted only by the usual stochastic unit probabilities (7.72).

This is our desired result. First note that it would vanish if $r(\mathbf{S},\boldsymbol{\xi}^\mu)$ were the same for all output configurations $\mathbf{S}$ that had appreciable probability $P(\mathbf{S}|\mathbf{w},\boldsymbol{\xi}^\mu)$, because $\langle[S_i^\mu - \langle S_i^\mu\rangle]\rangle = 0$. This is what happens when we reach a correct solution—we almost always get $r^\mu = +1$, so of course $\langle r \rangle^\mu$ has a local maximum there. Now comparing with (7.79) we see that we may write

$$\langle \Delta w_{ij}\rangle^\mu = \frac{\eta}{\beta}\frac{\partial \langle r\rangle^\mu}{\partial w_{ij}} \qquad (7.85)$$

so that the *average* weight update in response to input pattern μ is in the upward gradient direction of our (negative) cost function $\langle r \rangle^\mu$. Finally we can average this result over all patterns μ and hence write

$$\langle \Delta w_{ij}\rangle = \frac{\eta}{\beta}\frac{\partial \langle r\rangle}{\partial w_{ij}} \qquad (7.86)$$

where now the averages are across all patterns and outcomes.

The result suggests that the learning rule (7.79) will continue to increase the average reinforcement until a maximum is reached. It does not *prove* such convergence however, because it only tells us about the average behavior. In a particular case a sequence of unlucky chances might decrease $\langle r \rangle$ considerably, and might lead us to become stuck in a very poor configuration. The convergence of stochastic processes like this is very difficult to treat, and is not yet fully understood for the A_{RP} algorithm. Nevertheless the result is encouraging, and allows us once again to visualize the procedure in terms of gradient descent on a cost surface, or gradient ascent on an average reinforcement surface.

Equation (7.86) applies to the symmetric rule (7.79) and not to the original rule (7.77). It also relies on η being independent of r. Both these restrictions actually make the algorithm worse in practice, but seem to be needed to prove the result. In fact it may be that the better versions are *not* always moving uphill in $\langle r \rangle$, even on average, but are thus more free to explore alternatives and to find better solutions in the end. Indeed it is found that the algorithms to which the result applies tend to converge rapidly, but to rather poor local maxima [Williams and Peng, 1989].

Another case in which (7.86) applies is obtained by modifying our reinforcement signal from ± 1 to $0/1$. Then maximizing $\langle r \rangle$ is a *different* objective, in which failures are not penalized. Now (7.86) applies if we use the original A_{RP} rule (7.77), but set η^- to 0. This is sometimes called the associative reward-inaction rule, or A_{RI}.

Williams [1987, 1988b] has identified a wider class of algorithms that possess the gradient ascent property (7.86). His derivation is also applicable to class II problems with a stochastic environment, in which $\langle r \rangle$ would be an average over the environment's responses as well as over the stochastic choices of the output units. However the class of algorithms, which Williams calls REINFORCE algorithms, still does not include the full A_{RP} algorithm.

Models and Critics

Another approach to reinforcement learning involves modelling the environment with an auxiliary network, which can then be used to produce a target for each output of the main network instead of just a global signal r [Munro, 1987; Williams, 1988a; Werbos, 1987, 1988]. The general idea of a separate modelling network can also be extended to more complex reinforcement learning problems, including class III problems.

A suitable architecture is shown in Fig. 7.8. The lower network is the main one that produces an output to the environment in response to an input from it, as before. Again it normally employs stochastic output units to encourage exploration. The upper network is the modelling network. It monitors at its input both the input and the output of the main network, and thus sees everything that the environment does. It has a single continuous-valued output unit R, called the **evaluation unit**, which is intended to duplicate the reinforcement r produced by the environment. If the model is a good one we will have $R \approx r$ for each input-output pair that the model sees.

In class I and II problems it is easy to see how to train the modelling network. The environment provides the correct response r for each case encountered (at least on average), so we can use an ordinary supervised approach, such as backpropagation, to minimize $(r - R)^2$. So that the modelling network can see a broad

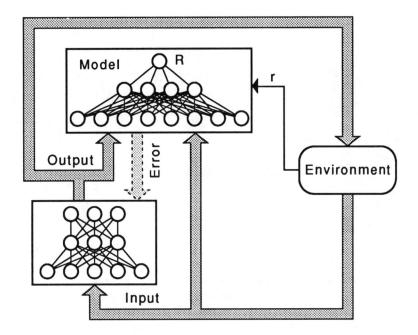

FIGURE 7.8 A reinforcement learning network which tries to model its environment.

sample of possible input-output pairs it may be expedient to turn up the temperature parameter $(T = 1/\beta)$ of the main network's stochastic output units while training the modelling network.

Once it is trained the modelling network can be used to train the main network. Now the aim is to maximize R, which we take to be a good model of r. This can also be done by back-propagation, through *both* networks. We do not change any weights in the modelling network, but propagate the appropriate δ's through it, starting from $\delta = -R$ at the evaluation unit. By the time these δ's propagate back to the main network we have an error signal for *each* unit instead of a global signal r, and can use ordinary supervised learning to adjust the weights.

Thus overall the whole scheme reduces the reinforcement learning problem to two stages of supervised learning with known targets. In principle at least we could do both stages at the same time, thinking of the two parts as a single cascaded network. However it is still essential to keep distinct the separate goals of the two subnetworks; it would be no good to train the whole network to minimize $(R - r)^2$ without any incentive towards maximizing r, and equally pointless to maximize R overall without regard to r. In practice though, it seems best to train the modelling network first, at least partially.

One advantage of the present approach is that the modelling network can smooth out the fluctuations of a stochastic environment in class II reinforcement

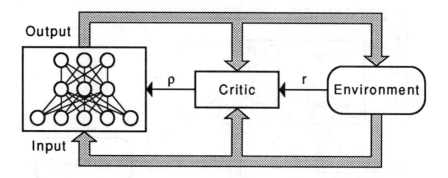

FIGURE 7.9 A reinforcement learning network with a critic. The main network might be a simple feed-forward network as sketched, or could itself include a modelling network as in Fig. 7.8.

learning problems. With the A_{RP} algorithm there is a tendency to chase meaningless fluctuations of r, but maximizing R should be less hazardous.

For class III problems there are various ways to go further, making the modelling network a better predictor of reinforcement. Actually it should try to estimate a weighted average of the future reinforcement, because the main network needs to maximize the cumulative reinforcement, not the instantaneous value. In a game, for example, the system might need to wait many moves for an eventual reinforcement. And even if continuing evaluation is provided by the environment, it may be appropriate to sacrifice short-term rewards for longer-term gains.

It may be worthwhile constructing a separate network (or algorithm) to perform this task of predicting reinforcement. As shown in Fig. 7.9 this **critic** receives the raw reinforcement signal r from the environment and feeds a processed signal ρ on to the main network. The ρ signal represents an evaluation of the *current* behavior of the main network, whereas r typically involves the past history. If the critic can produce a prediction R of r, we could for example use $\rho = r - R$, so the main network would be rewarded when r exceeded expectations, and penalized when it fell short; this is **reinforcement comparison** [Sutton, 1984].

The critic itself would normally be adaptive and would improve with experience; the name **adaptive heuristic critic** is sometimes used. We omit descriptions of appropriate algorithms, but note that they are often closely related to **dynamic programming**; see Sutton [1984, 1988] and Barto et al. [1991]. They are also related to models of classical conditioning in behavioral psychology [e.g., Tesauro, 1986; Sutton and Barto, 1991].

Barto et al. [1983] constructed a network consisting of an adaptive critic and a set of A_{RP} units that learned a problem of balancing a pole on a movable cart. The system was able to learn the task even though the "failure" reinforcement signal typically came long after the mistakes that produced it.

EIGHT

Unsupervised Hebbian Learning

8.1 Unsupervised Learning

This is the first of two chapters on unsupervised learning. In unsupervised learning there is no teacher. We still consider a network with both inputs and outputs, but there is no feedback from the environment to say what those outputs should be or whether they are correct. The network must discover for itself patterns, features, regularities, correlations, or categories in the input data and code for them in the output. The units and connections must thus display some degree of **self-organization**.

Unsupervised learning can only do anything useful when there is **redundancy** in the input data. Without redundancy it would be impossible to find any patterns or features in the data, which would necessarily seem like random noise. In this sense *redundancy provides knowledge* [Barlow, 1989]. More technically, the actual information content of the input data stream must be less than the maximum that could be carried on the same channel (the same input lines at the same rate); the difference is the redundancy.

The type of pattern that an unsupervised learning network detects in the input data depends on the architecture, and we will consider several cases. But it is interesting first to consider what sort of things such a network might tell us; what could the outputs be representing? There are a number of possibilities:

1. **Familiarity.** A single continuous-valued output could tell us how similar a new input pattern is to typical or average patterns seen in the past. The network would gradually learn what is typical.

2. **Principal Component Analysis.** Extending the previous case to several units involves constructing a multi-component basis, or set of axes, along which to measure similarity to previous examples. A common approach from

statistics, called principal component analysis, uses the leading eigenvector directions of the input patterns' correlation matrix.

3. **Clustering.** A set of binary-valued outputs, with only one on at a time, could tell us which of several categories an input pattern belongs to. The appropriate categories would have to be found by the network on the basis of the correlations in the input patterns. Each cluster of similar or nearby patterns would then be classified as a single output class.

4. **Prototyping.** The network might form categories as in the previous case, but then give us as output a prototype or exemplar from the appropriate class. It would then have the function of an associative memory, but the memories would be found directly from the input patterns, not imposed from outside.

5. **Encoding.** The output could be an encoded version of the input, in fewer bits, keeping as much relevant information as possible. This could be used for data compression prior to transmission over a limited-bandwidth channel, assuming that an inverse decoding network could also be constructed.

6. **Feature Mapping.** If the output units had a fixed geometrical arrangement (such as a two-dimensional array) with only one on at a time, they could map input patterns to different points in this arrangement. The idea would be to make a topographic map of the input, so that similar input patterns always triggered nearby output units. We would expect a global organization of the output units to emerge.

These cases are of course not necessarily distinct, and might also be combined in several ways. The encoding problem in particular could be performed using principal component analysis, or with clustering, which is often called **vector quantization** in this context. Principal component analysis could itself be used for **dimensionality reduction** of the data before performing clustering or feature mapping. This could avoid the "curse of dimensionality" usually encountered when looking for patterns in unknown data; a high-dimensional space with a modest number of samples is mostly empty.

It is worth noting that unsupervised learning may be useful even in situations where supervised learning would be possible:

■ Multi-layer back-propagation is extremely slow, in part because the best weight in one layer depends on all the other weights in all the other layers. This can be avoided to some extent by an unsupervised learning approach, or a hybrid approach, in which at least some layers train themselves just on the outputs from the immediately preceding layer.

■ Even after training a network with supervised learning, it may be advisable to allow some subsequent unsupervised learning so that the network can adapt to gradual changes or drift in its environment or sensor readings.

Unsupervised learning architectures are mostly fairly simple—the complications and subtleties come mainly from the learning rules. Most networks consist of only

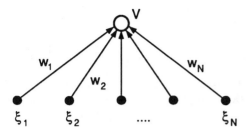

FIGURE 8.1 Architecture for simple Hebbian learning. The output unit is linear, so $V = \sum_j w_j \xi_j$.

a single layer. Most are essentially feed-forward, with the notable exception of Adaptive Resonance Theory. Except in the case of Feature Mapping, usually there are many fewer output units than inputs. The network architectures also tend to be more closely modelled after neurobiological structures than elsewhere in neural computation.

In most of the cases considered in this chapter and the next the architecture and learning rule come simply from intuitively plausible suggestions. However in some cases there is a well-defined quantity that is being maximized, such as the information content or variance of the output. These optimization approaches are closer to those of statisticians. There are in fact close connections between many of the networks discussed here and standard statistical techniques of pattern classification and analysis [Duda and Hart, 1973; Devijver and Kittler, 1982; Lippmann, 1987].

In this chapter we consider some techniques based on connections that learn using a modified Hebb rule. The outputs are continuous-valued and do *not* have a winner-take-all character, in contrast to most of the networks considered in the next chapter. Thus the purpose is not clustering or classification of patterns, but rather measuring familiarity or projecting onto the principal components of the input data. In a multi-layered network this can however lead to some remarkable feature extraction properties, as we will see in Section 8.4.

8.2 One Linear Unit

Let us assume that we have some input vectors ξ, with components ξ_i for $i = 1, 2,$ ..., N, drawn from some probability distribution $P(\xi)$. The components could be continuous-valued or binary-valued. At each time step an instance ξ is drawn from the distribution and applied to the network. After seeing enough such samples the network should learn to tell us—as its output—how well a particular input pattern conforms to the distribution. We will make this more precise shortly.

The case of a single *linear* output unit is simplest. The architecture is shown in Fig. 8.1 and we have simply

$$V = \sum_j w_j \xi_j = \mathbf{w}^T \xi = \xi^T \mathbf{w} \tag{8.1}$$

where **w** is the weight vector. We find it convenient in this chapter to use **matrix multiplication** notation (e.g., $\mathbf{w}^T\boldsymbol{\xi}$) rather than inner product notation (e.g., $\mathbf{w} \cdot \boldsymbol{\xi}$), writing $\mathbf{x}^T$ for the transpose (a row vector) of $\mathbf{x}$ (a column vector). All boldface symbols are vectors, except the matrix **C** introduced below.

With just one unit we clearly want the output V to become a scalar measure of familiarity. The more probable that a particular input $\boldsymbol{\xi}$ is, the larger the output $|V|$ should become, at least on average. It is therefore natural to try **plain Hebbian learning**

$$\Delta w_i = \eta V \xi_i \qquad (8.2)$$

where η controls the learning rate as usual. This strengthens the output in turn for each input presented, so frequent input patterns will have most influence in the long run, and will come to produce the largest output.

But there is a problem; the weights keep on growing without bound and learning never stops. It is nevertheless instructive to examine the rule (8.2) in more detail. Suppose for a moment that there *were* a stable equilibrium point for **w**. After sufficient learning the weight vector would remain in the neighborhood of the equilibrium point, just fluctuating around it (by an amount proportional to η) but not going anywhere on average. So at the equilibrium point we would expect the weight changes given by (8.2) to average to zero:

$$0 = \langle \Delta w_i \rangle = \langle V\xi_i \rangle = \left\langle \sum_j w_j \xi_j \xi_i \right\rangle = \sum_j C_{ij} w_j = \mathbf{C}\mathbf{w} \qquad (8.3)$$

where the angle brackets indicate an average over the input distribution $P(\boldsymbol{\xi})$ and we have defined the **correlation matrix C** by

$$C_{ij} \equiv \langle \xi_i \xi_j \rangle \qquad (8.4)$$

or

$$\mathbf{C} \equiv \langle \boldsymbol{\xi}\boldsymbol{\xi}^T \rangle . \qquad (8.5)$$

Several things should be noted about **C** before we proceed. First, it is not quite the **covariance matrix** of the input, which would be defined in terms of the means $\mu_i = \langle \xi_i \rangle$ as $\langle (\xi_i - \mu_i)(\xi_j - \mu_j) \rangle$. Secondly, **C** is *symmetric*, $C_{ij} = C_{ji}$, which implies that its eigenvalues are all real and its eigenvectors can be taken as orthogonal. Finally, because of the outer product form, **C** is *positive semi-definite*[1]; all its eigenvalues are positive or zero.

Now let us return to (8.3). It says that (at our hypothetical equilibrium point) **w** is an eigenvector of **C** with eigenvalue 0. But this could never be stable, because **C** necessarily has some positive eigenvalues; any fluctuation having a component along an eigenvector with positive eigenvalue would grow exponentially. We might suspect (correctly) that the direction with the largest eigenvalue $\lambda_{\max}$ of **C** would

---

[1]Proof: for any vector **x**, $\mathbf{x}^T\mathbf{C}\mathbf{x} = \mathbf{x}^T\langle \boldsymbol{\xi}\boldsymbol{\xi}^T \rangle \mathbf{x} = \langle \mathbf{x}^T\boldsymbol{\xi}\boldsymbol{\xi}^T\mathbf{x} \rangle = \langle (\boldsymbol{\xi}^T\mathbf{x})^2 \rangle \geq 0.$

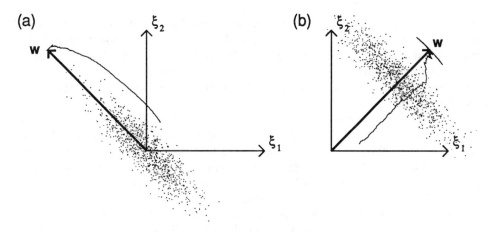

FIGURE 8.2 Examples of Oja's unsupervised learning rule. The dots show 1000 samples from each $P(\boldsymbol{\xi})$. The arrows represent the average weight vector $\mathbf{w}$ after many updates. The thin lines show a typical trajectory of $\mathbf{w}$ during training (for 2500 updates in (a); for 1000 updates in (b)). The parameter η was held at 0.1.

eventually become dominant, so that $\mathbf{w}$ would gradually approach an eigenvector corresponding to $\lambda_{\max}$, with a increasingly huge norm. But in any case $\mathbf{w}$ does not settle down; there are only unstable fixed points for the plain Hebbian learning procedure (8.2).

Oja's Rule

We can prevent the divergence of plain Hebbian learning by constraining the growth of the weight vector $\mathbf{w}$. There are several ways to do this, such as a simple renormalization $w_i' = \alpha w_i$ of all the weights after each update, choosing α so that $|\mathbf{w}'| = 1$. But Oja [1982] suggested a more clever approach. By modifying the Hebbian rule itself, he showed that it is possible to make the weight vector approach a constant length, $|\mathbf{w}| = 1$, without having to do any normalization by hand. Moreover, $\mathbf{w}$ does indeed approach an eigenvector of $\mathbf{C}$ with largest eigenvalue $\lambda_{\max}$. We will henceforth call this a **maximal eigenvector** for brevity.

Oja's rule corresponds to adding a weight decay proportional to V^2 to the plain Hebbian rule:

$$\Delta w_i = \eta V(\xi_i - V w_i). \tag{8.6}$$

Note that this looks like reverse delta-rule learning; Δw depends on the difference between the actual *input* and the back-propagated output.

It is not immediately obvious that (8.6) makes $\mathbf{w}$ approach unit length or tend to a maximal eigenvector, though we will prove these facts shortly. First let us see the rule in action. In each part of Fig. 8.2 we drew samples for $\boldsymbol{\xi}$ from a two-dimensional Gaussian distribution; $\boldsymbol{\xi}$ has two components ξ_1 and ξ_2, and there are two weights, w_1 and w_2. The weights were started from small random values and updated according to (8.6) after each sample. The thin lines show that $|\mathbf{w}|$ grew initially but reached a constant length (of 1) and then merely fluctuated on an arc of the circle $|\mathbf{w}| = 1$. The fluctuations did not die away and were much larger in case (b) than in case (a). Convergence to the unit circle was much faster in case (b) than in case (a).

The final average weight vectors $\mathbf{w}$ are shown by the heavy arrows. What do these tell us? Do they really make the output V represent the familiarity of a particular input $\boldsymbol{\xi}$? Well, yes and no; they do the best they can within the constraints of our architecture. Because we chose linear units, the output V is just the component of the input $\boldsymbol{\xi}$ along the $\mathbf{w}$ direction. In case (a), with zero-mean data, this is zero on average, whatever the direction of $\mathbf{w}$, but it is largest in magnitude for the direction found. In case (b), the average of V itself is maximized for the direction found. So in both cases the direction of $\mathbf{w}$ found by Oja's rule gives larger $|V|$'s on average than for any other direction, for points drawn from the original distribution. Points drawn from another distribution—"unfamiliar" points—would tend to give smaller values of $|V|$, unless they had larger magnitudes on average. Thus the network does develop a familiarity index for the distribution as a whole, though not necessarily for any particular sample $\boldsymbol{\xi}$; in case (a), for instance, the most probable (and mean) $\boldsymbol{\xi}$ is at the origin, which gives $V = 0$.

In fact Oja's rule *chooses the direction of* $\mathbf{w}$ *to maximize* $\langle V^2 \rangle$. This confirms, and makes more precise, the above observations based on Fig. 8.2. For zero-mean data, such as case (a), it corresponds to **variance maximization** at the output, and also to finding a principal component, as discussed in Section 8.3.

Theory of Oja's Rule ⋆

We have left ourselves with a number of claims to prove. We have said that Oja's rule converges to a weight vector $\mathbf{w}$ with the following properties:

1. Unit length: $|\mathbf{w}| = 1$, or $\sum_i w_i^2 = 1$.
2. Eigenvector direction: $\mathbf{w}$ lies in a maximal eigenvector direction of $\mathbf{C}$.
3. Variance maximization: $\mathbf{w}$ lies in a direction that maximizes $\langle V^2 \rangle$.

Let us first show that property 3 is a simple consequence of property 2. Using (8.1) and (8.5) we can write

$$\langle V^2 \rangle = \langle (\mathbf{w}^T \boldsymbol{\xi})^2 \rangle = \langle \mathbf{w}^T \boldsymbol{\xi} \boldsymbol{\xi}^T \mathbf{w} \rangle = \mathbf{w}^T \mathbf{C} \mathbf{w}. \tag{8.7}$$

Now for fixed $|\mathbf{w}|$ and a symmetric matrix $\mathbf{C}$, it is a standard result that the quadratic form $\mathbf{w}^T \mathbf{C} \mathbf{w}$ is maximized when $\mathbf{w}$ is along a maximal eigenvector direction of $\mathbf{C}$. So this direction also maximizes $\langle V^2 \rangle$, as claimed.

[The standard result can be proved by diagonalizing the quadratic form, writing it as

$$\mathbf{w}^T \mathbf{C} \mathbf{w} = \sum_\alpha \lambda^\alpha w_\alpha^2 \tag{8.8}$$

where the λ^α's are the eigenvalues of **C** and the w_α's are the components of **w** along the corresponding eigenvectors. These eigenvectors are orthogonal and $|\mathbf{w}| = 1$, so $\sum_\alpha w_\alpha^2 = 1$. To maximize (8.8) under this constraint we clearly have to put all the weight into the term with the largest λ^α, which is $\lambda_{\max}$ by definition. Thus **w** must be along a maximal eigenvector. In the case of a degenerate largest eigenvalue any direction in the space spanned by the corresponding eigenvectors will suffice.]

To prove properties 1 and 2 we must return to Oja's rule (8.6) itself. Just as in (8.3) for the plain Hebbian case, we expect that the average weight change will vanish when an equilibrium has been reached:

$$
\begin{aligned}
0 = \langle \Delta w_i \rangle &= \langle V\xi_i - V^2 w_i \rangle \\
&= \left\langle \sum_j w_j \xi_j \xi_i - \sum_{jk} w_j \xi_j w_k \xi_k w_i \right\rangle \\
&= \sum_j C_{ij} w_j - \left[\sum_{jk} w_j C_{jk} w_k \right] w_i
\end{aligned}
\tag{8.9}
$$

or

$$
0 = \langle \Delta \mathbf{w} \rangle = \mathbf{C}\mathbf{w} - [\mathbf{w}^T \mathbf{C} \mathbf{w}]\mathbf{w}.
\tag{8.10}
$$

Thus an equilibrium vector must obey

$$
\mathbf{C}\mathbf{w} = \lambda \mathbf{w}
\tag{8.11}
$$

with

$$
\lambda = \mathbf{w}^T \mathbf{C} \mathbf{w} = \mathbf{w}^T \lambda \mathbf{w} = \lambda |\mathbf{w}|^2.
\tag{8.12}
$$

Equation (8.11) shows clearly that an equilibrium **w** must be an eigenvector of **C**, and (8.12) proves that $|\mathbf{w}| = 1$. All that remains is to show that $\lambda = \lambda_{\max}$.

Any normalized eigenvector of **C** would satisfy (8.10). But only one belonging to $\lambda_{\max}$ is stable. To show this, let us take **w** close to a particular normalized eigenvector $\mathbf{c}^\alpha$ of **C**

$$
\mathbf{w} = \mathbf{c}^\alpha + \varepsilon
\tag{8.13}
$$

with $\mathbf{C}\mathbf{c}^\alpha = \lambda^\alpha \mathbf{c}^\alpha$ and $|\mathbf{c}^\alpha| = 1$. Then, using (8.9), the average subsequent change of ε to order ε is

$$
\begin{aligned}
\langle \Delta \varepsilon \rangle = \langle \Delta \mathbf{w} \rangle &= \mathbf{C}(\mathbf{c}^\alpha + \varepsilon) - [((\mathbf{c}^\alpha)^T + \varepsilon^T)\mathbf{C}(\mathbf{c}^\alpha + \varepsilon)](\mathbf{c}^\alpha + \varepsilon) \\
&= \lambda^\alpha \mathbf{c}^\alpha + \mathbf{C}\varepsilon - [(\mathbf{c}^\alpha)^T \mathbf{C}\mathbf{c}^\alpha]\mathbf{c}^\alpha - [\varepsilon^T \mathbf{C}\mathbf{c}^\alpha]\mathbf{c}^\alpha \\
&\quad - [(\mathbf{c}^\alpha)^T \mathbf{C}\varepsilon]\mathbf{c}^\alpha - [(\mathbf{c}^\alpha)^T \mathbf{C}\mathbf{c}^\alpha]\varepsilon + O(\varepsilon^2) \\
&= \mathbf{C}\varepsilon - 2\lambda^\alpha [\varepsilon^T \mathbf{c}^\alpha]\mathbf{c}^\alpha - \lambda^\alpha \varepsilon + O(\varepsilon^2).
\end{aligned}
\tag{8.14}
$$

Now take the component of this along the direction of another normalized eigenvector $\mathbf{c}^\beta$ of **C** by multiplying by $(\mathbf{c}^\beta)^T$ on the left (ignoring the $O(\varepsilon^2)$ terms):

$$
\begin{aligned}
(\mathbf{c}^\beta)^T \langle \Delta \varepsilon \rangle &= \lambda^\beta (\mathbf{c}^\beta)^T \varepsilon - 2\lambda^\alpha [\varepsilon^T \mathbf{c}^\alpha]\delta_{\alpha\beta} - \lambda^\alpha (\mathbf{c}^\beta)^T \varepsilon \\
&= [\lambda^\beta - \lambda^\alpha - 2\lambda^\alpha \delta_{\alpha\beta}](\mathbf{c}^\beta)^T \varepsilon.
\end{aligned}
\tag{8.15}
$$

Note that we used $(\mathbf{c}^\beta)^T \mathbf{c}^\alpha = \delta_{\alpha\beta}$ by orthonormality of the eigenvectors. Equation (8.15) is just what we need; for $\beta \neq \alpha$ it says that the component of ε along $\mathbf{c}^\beta$ will grow—and hence the solution will be unstable—if $\lambda^\beta > \lambda^\alpha$. So if λ^α is not the largest eigenvalue $\lambda_{\max}$ there will always be an unstable direction. On the other hand, an eigenvector corresponding to $\lambda_{\max}$ *is* stable in all directions, including the $\mathbf{c}^\alpha$ direction itself. This completes the proofs of our claims.

It is worth remarking that we have not really proved convergence to the solution, only that it is on average a stable fixed point of Oja's rule. In a nonlinear stochastic dynamical system like this there are several things that might conceivably happen besides convergence to a fixed point. We might see quasi-cyclic behavior, or we might simply see the system continuing to fluctuate, just like a thermodynamic system at $T > 0$ (above a phase transition perhaps, so the typical state could be quite unlike the ground state). We could consider quenching the fluctuations by steadily reducing η to zero, but if we did this too rapidly the system might get stuck in the wrong place and never get to the desired solution. More powerful techniques, such as **stochastic approximation theory**, are needed to prove convergence in these systems [see e.g., Kushner and Clark, 1978]. Such proofs have in fact been constructed for Oja's rule [Oja and Karhunen, 1985].

Other Rules

Oja's rule (8.6) is not the only way one can modify the plain Hebbian rule (8.2) to keep the weights bounded. Linsker [1986, 1988] uses simple clipping; the individual weights w_i are constrained to obey $w_- \leq w_i \leq w_+$. Yuille et al. [1989] suggest a rule

$$\Delta w_i = \eta(V\xi_i - w_i|\mathbf{w}|^2) \tag{8.16}$$

which makes $\mathbf{w}$ converge to the same maximal eigenvector direction as Oja's rule, but with $|\mathbf{w}| = \sqrt{\lambda_{\max}}$ instead of $|\mathbf{w}| = 1$. It has the disadvantage over Oja's rule of being nonlocal—to update w_i we need information about other w_j's—but the mathematical advantage that there is an associated **cost function**

$$E = -\frac{1}{2}\sum_{ij} C_{ij} w_i w_j + \frac{1}{4}\left(\sum_i w_i^2\right)^2 = -\frac{1}{2}\mathbf{w}^T \mathbf{C} \mathbf{w} - \frac{1}{4}|\mathbf{w}|^4 . \tag{8.17}$$

The average effect $\langle \Delta w_i \rangle$ of (8.16) corresponds exactly to gradient descent on this cost surface.

Pearlmutter and Hinton [1986] also derive a gradient-descent learning rule, based on maximizing the information content of the output. It is applicable to nonlinear as well as linear units.

8.3 Principal Component Analysis

A common method from statistics for analyzing data is **principal component**

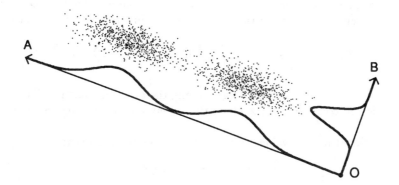

FIGURE 8.3 Illustration of principal component analysis. OA is the first principal component direction of the distribution that generated the cloud of points. The projection onto OA shows up more structure than the projection onto another direction OB. After Linsker [1988].

analysis (PCA) [see e.g., Jolliffe, 1986]. In communication theory it is known as the Karhunen-Loève transform. It is also closely related to least squares methods, factor analysis, singular value decomposition, and matched filtering. Linsker [1988] notes that performing principal component analysis is equivalent to maximizing the information content of the output signal in situations where that has a Gaussian distribution.

The aim is to find a set of M orthogonal vectors in data space that account for as much as possible of the data's variance. Projecting the data from their original N-dimensional space onto the M-dimensional subspace spanned by these vectors then performs a **dimensionality reduction** that often retains most of the intrinsic information in the data. Typically $M \ll N$, making the reduced data much easier to handle when, for example, subsequently searching for clusters. Figure 8.3 shows by example that clusters are more likely to be distinguished by projecting onto a high-variance direction than onto a low-variance one.

Specifically the first principal component is taken to be along the direction with the maximum variance. The second principal component is constrained to lie in the subspace perpendicular to the first. Within that subspace it is taken along the direction with the maximum variance. Then the third principal component is taken in the maximum variance direction in the subspace perpendicular to the first two, and so on.

In general it can be shown that the kth principal component direction is along an eigenvector direction belonging to the kth largest eigenvalue of the full **covariance matrix** $\langle (\xi_i - \mu_i)(\xi_j - \mu_j) \rangle$, where $\mu_i = \langle \xi_i \rangle$. For zero-mean data this reduces to the corresponding eigenvectors of the correlation matrix $\mathbf{C}$ defined by (8.5), and *we will restrict our attention to this case*. We have already seen that the unconstrained maximum variance direction—the first principal component—corresponds to a maximal eigenvector of **C**. To prove the corresponding result for the kth principal component, we write the variance along the direction of a unit vector **x** as

$$\sigma_{\mathbf{x}}^2 = \langle(\xi^T\mathbf{x})^2\rangle = \langle\mathbf{x}^T\xi\xi^T\mathbf{x}\rangle = \mathbf{x}^T\mathbf{C}\mathbf{x} = \sum_{\alpha}\lambda^{\alpha}x_{\alpha}^2. \tag{8.18}$$

The last expression results from diagonalizing the quadratic form $\mathbf{x}^T\mathbf{C}\mathbf{x}$, just as in (8.8). x_{α} is the component of **x** along the eigenvector $\mathbf{c}^{\alpha}$ belonging to eigenvalue λ^{α} of **C**.

Let us take the eigenvalues to be in decreasing order, so that

$$\lambda^1 \geq \lambda^2 \geq \ldots \geq \lambda^N \tag{8.19}$$

with $\lambda^1 = \lambda_{\max}$. Now we proceed by mathematical induction and assume that principal components 1 to $k-1$ are along the first $k-1$ eigenvector directions. The kth principal component is constrained to be perpendicular to these directions, so we must have $x_1 \ldots x_{k-1} = 0$. Maximizing $\sigma_{\mathbf{x}}^2$ subject to this condition, with $|\mathbf{x}| = 1$ (and thus $\sum_{\alpha} x_{\alpha}^2 = 1$), clearly results in

$$x_j = \begin{cases} \pm 1 & \text{if } j = k\,; \\ 0 & \text{otherwise.} \end{cases} \tag{8.20}$$

Thus the kth principal component is along the kth eigenvector, as claimed. Moreover (8.18) shows that the variance $\sigma_{\mathbf{x}}^2$ itself is equal to λ^k when **x** is along the kth principal component direction.

One-Layer Feed-Forward Networks

Oja's rule (8.6) finds a unit weight vector which maximizes the mean square output $\langle V^2\rangle$. For zero-mean data this is just the first principal component. It would however be desirable to have an M-output network that extracts the first M principal components. Sanger [1989a] and Oja [1989] have both designed one-layer feed-forward networks that do this. Other network architectures for principal component analysis will be discussed later.

The networks are linear, with the ith output V_i given by

$$V_i = \sum_j w_{ij}\xi_j = \mathbf{w}_i^T\xi = \xi^T\mathbf{w}_i. \tag{8.21}$$

Note that the vectors here are N-dimensional (i.e., in the *input* space); $\mathbf{w}_i$ is the weight vector for the ith output. Sanger's approach can also be extended to nonlinear units [Sanger, 1989b], but we discuss only the linear case.

Sanger's learning rule is

$$\Delta w_{ij} = \eta V_i\left(\xi_j - \sum_{k=1}^{i} V_k w_{kj}\right) \tag{8.22}$$

and Oja's M-unit rule is

$$\Delta w_{ij} = \eta V_i \left(\xi_j - \sum_{k=1}^{N} V_k w_{kj} \right). \tag{8.23}$$

The only difference is in the upper limit of the summation. Both rules reduce to Oja's 1-unit rule (8.6) for the $M = 1$ single output case. Sanger's rule always reduces to Oja's 1-unit rule for the first unit $i = 1$, so we already know that that unit will find the first principal component of the input data.[2]

It turns out that in both cases the $\mathbf{w}_i$ vectors converge to orthogonal unit vectors, $\mathbf{w}_i^T \mathbf{w}_j = \delta_{ij}$. For Sanger's rule the weight vectors become exactly the first M principal component directions, in order:

$$\mathbf{w}_i \to \pm \mathbf{c}^i \tag{8.24}$$

where $\mathbf{c}^i$ is a normalized eigenvector of the correlation matrix $\mathbf{C}$ belonging to the ith largest eigenvalue λ^i; we take the eigenvalues to be in decreasing order, as in (8.19). For Oja's M-unit rule the M weight vectors converge to span the same subspace as these first M eigenvectors, but do not find the eigenvector directions themselves.

In both cases the outputs project an input vector $\boldsymbol{\xi}$ onto the space of the first M principal components. Sanger's rule is usually more useful in applications because it extracts the principal components individually in order, and gives a reproducible result (up to sign differences) on a given data set if the eigenvalues are nondegenerate. It performs exactly the Karhunen-Loève transform. Different outputs are statistically uncorrelated and their variance decreases steadily with increasing i. Thus, in applications to data compression and encoding, fewer bits are needed for later outputs. It may be appropriate to look at the variance of a given output (which is just the value of the corresponding eigenvalue) as a measure of the usefulness of that output, perhaps taking only the first few outputs down to a variance cutoff.

Oja's M-unit rule gives weight vectors which, while spanning the right subspace, differ individually from trial to trial. They depend on the initial conditions and on the particular data samples seen during learning. On average the variance of each output is the same; this may be useful in some applications (such as in one layer of a multilayer network) where one wants to keep the information spread uniformly across the units. Furthermore, if any algorithm of this sort is implemented in real brains, it would probably look more like Oja's than Sanger's: there is no obvious advantage for an animal in having information sorted out into individual principal components.

Neither of these new learning rules (8.22) and (8.23) is local. Updating weight w_{ij} requires more information than is available at input j and output i. However,

---

[2]Here, and henceforth, we consider only zero-mean data. Strictly speaking, the networks find the eigenvectors of the correlation matrix $\langle \xi_i \xi_j \rangle$, but the principal components are the eigenvectors of the full covariance matrix $\langle (\xi_i - \mu_i)(\xi_j - \mu_j) \rangle$. For zero-mean data there is no difference.

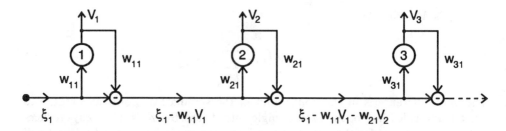

FIGURE 8.4 Network to implement Sanger's unsupervised learning rule. Only one input line is shown. The weighted output V_i of unit i is subtracted from the input before it reaches unit $i + 1$. Note that each weight appears twice; the two values must be kept equal.

Sanger suggested a reformulation of his rule which does allow a local implementation. We simply separate out the $k = i$ term in (8.22)

$$\Delta w_{ij} = \eta V_i \left[\left(\xi_j - \sum_{k=1}^{i-1} V_k w_{kj} \right) - V_i w_{ij} \right] \tag{8.25}$$

and observe that this is just the same as Oja's 1-unit rule for unit i except that the input ξ_j has been replaced by $\xi_j - \sum_{k=1}^{i-1} V_k w_{kj}$. So with inputs modified in this way we can use Oja's original rule, which *is* local. The modified inputs can be calculated by progressively decrementing the total inputs, as indicated in Fig. 8.4.

Theory of Sanger's Rule ⋆

We examine only Sanger's rule in detail; Oja's M-unit rule has been analyzed by Krogh and Hertz [1990]. Our aim is to demonstrate the result (8.24).

First let us substitute (8.21) into (8.22) and average:

$$\langle \Delta w_{ij} \rangle / \eta = \left\langle \sum_p w_{ip} \xi_p \xi_j - \sum_p w_{ip} \xi_p \sum_{k=1}^{i} \sum_q w_{kq} \xi_q w_{kj} \right\rangle$$

$$= \sum_p w_{ip} C_{pj} - \sum_{k=1}^{i} \left[\sum_{pq} w_{kq} C_{pq} w_{ip} \right] w_{kj} \tag{8.26}$$

or

$$\langle \Delta \mathbf{w}_i \rangle / \eta = \mathbf{C} \mathbf{w}_i - \sum_{k=1}^{i-1} [\mathbf{w}_k^T \mathbf{C} \mathbf{w}_i] \mathbf{w}_k - [\mathbf{w}_i^T \mathbf{C} \mathbf{w}_i] \mathbf{w}_i \tag{8.27}$$

where we have separated out the $k = i$ term from the summation. Now let us proceed by mathematical induction and assume that weight vectors $1, 2, \ldots, i-1$ have already found their appropriate eigenvectors, so $\mathbf{w}_k = \pm \mathbf{c}^k$ for $k < i$. The first two terms on the right-hand side of (8.27) then become the projection of $\mathbf{Cw}_i$ onto the subspace orthogonal to these $i - 1$ eigenvectors; recall (e.g., from the Gram-Schmidt orthogonalization procedure) that $\mathbf{x} - (\mathbf{y}^T\mathbf{x})\mathbf{y}$ is the projection of $\mathbf{x}$ perpendicular to a unit vector $\mathbf{y}$. Thus we have

$$\langle\Delta\mathbf{w}_i\rangle/\eta = (\mathbf{Cw}_i)^{\perp} - [\mathbf{w}_i^T\mathbf{Cw}_i]\mathbf{w}_i \qquad (8.28)$$

where $(\mathbf{Cw}_i)^{\perp}$ means the projection of $\mathbf{Cw}_i$ into the subspace orthogonal to the first $i-1$ eigenvectors. Note that $(\mathbf{Cw}_i)^{\perp}$ is equal to $\mathbf{Cw}_i^{\perp}$ because $\mathbf{C}$ preserves this subspace.

Suppose now that $\mathbf{w}_i$ has a component *not* in this subspace. For that component the first term on the right-hand side of (8.28) gives nothing, and the second causes a decay towards zero (recall that $\mathbf{w}_i^T\mathbf{Cw}_i$ is a positive scalar). Thus $\mathbf{w}_i$ relaxes towards the subspace. But when restricted to the subspace the whole equation (8.28) becomes just Oja's 1-unit rule (8.6) for unit i, and so leads to the maximal eigenvector in the subspace, which is $\pm\mathbf{c}^i$ with eigenvalue λ_i. This completes the induction. In fact the weight vectors $\mathbf{w}_i$ approach their final values simultaneously, not one at a time, but our argument still applies to the endpoint.

Other Networks

The architectures considered so far for principal component analysis have all been one-layer feed-forward networks. Other networks, with more layers or with lateral connections, can also perform it, and may have some advantages.

We have already met the **self-supervised back-propagation** approach, on page 136. A two-layer linear network with N inputs, N outputs, and $M < N$ hidden units is trained using back-propagation so that the outputs are as close as possible to the inputs in the training set. This forces the network to encode the N-dimensional data as closely as possible in the M dimensions of the hidden layer, and it should not be surprising to find that the hidden units end up projecting onto the subspace of the first M principal components. The network produces essentially the same result as Oja's M-unit rule, with equal variance on average on each of the hidden units. The dynamical approach to the solution is also similar, though not identical [Sanger, 1989a].

Another approach is to use a one-layer network, but have trainable lateral connections between the M output units V_i. One architecture is shown in Fig. 8.5 [Rubner and Tavan, 1989; Rubner and Schulten, 1990]. Note that a lateral connection u_{ij} from V_j to V_i is present only if $i > j$. The ordinary weights w_{ik} from the inputs are trained with a plain Hebbian rule (8.2), followed by renormalization of each weight vector to unit length. But the lateral weights employ **anti-Hebbian learning**, equivalent to a Hebb rule with negative η:

$$\Delta u_{ij} = -\gamma V_i V_j \,. \qquad (8.29)$$

Clearly the first output unit V_1 extracts the first principal component, just as in Oja's 1-unit rule and Sanger's rule. The second unit would do the same but for its

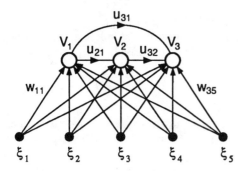

FIGURE 8.5 A network for principal component analysis. The lateral u_{ij} connections between the output units use *anti*-Hebbian learning.

lateral connection u_{21} from unit 1, which tries through (8.29) to decorrelate it with V_1. This is very like the effect of Sanger's rule, and leads to the same result; the second unit extracts the second principal component. And so on; the network learns to extract the first M principal components in order, just as in Sanger's network. The lateral weights end up going to zero, but are still required to be present for stability against fluctuations. A very similar proposal was made by Földiák [1989] using full lateral connections in the output layer; there was no $i > j$ restriction.

8.4 Self-Organizing Feature Extraction

Hebbian learning has been applied in a number of ways to producing **feature detectors** in a network whose input is a two-dimensional pixel array or "retina." Usually there is a well-defined set of possible inputs—such as letter patterns, or bars at certain angles—and each output cell is expected to become most sensitive to *one* of those inputs, with different output cells choosing different input patterns. There may be as many or more output units as input units; the aim is *not* to reduce the input information, but to transform it.

One can usefully define the **selectivity** of a particular output O_i as [Bienenstock et al., 1982]

$$\text{Selectivity}_i = 1 - \frac{\langle O_i \rangle}{\max O_i} \tag{8.30}$$

where the average $\langle O_i \rangle$ and the max are both taken over all the possible inputs. The selectivity approaches 1 if the output unit only responds appreciably to a single input (or a narrow range of inputs in a continuous case), and goes to zero if it responds equally to all inputs.

The problem is to define an architecture and learning rule in which the outputs go from low initial selectivity to high final selectivity. Moreover one wants different output units to become optimally sensitive to different input patterns, with some output unit or units approximately matched to every input pattern. Finally, if the output units are arranged in a geometrical way such as two-dimensional array, one would like similar input patterns to give maximum response at nearby output units; this is a form of feature mapping, which is discussed more fully in Section 9.4.

Most investigation has been with the visual cortex in mind. Experimental evidence shows the existence (e.g., in area 17 or V1) of neurons that respond preferentially to bars or edges in a particular orientation in the visual field [Hubel and Wiesel, 1962]. These are largely organized into **orientational columns** perpendicular to the cortical surface; neurons in a column respond maximally to approximately the same orientation.

After the work of Blakemore and Cooper [1970] on the imprinting of orientational preferences in kittens brought up in unusual environments, it was commonly thought that the whole orientational column structure is learned postnatally. Early models therefore employed input patterns of lines or edges, such as might be encountered postnatally, and showed how these could lead to orientational selectivity [von der Malsburg, 1973; Pérez et al., 1975; Nass and Cooper, 1975; Bienenstock et al., 1982]. Now it is known that some orientation specificity is present prenatally, so such models cannot be the whole story [von der Malsburg and Cowan, 1982]. They may however be appropriate for later optimization and stabilization of the pattern, which occurs during a critical developmental period after birth.

Recently, Linsker [1986] proposed a model of self-organization in the visual system that does not required structured input. It uses a version of Hebbian learning in a *layered* network. His work is summarized in Linsker [1988], a paper that also treats principal components, maximum information, etc., and is highly recommended.

Linsker's network resembles the visual system, with an input retina feeding on to a number of layers corresponding to the layers of the visual cortex. The units of the network are *linear* and are organized into two-dimensional layers indexed A (input), B, C, and so on. There are feed-forward connections between layers, with each unit receiving inputs only from a neighborhood in the previous layer, the **receptive field**. These limited receptive fields are crucial; they let units respond to spatial correlations in the input or the previous layer. Figure 8.6 shows the arrangement.

Note that this is the first time we have encountered a multi-layer unsupervised learning network. Because the units are linear the final output in principle could be found by a network with just one linear layer—a product of linear transformations is a linear transformation. But this does not apply to the unsupervised learning rule itself, which *can* benefit from multiple layers.

If a particular unit receives input from K units numbered $j = 1, 2, \ldots, K$ its output O is

$$O = a + \sum_{j=1}^{K} w_j V_j \tag{8.31}$$

where V_j is either the input pattern ξ_j (if the unit is in layer B) or the output of a previous layer. The threshold term a could be omitted.

Linsker used a modified Hebbian learning rule

$$\Delta w_i = \eta(V_i O + b V_i + c O + d) \tag{8.32}$$

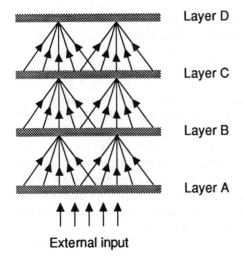

Layer D

Layer C

Layer B

Layer A

External input

FIGURE 8.6 Architecture of Linsker's multi-layer Hebbian learning network, showing the receptive fields of a few units.

where b–d are parameters which may be tuned to produce various types of behavior. The weights were clipped to the range $w_- \leq w_i \leq w_+$ to prevent them growing indefinitely. Explicit clipping can be avoided through the use of alternative rules such as (8.16) [Kammen and Yuille, 1988; Yuille et al., 1989]. Or, going more in the direction of biological realism, a mixture of purely excitatory $(0 \leq w_i \leq w_+)$ and purely inhibitory $(w_- \leq w_i \leq 0)$ connections can be used instead, and leads to the same results [Linsker, 1986].

Let us find the average weight change $\langle \Delta w_i \rangle$. We assume that all the inputs V_i have the same mean $\overline{V}$ and put $V_i = \overline{V} + v_i$. Then (8.32) becomes

$$
\begin{aligned}
\langle \Delta w_i \rangle &= \eta \left[\left\langle (\overline{V} + v_i) \sum_j [a + w_j (\overline{V} + v_j)] \right\rangle + b\overline{V} + c \sum_j (a + w_j \overline{V}) + d \right] \\
&= \eta \left[\sum_j C_{ij} w_j + \lambda \left(\mu - \sum_j w_j \right) \right]
\end{aligned}
\tag{8.33}
$$

where the constants λ and μ are combinations of the old constants a–d and $\overline{V}$, and C_{ij} is the $K \times K$ covariance matrix $\langle v_i v_j \rangle$ of the inputs to our chosen unit.

Equation (8.33) is the form of the rule actually used. It is easy to verify that it is equivalent to the average of gradient descent learning, $\Delta w_i = -\eta \partial E / \partial w_i$, on the **cost function**

$$
E = -\frac{1}{2} \mathbf{w}^T \mathbf{C} \mathbf{w} + \frac{\lambda}{2} \left(\mu - \sum_j w_j \right)^2.
\tag{8.34}
$$

In the first term, $\mathbf{w}^T \mathbf{C} \mathbf{w}$ is just the variance of the output O, by extension of (8.7) to the full covariance matrix. The second term may be regarded as a Lagrange multiplier or penalty term to impose the constraint $\sum_j w_j = \mu$. So Linsker's rule tries to maximize the output variance subject to this constraint and to the clipping constraint $w_- \leq w_i \leq w_+$. This may be compared to Oja's rule (8.6), which maximizes the same output variance subject to $\sum_j w_j^2 = 1$, and does not need a clipping constraint.

An equilibrium state of Linsker's rule will *not* normally have the right-hand side of (8.33) equal to zero, which would require $(1, 1, \ldots, 1)$ to be an eigenvector of **C**. Instead, most of the weights w_i will be pinned at one of their boundary values; either at w_- with $\langle \Delta w \rangle \leq 0$ or at w_+ with $\langle \Delta w \rangle \geq 0$. In fact all, or all but one, of the weights *must* be at a boundary value; any weight vector with two or more components not at a boundary is unstable under (8.33). To prove this, suppose the contrary and consider a perturbation $\mathbf{w}' = \mathbf{w} + \boldsymbol{\varepsilon}$ away from the supposed equilibrium point. In particular, choose $\boldsymbol{\varepsilon}$ so that $\sum_j \varepsilon_j = 0$ (with $\varepsilon_i = 0$ if w_i is at a boundary). Then (8.33) gives simply

$$\langle \Delta \boldsymbol{\varepsilon} \rangle = \eta \mathbf{C} \boldsymbol{\varepsilon} \qquad (8.35)$$

which causes $|\boldsymbol{\varepsilon}|$ to grow, because **C** is positive definite even when restricted to the non-boundary rows and columns. So the chosen point cannot be an equilibrium point, which proves the claim.

The most interesting aspects of Linsker's network arise from the spatial geometry of the layers and the receptive fields. We will approach them in a moment by describing Linsker's simulation results. Theoretical analysis, in detail beyond the level of this book, has been provided by Kammen and Yuille [1988], Yuille et al. [1989], and MacKay and Miller [1990], each using slightly modified rules and a **continuum approximation**. In the continuum approximation the weights w_i become a function $w(x, y)$ or $w(\mathbf{r})$ of position in the plane. The covariance matrix $\mathbf{C}_{ij}$ becomes a two-point correlation function $C(\mathbf{r}, \mathbf{s})$ which takes the form $C(\mathbf{r} - \mathbf{s})$ if there is translational invariance. A Fourier transform to wave-vector space makes the analysis easier. The weight patterns that arise can also be described in terms of a set of functions much like those encountered in the quantum mechanics of atoms (1s, 2s, 2p, etc.).

Simulation Results

Linsker first trained the connections between the input layer A and the first hidden layer B, then the next layer of connections from B to C, and so forth. Such sequential training may in general be important for teaching successively higher-level "concepts" to a layered unsupervised network, building gradually on elements learned earlier. This may require a teacher who can monitor the progress of learning within the network, and selectively activate or inhibit learning at different levels [Silverman and Noetzel, 1988]. However in Linsker's case the sequential approach was mainly for convenience; only one layer was simulated at a time, using an input covariance matrix calculated from the output properties of the previous layer.

The simulations used the equations (8.33) for the average $\langle \Delta w_i \rangle$, not the original learning rule (8.32) for Δw_i itself. Thus only the appropriate input covariance

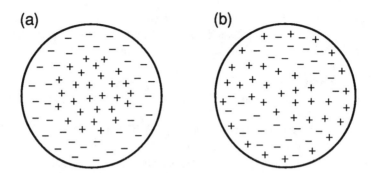

FIGURE 8.7 Sketch of positive and negative connection strengths within the receptive fields of units in Linsker's network. (a) A center-surround cell in layer C. (b) A bilobed orientation-selective cell in layer G. After Linsker [1986].

matrix C was needed to calculate how a particular unit would behave. Once the resulting connection strengths w_i feeding a whole layer had been calculated, the covariance matrices for that layer could be found from those of the preceding layer (giving a linear transformation of the matrices). Then the next layer could be treated in the same way.

Linsker took independent random noise as input in layer A, making the covariance matrix proportional to the identity matrix, the same for every site in layer B. The resulting weights from layer A to layer B depend on the parameters λ and μ, and on the size of the receptive fields. For a range of these parameters it was found that all the weights saturated to w_+, so the units in layer B simply averaged the input activity over their receptive fields. This made the B-units highly correlated because their receptive fields overlapped one another; a high activity in one of B_1's inputs would typically be seen also by many of B_1's neighbors.

As a result of this neighbor correlation in B, the units of layer C developed into **center-surround cells.** They responded maximally either to a bright spot in the center of their receptive field surrounded by a dark area or to a dark spot on a bright background. In the quantum mechanical analogy this is a "2s" function [MacKay and Miller, 1990]. Figure 8.7(a) illustrates the pattern of connection strengths found for one such unit; since most connections are pinned at either w_- or w_+ we show their values simply by +'s and −'s.

Different units of layer C had a Mexican hat covariance function; nearby units were positively correlated while further away there was a ring of negative correlation. Note that this is a correlation that has evolved through learning, *not* imposed by the terms of the model. Using the parameter values chosen by Linsker, this correlation gave rise in the following few layers (D–F) to center-surround units with sharper and sharper Mexican hat correlations; the negative trough became deeper and deeper.

For layer G Linsker changed the parameters, increasing the radius of the receptive field. This produced a variety of possible weight patterns, depending on the parameters, many of which were no longer circularly symmetric. Units were found with alternating bands of positive and negative connections, or with two or three islands of inhibition around the center of an otherwise excitatory field; Fig. 8.7(b) shows such a "bilobed" unit. These units had thus indeed become **orientation-selective cells**, responding maximally for example to a bright edge or bar in a particular orientation against a dark background.

The emergence of orientationally asymmetric units in a system with symmetric architecture and parameters is a **broken symmetry** phenomenon, and has been examined in those terms by Kammen and Yuille [1988]. The orientation selected by a given unit is rather arbitrary and varies from unit to unit.

Up to now we have needed no lateral connections within each layer. But now they can help us to obtain a smooth variation of orientational preference across the output layer. The addition of local excitatory connections within layer G— local lateral excitation—was found to produce banded patterns of approximately co-oriented cells, strikingly similar to some patterns of orientational columns found biologically [Linsker, 1986]. This is an example of feature mapping, which will be discussed in greater detail in the following chapter. Alternatively, lateral *inhibitory* connections between pairs of neighboring units can give rise to "quadrature pairs" of cells with spatial receptive field patterns that are approximately 90° out of phase [Yuille et al., 1989].

In the mammalian visual system center-surround cells are found as early as in the retina itself, and orientation-selective cells (including quadrature pairs) are found in the visual cortex. Considering the complexity of the visual system, including feedback and nonlinearity, it is remarkable that such a simple model can develop a similar structure. Of course this should not be taken to imply that feature detecting cells in retina and visual cortex develop just as in the model; there are many alternative ideas as to how it might occur [see e.g., Rose and Dobson, 1985]. However it does show that relatively simple mechanisms of Hebbian learning *could* produce such structures, without either visual input or detailed genetic programming.

NINE

Unsupervised Competitive Learning

In the preceding chapter we studied unsupervised learning approaches—all based on Hebbian learning—in which multiple output units are often active together. We now turn to **competitive learning** in which only *one* output unit, or only one per group, is on at a time. The output units *compete* for being the one to fire, and are therefore often called **winner-take-all** units. They are also sometimes called **grandmother cells** (cf. page 143).

The aim of such networks is to **cluster** or **categorize** the input data. Similar inputs should be classified as being in the same category, and so should fire the same output unit. The classes must be found by the network itself from the correlations of the input data; in the language of statistics we must deal with *unlabelled* data.

Categorization has obvious uses for any artificially intelligent machine trying to understand its world. More immediately, it can be used for data encoding and compression through **vector quantization**, in which an input data vector is replaced by the index of the output unit that it fires. It also has applications to function approximation, image processing, statistical analysis, and combinatorial optimization.

It is worth mentioning at the outset that grandmother-cell representations have some rather generic disadvantages over distributed representations [Caudill, 1989]:

- They need one output cell (and associated connections) for every category involved. Thus N units can represent only N categories, compared to 2^N for a binary code. Actually K-out-of-N codes may prove best for network computation, but are not yet well understood or exploited [Hecht-Nielsen, 1987a].

- They are not robust to degradation or failure. If one output unit fails, then we lose that whole category. Of course the same sort of problem arises in most *digital* circuitry, but we have come to expect better of neural networks.

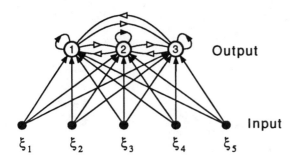

Output

Input

FIGURE 9.1 A simple competitive learning network. The connections shown with open arrows are inhibitory; the rest are excitatory.

- They cannot represent hierarchical knowledge. Two input patterns are either lumped together or not; there is no way to have categories within categories. Adding further layers to a network clearly does not help unless we relax the single-winner condition.

A closely related topic is **feature mapping**, which we discuss in Sections 9.4–9.6. Feature mapping is distinguished by the development of significant spatial organization in the output layer, whereas in simple competitive learning there is no particular geometrical relationship between the output units. Nevertheless the history and concepts of the two ideas are closely intertwined. Indeed, because feature mapping is so closely associated with the name of Kohonen, even simple competitive learning units are sometimes called Kohonen units. A history of either field should include the seminal work of Rosenblatt [1962], von der Malsburg [1973], Fukushima [1975, 1980], Grossberg [1976a, b], Kohonen [1982], and Rumelhart and Zipser [1985].

9.1 Simple Competitive Learning

In the simplest competitive learning networks there is a single layer of output units O_i, each fully connected to a set of inputs ξ_j via excitatory connections $w_{ij} \geq 0$. Figure 9.1 shows the architecture. We consider mainly binary 0/1 inputs and outputs. Only one of the output units, called the **winner**, can fire at a time. The winner is normally the unit with the largest net input

$$h_i = \sum_j w_{ij}\xi_j = \mathbf{w}_i \cdot \boldsymbol{\xi} \tag{9.1}$$

for the current input vector $\boldsymbol{\xi}$. Thus

$$\mathbf{w}_{i^*} \cdot \boldsymbol{\xi} \geq \mathbf{w}_i \cdot \boldsymbol{\xi} \qquad \text{(for all } i\text{)} \tag{9.2}$$

defines the winning unit i^* with $O_{i^*} = 1$. If the weights for each unit are normalized, so that (say) $|\mathbf{w}_i| = 1$ for all i, then (9.2) is equivalent to

$$|\mathbf{w}_{i^*} - \boldsymbol{\xi}| \leq |\mathbf{w}_i - \boldsymbol{\xi}| \qquad \text{(for all } i\text{).} \tag{9.3}$$

This says that the winner is the unit with normalized weight vector $\mathbf{w}$ closest to the input vector $\boldsymbol{\xi}$.

It does not much matter how the winner-take-all character of the output is implemented. In a computer simulation one can merely search for the maximum h_i. In a real network it is possible to implement a set of winner-take-all units with **lateral inhibition**; each unit inhibits the others, as shown in Fig. 9.1. A self-excitatory connection is also required, and the lateral weights and nonlinear activation function must be chosen correctly to ensure that only one output is chosen and that oscillation is avoided. For details see Grossberg [1976a, 1980]. Other schemes for winner-take-all networks have been proposed by Feldman and Ballard [1982], Lippmann [1987], and Winters and Rose [1989].

A winner-take-all network implements a pattern classifier using the criterion (9.2) or (9.3). The problem now is how to get it to find clusters in the input data and choose the weight vectors $\mathbf{w}_i$ accordingly.

We start with small random values for the weights; it is important that any symmetry be broken. Then a set of input patterns $\boldsymbol{\xi}^\mu$ is applied to the network in turn or in random order. Alternatively the inputs might be drawn independently from a distribution $P(\boldsymbol{\xi})$. For each input we find the winner i^* among the outputs and then update the weights $w_{i^* j}$ *for the winning unit only* to make the $\mathbf{w}_{i^*}$ vector closer to the current input vector $\boldsymbol{\xi}^\mu$. This makes the winning unit more likely to win on that input in the future. The most obvious way to do this would be

$$\Delta w_{i^* j} = \eta \xi_j^\mu \tag{9.4}$$

but this is no good by itself because it makes the weights grow without bound, and one unit comes to dominate the competition for all inputs.

One way to correct (9.4) is to follow it with a normalization step $w'_{i^* j} = \alpha w_{i^* j}$ (for all j), choosing α so that $\sum_j w'_{i^* j} = 1$ or $\sum_j (w'_{i^* j})^2 = 1$ (i.e., $|\mathbf{w}'_{i^*}| = 1$). It is easy to show that the first of these choices corresponds to an overall rule

$$\Delta w_{i^* j} = \eta' \left(\frac{\xi_j^\mu}{\sum_j \xi_j^\mu} - w_{i^* j} \right) \tag{9.5}$$

in which the first term is a normalized version of the input.

Another solution, which we will henceforth refer to as the **standard competitive learning rule**, is to use

$$\Delta w_{i^* j} = \eta (\xi_j^\mu - w_{i^* j}) \tag{9.6}$$

which moves $\mathbf{w}_{i^*}$ directly towards $\boldsymbol{\xi}^\mu$. Of course (9.6) is equivalent to (9.5) if the input is already normalized, and indeed (9.6) works best for pre-normalized inputs.

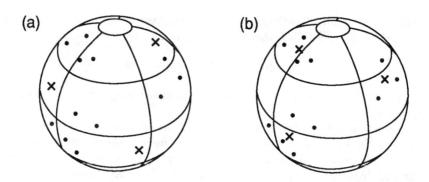

FIGURE 9.2 Competitive learning. The dots represent the input vectors, the crosses represent the weights for each of three units. (a) Before learning. (b) After learning.

Grossberg [1976a] shows how to normalize the input appropriately using an extra input layer of units.

Note that these learning rules are for the winner only. Because $O_{i^*} = 1$ and $O_i = 0$ for $i \neq i^*$ by definition of the winner, we can write (9.6) in the form

$$\Delta w_{ij} = \eta O_i (\xi_j^\mu - w_{ij}) \tag{9.7}$$

which is valid for all i and looks more like a Hebb rule with a decay term. In fact it is identical to Sanger's rule (8.22) and to Oja's M-unit rule (8.23) when they are specialized to the single winner case. It is also an appropriate generalization of competitive learning for multiple continuous-valued winners [Grossberg, 1976a].

A geometric analogy is very useful for understanding the consequences of competitive learning rules. Consider a case with 3 inputs so that each input pattern is a three-dimensional vector $\boldsymbol{\xi}^\mu = (\xi_1^\mu, \xi_2^\mu, \xi_3^\mu)$. For binary ξ_j^μ this vector lies on one of the vertices of a cube, but in this low dimension it is more convenient to think in terms of continuous variables. We can represent the direction of each $\boldsymbol{\xi}^\mu$ by a point on the surface of a unit sphere, as shown by the dots in Fig. 9.2. The direction of the weight vector $\mathbf{w}^i = (w_{i1}, w_{i2}, w_{i3})$ for each output unit i also corresponds to a point on the same sphere; the weights for 3 output units are depicted by crosses in the figure. These vectors are normalized to unit length (and therefore lie on the surface of the sphere) if we use the $|\mathbf{w}_i| = 1$ normalization. Diagram (a) shows a possible initial state and diagram (b) a typical final state; the output units have each discovered a cluster of inputs and have moved to its center of gravity. This is the principal function of a competitive learning network—it discovers clusters of overlapping input vectors.

We can see how this works geometrically. First, (9.2) defines the winner for a given input as the output unit with the greatest value of $\mathbf{w}_i \cdot \boldsymbol{\xi}$, which means the smallest angle θ to the $\boldsymbol{\xi}$ direction (if $|\mathbf{w}_i| = 1$). So the winner of a given dot is the closest cross. Secondly, the winning $\mathbf{w}_i$ is modified by (9.5) or (9.6), which moves it towards the active input vector. The patterns thus compete for output units, continually trying to bring the nearest one closer. The final state shown in Fig. 9.2(b) is a stable attractor if the inputs are activated about equally often.

One problem is that units with $\mathbf{w}_i$ far from any input vector may never win, and therefore never learn. These are sometimes called **dead units**. They may actually be desirable if the future might bring different input patterns, but otherwise can be prevented in several ways:

1. We can initialize the weights to samples from the input itself, thus ensuring that they are all in the right domain.

2. We can update the weights of all the losers as well as those of the winners, but with a much smaller η [Rumelhart and Zipser, 1985; see also Grossberg, 1987b]. Then a unit that has always been losing will gradually move towards the average input direction until eventually it succeeds in winning a competition. This has been called **leaky learning**.

3. If the units are arranged in a geometrical way, such as in a two-dimensional array, we can update the weights of *neighboring* losers as well as those of the winners. This is the essence of Kohonen feature mapping, discussed in detail in Section 9.4.

4. We can turn the input patterns on gradually, using $\alpha \xi^\mu + (1-\alpha)\mathbf{v}$, where $\mathbf{v}$ is some constant vector to which all the weight vectors $\mathbf{w}_i$ are initialized. As we turn α up gradually from 0 to 1 the pattern vectors move away from $\mathbf{v}$ and take weight vectors with them [Hecht-Nielsen, 1987b].

5. We can subtract a bias (or threshold) term μ_i from h_i in (9.1) and adjust the thresholds to make it easier for frequently losing units to win. Units that win often should raise their μ_i's, while the losers should lower them [Grossberg, 1976b; Bienenstock et al., 1982]. Stabilizing such a scheme is a little tricky, but can be done; indeed one can force M output units to win $1/M$th of the time on average [DeSieno, 1988]. The mechanism is sometimes called a **conscience**; frequent winners are supposed to feel guilty and so reduce their winning rate.

6. We can smear the pattern vectors with the addition of noise, using a distribution with a long tail so that there is some positive probability for any ξ [Szu, 1986; Hueter, 1988].

Cost Functions and Convergence

It would be nice to prove that competitive learning converges to the "best" solution of a given problem. But the best solution of a general clustering problem is not very clearly defined, and there exist various different criteria in the statistics literature [see e.g., Krishnaiah and Kanal, 1982 or Duda and Hart, 1973]. In practice most authors have defined an algorithm and then looked after the fact (if at all) at what it optimized.

Of most interest is the standard rule (9.6), for which there is an associated cost (Lyapunov) function given by [Ritter and Schulten, 1988c]

$$E\{w_{ij}\} = \frac{1}{2}\sum_{ij\mu} M_i^\mu (\xi_j^\mu - w_{ij})^2 = \frac{1}{2}\sum_\mu |\boldsymbol{\xi}^\mu - \mathbf{w}_{i*}|^2. \qquad (9.8)$$

Here M_i^μ is the **cluster membership matrix** which specifies whether or not input pattern $\boldsymbol{\xi}^\mu$ activates unit i as winner:

$$M_i^\mu = \begin{cases} 1 & \text{if } i = i^*(\mu); \\ 0 & \text{otherwise.} \end{cases} \qquad (9.9)$$

Note that the winner i^* is a function of the input μ, and of all the weights w_{ij}, in both (9.8) and (9.9). Thus the membership matrix in general will change in the course of learning.

Gradient descent on the cost function (9.8) yields

$$\langle \Delta w_{ij} \rangle = -\eta \frac{\partial E}{\partial w_{ij}} = \eta \sum_\mu M_i^\mu (\xi_j^\mu - w_{ij}) \qquad (9.10)$$

which is just the sum of the standard rule (9.6) over all the patterns μ for which i is the winner.[1] Thus on average (for small enough η) the standard rule decreases (9.8) until we reach a local minimum. This works even though (9.8) is only piecewise differentiable.

This result is, however, somewhat deceptive for two reasons. Firstly, there are typically very many local minima, corresponding to different clusterings of the data. Nevertheless the cost function does let us rank different clusterings in order of desirability, regarding lower cost ones as better. Stochastic noise (e.g., from the input presentation order) may be able like simulated annealing to kick us out of the higher minima and towards progressively lower ones, but there is no guarantee of finding the global minimum.

Secondly, we have averaged across the different patterns. We could actually use (9.10) directly, and update in **batch mode** by accumulating the changes Δw_{ij} given by each of a finite set of patterns $\boldsymbol{\xi}^\mu$ before actually updating the weights. Then our argument would be watertight and we would necessarily reach a stable equilibrium (though still only a local minimum). In fact that procedure corresponds exactly to the classical **k-means clustering** algorithm [e.g., Krishnaiah and Kanal, 1982]. But with incremental updates—after each pattern μ—continuing changes in the $i^*(\mu)$ classification can occur. With a regular cyclic presentation of patterns one can even produce examples in which the winners are permuted in every cycle [Grossberg, 1976a]. These problems can be reduced to some extent by doing the weight adjustments with "momentum" as in (6.24), effectively performing a

---

[1] We assume that the patterns are weighted equally. In general a probability P^μ could be inserted into (9.8) and (9.10).

weighted moving average over a set of recent patterns that resulted in the same winner.

Only in the case of sufficiently sparse patterns can one prove stability and convergence theorems for incremental updating [Grossberg, 1987b, and references therein]. The patterns are sparse enough if, for example, we can find a set of clusters so that the minimum overlap $\boldsymbol{\xi}^\mu \cdot \boldsymbol{\xi}^\nu$ within a cluster exceeds the maximum overlap between that cluster and any other.

In practice, both to prove theorems (using e.g., stochastic approximation methods) and to classify real data, it helps to decrease the learning rate η during learning. An initial large value encourages wide exploration, while later on a smaller value suppresses further winner changes and allows refinement of the weights. It is common to use either $\eta(t) = \eta_0 t^{-\alpha}$ (with $\alpha \le 1$) or $\eta(t) = \eta_0(1 - \alpha t)$.

It is possible to *start* with a cost function and derive a learning rule from it, as we have seen elsewhere. An interesting example was suggested by Bachmann et al. [1987] who replaced the quadratic form in (9.8) by a different power law:

$$E\{w_{ij}\} \;=\; -\frac{1}{p}\sum_{ij\mu} Q_i^\mu |\xi_j^\mu - w_{ij}|^{-p}. \tag{9.11}$$

They also let the winning weight vector be repelled by input vectors in other clusters, in addition to being attracted by those in its own cluster, taking

$$Q_i^\mu = 2M_i^\mu - 1. \tag{9.12}$$

For $p = N - 2$ the motion[2] of the weight vectors in the N-dimensional input space is like that of electrostatic point charges attracted toward stationary charges fixed at the ξ_j^μ's of the winning cluster, and repelled by charges at other ξ_j^μ's. This kind of model is therefore called a **Coulomb energy network** [Scofield, 1988].

9.2 Examples and Applications of Competitive Learning

Graph Bipartitioning

As a simple illustration of competitive learning we consider the graph bipartitioning problem defined in Section 4.3; we want to divide a given graph into two equal parts with as few links as possible between the parts. Let us have one binary 0/1 input for each vertex of the graph and two output units, one to indicate each partition. Then, after the problem has been solved, we should be able to turn on one vertex and have the output tell us the partition to which it belongs.

---

[2] As in almost all the physical analogies in this book, the motion we think of here is overdamped; we should imagine it as taking place in a very viscous fluid.

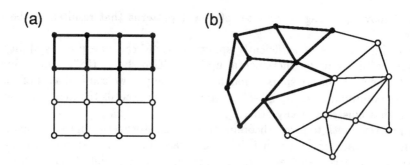

FIGURE 9.3 Application of competitive learning to graph bipartitioning. There are two output units, L and R, and one input for each vertex. Vertices are shown solid if L wins when that vertex is on alone (monopole stimulus), and open if R wins. Edges are shown heavy if L wins when the vertices at both ends are both on (dipole stimulus), and light if R wins. After Rumelhart and Zipser [1985].

How can we hope to produce such a network? Clearly we have to tell it about the *edges* of the graph. A good way to do this is to represent each edge by a **dipole stimulus** in which the two vertices belonging to the edge are turned on together. If we use this set of patterns (one for each edge) as our input data ξ_j^μ, the network should learn to divide them into two roughly equal halves with minimum overlap. This does not quite correspond to the graph bipartitioning problem, for which we want equal numbers of *vertices* in each partition, not necessarily equal number of *edges*. However, it is often close enough to be interesting, and serves as a useful illustration [McClelland and Rumelhart, 1988].

Figure 9.3 shows two examples. In (a) the graph is a regular grid, examined in detail by Rumelhart and Zipser [1985] using the rule (9.5). The figure shows a typical result after learning, indicating which of the two output units fires in response to monopole stimuli (one vertex) and to dipole stimuli (two vertices on one edge). The network has found one of the two optimum solutions to the bipartitioning problem, and has divided the dipole stimuli roughly equally. However, it sometimes finds a solution which divides the graph diagonally, which is far from optimal for the bipartitioning problem.

In (b) we show the result of applying the rule (9.6) to the bipartitioning problem of Fig. 4.6. We accumulated the changes Δw_{ij} for each of the 28 dipole stimuli (edges) before actually updating the weights. Starting the weights from uniform random values in the range $[0, 1]$ we found convergence to an optimum solution in approximately 30% of our trials, using $\eta = 0.1$. Convergence usually occurred within 10 weight updates. The typical solution shown has not only solved the bipartitioning problem for the vertices, but has also divided the edges exactly in two.

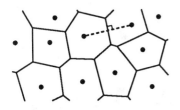

FIGURE 9.4 Voronoi tessellation. The space is divided into polyhedral regions according to which of the prototype vectors (dots) is closest. The boundaries are perpendicular bisector planes of lines joining pairs of neighboring prototype vectors.

Vector Quantization

Probably the most important application of competitive learning is **vector quantization** for data compression. Vector quantization is used both for storage and for transmission of speech and image data. Gray [1984] and Nasrabadi and King [1988] provide reviews of the general problem and of traditional algorithms.

The idea is to categorize a given set or distribution of input vectors ξ^μ into M classes, and then represent any vector just by the class into which it falls. The vector components ξ_j^μ are usually continuous-valued. We can just transmit or store the index of the appropriate class, rather than the input vector itself, once a set of classes, or a **codebook**, has been agreed on. Normally the classes are defined by a set of M **prototype vectors**, and we find the class of a given input by finding the *nearest* prototype vector using the ordinary (Euclidean) metric. This divides up the input space into a **Voronoi tessellation** (or **Dirichlet tessellation**), illustrated in Fig. 9.4

The translation to a competitive learning network is immediate. When an input vector ξ^μ is applied at the network input, the winning output tells us the appropriate class. The weight vectors $\mathbf{w}_i$ are the prototype vectors, and we find the winner using (9.3):

$$|\mathbf{w}_{i^*} - \xi| \le |\mathbf{w}_i - \xi| \qquad \text{(for all } i\text{)}. \tag{9.13}$$

Note that this is not equivalent to maximizing $\mathbf{w}_i \cdot \xi$ unless the weight vectors are normalized.

We can find an appropriate set of prototype vectors (i.e., weights) with the standard competitive learning algorithm (9.6). When exposed to sample data the weights change to divide up the input space into Voronoi polyhedra containing roughly equal numbers of sample points; they provide a discretized map of the input probability $P(\xi)$. After training we can freeze the weights to establish a static codebook.

Figure 9.5 shows two simple examples. In each case we defined a probability distribution $P(\xi)$ for the two-dimensional input vectors $\xi = (\xi_x, \xi_y)$. In (a) the two output units correctly classified the input into its two clusters. For the L-shaped distribution in (b) the 10 units always divided up the input distribution fairly evenly. More input samples gave a more precise division, symmetric about the diagonal.

In practical applications for data compression it is essential to add additional mechanisms to avoid dead units and to ensure an equitable distribution of units in the pattern space. This has been done using Kohonen feature mapping [Nasrabadi and Feng, 1988; Naylor and Li, 1988] and by the use of a conscience mechanism

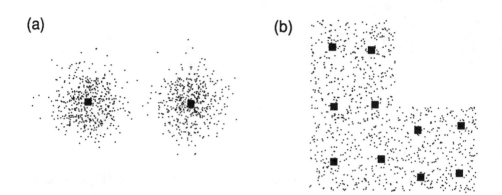

FIGURE 9.5 Examples of vector quantization. In each case the 1000 dots are samples from an input probability distribution. The black squares show the weight vectors of each unit after training on these data. A unit wins the competition when its weight vector is closest to the input vector, so the weight vectors shown define a Voronoi tessellation of the plane.

[Ahalt et al., 1990]. The conscience mechanism appears best, and produces near-optimal results.

Kohonen [1989] has also suggested a *supervised* version of vector quantization called **learning vector quantization** (LVQ). In a supervised case the classes are predefined and we have a body of *labelled* sample data; each sample input vector ξ is tagged with its correct class. We may usefully have several prototype vectors per class. The winning rule (9.13) is unmodified but the update rule depends on whether the class of the winner is correct or incorrect:

$$\Delta w_{i\cdot j} = \begin{cases} +\eta(\xi_j^\mu - w_{i\cdot j}) & \text{if class is correct;} \\ -\eta(\xi_j^\mu - w_{i\cdot j}) & \text{if class is incorrect.} \end{cases} \tag{9.14}$$

In the first case the rule is the standard one, but in the second case the weight vector is moved in the opposite direction, away from the sample vector. This minimizes the number of misclassifications.

An improved algorithm called LVQ2, closer in effect to Bayes decision theory, has also been suggested by Kohonen [1989]. The learning rule is only applied if:

1. the input vector ξ is misclassified by the winning unit i^*;

2. the next-nearest neighbor i' has the correct class; and

3. the input vector ξ is sufficiently close to the decision boundary (perpendicular bisector plane) between $\mathbf{w}_{i\cdot}$ and $\mathbf{w}_{i'}$.

Then *both* $\mathbf{w}_{i\cdot}$ and $\mathbf{w}_{i'}$ are updated, using the *incorrect* and *correct* cases of (9.14) respectively. This rule has shown excellent performance in studies on statistical and speech data [Kohonen et al., 1988].

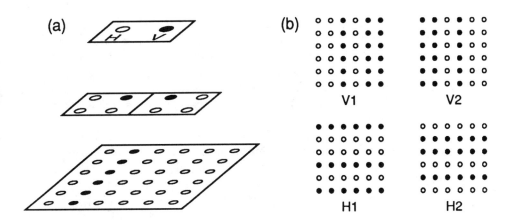

FIGURE 9.6 The horizontal/vertical line problem of Rumelhart and Zipser [1985]. (a) Architecture of the network. Not shown are teacher inputs nor connections; each layer is fully connected to the next. The input layer is a 6 × 6 pixel array on which line stimuli are presented. The intermediate layer has two groups of four winner-take-all units. The output layer has two winner-take-all units, signalling horizontal or vertical input lines. (b) Typical sensitivity of the four units in one intermediate layer group; one fires for each of the 12 line stimuli.

Multi-Layer Networks

Our competitive learning models so far have had a single layer in which exactly one unit wins for each input pattern. To generalize this to a multi-layer network we need to allow several winners per layer so that there are interesting patterns for subsequent layers to analyze. In this way we can hope to detect successively higher order patterns in the data.

One approach is to divide up each layer into groups, in *each* of which one unit wins [Rumelhart and Zipser, 1985]. Even if the groups in a given layer are structurally identical, they may well cluster the data in different ways (starting from random connection strengths) and hence add to the available information.

As an example, consider the horizontal/vertical line problem studied by Rumelhart and Zipser [1985]. The inputs are arranged in a 6 × 6 array that is fully connected to the remaining network (which therefore has no architectural information about the two-dimensional layout). Input stimuli consist of excitation of a whole horizontal or vertical line of the inputs. We want to have a pair of output units which learn to fire on horizontal and vertical lines respectively.

With a single output layer this task is impossible. If the desired vertical feature detector is to win for any of the 6 vertical lines, by symmetry it will have equal weight from each of the 6 inputs in each of the 6 vertical lines. But this is all 36 units! Even with a teacher it cannot work.

The simplest architecture that *did* work required two layers, and is shown in Figure 9.6(a). An extra teaching input was needed initially to say whether the line was horizontal or vertical, but could be omitted after training. In each group of the intermediate layer the four units learn to respond to three vertical, or three horizontal lines. The output layer learns the correlations between the two groups in the intermediate layer, which normally leads to discrimination between horizontal and vertical input lines. This fails if both groups in the intermediate layer happen to divide up the lines $(6 \rightarrow 3 + 3)$ in the same way, but this is rare and could be cured using more than two groups.

Another approach to multi-layer architecture is to impose mutual inhibition between units only when they are within a certain distance d of each other (we assume that the units are arranged geometrically, usually within a two-dimensional layer). Then no two units within d can fire together, though further units may. Each unit has a **vicinity area** of radius d which is its local group, but groups can overlap. This is perhaps more appropriate for biological modelling than the discrete groups considered above.

Fukushima [1975, 1980] employed this idea in his multi-layer unsupervised learning networks called the **cognitron** and **neocognitron**.[3] Up to eight layers were used, with a very specific architecture of interconnections between layers, using limited receptive fields and different functional groups. The network was able to learn to differentiate different letters (presented as patterns on a 16 × 16 pixel array) even with the imposition of considerable translation, scaling, or distortion.

9.3 Adaptive Resonance Theory

As noted earlier (page 222), there is no guarantee of stability of the categories formed by by competitive learning. Even if we continue to present a finite fixed sequence of patterns, the category (i.e., winning unit) of a particular pattern may continue changing endlessly. One way of preventing this is to reduce the learning rate η gradually to zero, thus freezing the learned categories. But then the network loses its plasticity, or ability to react to any new data. It is not easy to have both stability and plasticity; this is Grossberg's **stability-plasticity dilemma**.

To make a real-time learning system that can exist and continue adapting in a nonstationary world we must clearly deal with this dilemma. But there is also another related stability issue: how many output units should we utilize? If we keep that number fixed and avoid dead units, then enforcing long-term stability of the learned categories means that there will be no units available for new patterns subsequently encountered. On the other hand, if we somehow have an inexhaustible supply of output units we must avoid paving the input space ever more finely, since that is not categorization at all. A good solution would be to have a finite (or infinite) supply of output units, but *not use them until needed*. Then at any time there would be some output units in use, stably detecting the known categories,

---

[3]Later versions of the neocognitron [Fukushima et al., 1983] used supervised learning.

and some more waiting in the wings until needed. If we used up the whole supply we should probably stop adapting; stability should come above plasticity when we are at full capacity.

Carpenter and Grossberg [1987a, b, 1988] have developed networks called ART1 and ART2 that behave in just this way. These networks overcome the stability-plasticity dilemma by accepting and adapting the stored prototype of a category *only* when the input is sufficiently similar to it. "ART" stands for **adaptive resonance theory**; the input and stored prototype are said to *resonate* when they are sufficiently similar. When an input pattern is not sufficiently similar to *any* existing prototype a new category is formed, with the input pattern as the prototype, using a previously uncommitted output unit. If there are no such uncommitted units left, then a novel input gives no response.

The meaning of *sufficiently similar* above is dependent on a **vigilance parameter** ρ, with $0 < \rho \leq 1$. If ρ is large the similarity condition becomes very stringent, so many finely divided categories are formed. On the other hand a small ρ gives a coarse categorization. The vigilance level can be changed during learning; increasing it can prompt subdivision of existing categories.

ART1 is designed for binary 0/1 inputs, whereas ART2 is for continuous-valued inputs. We focus exclusively on ART1, the simpler case. We suggest Moore [1988] for further reading, as well as Carpenter and Grossberg [1988] for an overview, and Carpenter and Grossberg [1987a, b] for the details of ART1 and ART2 respectively.

The ART1 algorithm

It is easiest to present ART1 as an algorithm before describing the network implementation. Let us take input vectors ξ and stored prototype vectors $\mathbf{w}_i$, both with N binary 0/1 components. i indexes the output units, or categories, each of which can be *enabled* or *disabled*. We start with $\mathbf{w}_i = \mathbf{1}$ for all i, where $\mathbf{1}$ is the vector of all ones; this will represent an uncommitted state, not a category. Then the algorithm upon presentation of a new input pattern ξ is as follows:

1. Enable all the output units.

2. Find the winner i^* among all the enabled output units (exit if there are none left). The winner is defined as the one for which $\overline{\mathbf{w}}_i \cdot \xi$ is largest, where $\overline{\mathbf{w}}_i$ is a normalized version of $\mathbf{w}_i$. The normalization is given by

$$\overline{\mathbf{w}}_i = \frac{\mathbf{w}_i}{\varepsilon + \sum_j w_{ji}} \tag{9.15}$$

where w_{ji} is the jth component of $\mathbf{w}_i$. The small number ε is included to break ties, selecting the longer of two $\mathbf{w}_i$'s which both have all their bits in ξ. Note that an uncommitted unit wins if there is no better choice.

3. Test whether the match between ξ and $\mathbf{w}_{i^*}$ is good enough by computing the ratio

$$r = \frac{\mathbf{w}_{i^*} \cdot \xi}{\sum_j \xi_j} . \tag{9.16}$$

This is the fraction of bits in $\boldsymbol{\xi}$ that are also in $\mathbf{w}_{i^*}$. If $r \geq \rho$, where ρ is the vigilance parameter, there is resonance; go to step 4. Otherwise if $r < \rho$ the prototype vector $\mathbf{w}_{i^*}$ is rejected; disable unit i^* and go back to step 2.

4. Adjust the winning vector $\mathbf{w}_{i^*}$ by deleting any bits in it that are not also in $\boldsymbol{\xi}$. This is logical AND operation, and is referred to as **masking** the input.

This algorithm can terminate in one of three ways. If we find a matching prototype vector we adjust it (if necessary) in step 4 and output that category i^*. If we find no suitable prototype vector from among the previous categories, then one of the uncommitted vectors is selected and made equal to the input $\boldsymbol{\xi}$ in step 4; again we output the appropriate (new) category i^*. Finally, if there are no matches and no uncommitted vectors we end up with all units disabled, and hence no output.

It should now be clear how this algorithm solves the various problems we raised. It continues to have plasticity until all the output units are used up. It also has stability; a detailed analysis shows that all weight changes cease after a *finite* number of presentations of any fixed set of inputs. This comes essentially from the fact that the adaptation rule, step 4, can only remove bits from the prototype vector, never add any. Thus a given prototype vector can never cycle back to a previous value.

Note that the loop from step 3 back to step 2 constitutes a **search** through the prototype vectors, looking at the closest, next closest, etc., by the maximum $\overline{\mathbf{w}}_i \cdot \boldsymbol{\xi}$ criterion until one is found that satisfies the $r \geq \rho$ criterion. These criteria are different, so going further away by the first measure may actually bring us closer by the second. The first measure is concerned with the fraction of the bits in $\mathbf{w}_i$ that are also in $\boldsymbol{\xi}$, whereas r is the fraction of the bits in $\boldsymbol{\xi}$ that are also in $\mathbf{w}_i$. Of course this search is comparatively slow (and more like a conventional algorithm), but it only occurs before stability is reached for a given input set. After that each category is found on the first attempt and there is never a jump back from step 3.

Network Implementation of ART1

Carpenter and Grossberg designed the ART1 network using previously developed building blocks that were based on biologically reasonable assumptions. The selection of a winner, the input layer, the weight changes, and the enable/disable mechanism can all be described by realizable circuits governed by differential equations. There are at least three time scales involved: the relaxation time of the winner-take-all circuit; the cycling time of the search process; and the rate of weight update.

We describe a simpler version, taking the winner-take-all circuit for granted and simplifying certain other features. Figure 9.7 shows this reduced version. There are two layers, with units V_j in the input layer and O_i in the output layer fully connected in both directions. The forward weights $\overline{w}_{ij}$ are normalized copies of the backward weights w_{ji}, according to (9.15). Note that the w_{ji} are each 0 or 1, as are ξ_j, V_j, O_i, A, and R.

The output layer consists of winner-take-all units; only the unit with the largest net input $\sum_j \overline{w}_{ij} V_j$ among all enabled units has $O_i = 1$. If the "reset" signal R is

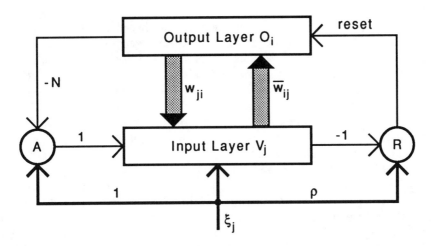

FIGURE 9.7 The ART1 network. The large shaded arrows represent connections between all pairs of output units O_i and input units V_j.

turned on while a winner is active, that unit is disabled and removed from future competitions. All units can be re-enabled by another signal that is not shown.

The input units V_j are designed so that

$$V_j = \begin{cases} \xi_j & \text{if no output } O_i \text{ is on;} \\ \xi_j \wedge \sum_i w_{ji} O_i & \text{otherwise} \end{cases} \qquad (9.17)$$

where $\wedge$ means logical AND. For technical reasons, and to allow some other functions that we omit, this is done using an auxiliary unit A, which is on ($A = 1$) if any input is on but no output is, and off ($A = 0$) otherwise. A could be a 0/1 threshold unit with total input $\sum_j \xi_j - N \sum_i O_i$ and threshold 0.5, as suggested by the connection weights shown in the figure. The input units receive total input

$$h_j = \xi_j + \sum_i w_{ji} O_i + A \qquad (9.18)$$

and fire ($V_j = 1$) if this exceeds a threshold of 1.5. This is equivalent to (9.17), and is referred to as the 2/3 rule; two out of its three inputs ξ_j, $\sum_i w_{ji} O_i$, and A must be on for a unit V_j to fire.

The disabling or "reset" signal is generated when r from (9.16) is less than ρ. This can be accomplished with a 0/1 threshold unit R that receives input $\rho \sum_j \xi_j - \sum_j V_j$ and has threshold 0, as shown by the connection weights in the figure.

Finally the backward weights are updated slowly according to

$$\frac{dw_{ji}}{dt} = \eta O_i (V_j - w_{ji}) \qquad (9.19)$$

so that the prototype vector $w_{i^* j}$ for the winner i^* becomes equal to the masked input V_j after resonance has occurred. The forward weights have a slightly more complicated learning rule which leads to a similar, but normalized, result.

Given the details that we have omitted, this network runs entirely autonomously. It does not need any external sequencing or control signals, and can cope stably with an infinite stream of input data. It has fast access of well-known categories, automatic search for less-known categories, creation of new categories when necessary, and refusal to respond to unknown inputs when its capacity is exhausted. And its architecture is entirely parallel.

In practice ART1 networks are somewhat tricky to adjust, and are very sensitive to noise in the input data. If random bits are sometimes missing from the input patterns, then the stored prototype vectors can be gradually degraded by the one-way masking form of the adaptation rule. ART1 networks are also rather inefficient in their storage requirements; we need one unit and $2N$ modifiable connections for each category, and many fixed connections for the A unit, R unit, and lateral inhibition. They also share the limitations of a grandmothering approach common to most competitive learning schemes (see page 217). Some of these problems are solved in the ART2 network.

9.4 Feature Mapping

Up to now we have paid little attention to the geometrical arrangement of our competitive output units. If however we place them in a line or a plane, we may consider networks in which the *location* of the winning output unit conveys some information, with *nearby outputs corresponding to nearby input patterns*. If ξ^1 and ξ^2 are two input vectors, and $\mathbf{r}^1$ and $\mathbf{r}^2$ are the locations of the corresponding winning outputs, then $\mathbf{r}^1$ and $\mathbf{r}^2$ should get closer and closer, eventually coinciding, as ξ^1 and ξ^2 are made more and more similar. Additionally we should *not* find $\mathbf{r}^1 = \mathbf{r}^2$ unless the patterns ξ^1 and ξ^2 *are* similar. A network that performs such a mapping is called a **feature map**.

Technically what we are asking for is a **topology preserving map** from the space of possible inputs to the line or plane of the output units. A topology preserving map, or **topographic map**, is essentially a mapping that preserves neighborhood relations. The idea may seem almost trivial, but it is not. This is best seen by realizing that such a map is *not* necessarily possible between a given pair of spaces. For example we cannot map a circle onto a line without a discontinuity somewhere.

We have not yet specified the nature of the inputs, beyond assuming a well-defined distance (metric) between pairs of input patterns. There are actually two cases commonly considered, as exemplified by the two architectures of Fig. 9.8. In both cases there is a geometrical array of outputs, which we show as two-dimensional, but the form of the inputs is quite different.

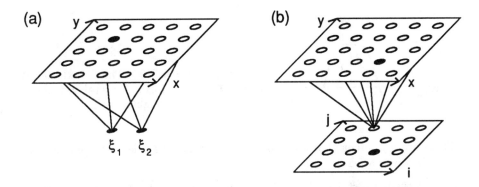

FIGURE 9.8 Two types of feature mapping architecture. (a) is the conventional one, with continuous-valued inputs ξ_1, ξ_2. (b) is of biological interest in mapping from, e.g., retina to cortex. Layers are fully connected, though only a few connections are shown.

In the first case there are a few continuous-valued inputs, such as ξ_1 and ξ_2 shown in Fig. 9.8(a). We expect a map onto the output space (x, y) from the two-dimensional region of (ξ_1, ξ_2) space in which inputs occur; this might or might not be a square domain. Another natural situation in this class would be a single input ξ_1 mapped to a linear array (line) of outputs. We could also have, for instance, three inputs ξ_1–ξ_3 mapped topologically onto a two-dimensional plane if the actual input patterns all fell into a two-dimensional subspace of the three-dimensional ξ space.

The second case, Fig. 9.8(b), is studied less often but has more biological importance. The inputs ξ_{ij} themselves are arranged in a two-dimensional array that defines the input space (i, j). In the simplest case the input patterns have only one input turned on at a time, thus defining a single point in this space. The problem is to learn a continuous mapping from (i, j) onto (x, y), roughly between points "above" one another in the two spaces. The significance of the problem is the frequent occurrence of such topologically correct mappings in the brain, including connections from the sense organs (eye, ear, skin) to the cortex, and connections between different areas of the cortex. The **retinotopic map** from the retina to the visual cortex (in higher vertebrates) or optic tectum (in lower vertebrates) is the most studied example of a two-dimensional map, though the **somatosensory map** from the skin onto the somatosensory cortex (where there is an image of the body surface) is also important. The **tonotopic map** from the ear onto the auditory cortex forms a one-dimensional example, with sounds of different frequencies laid out in order on the cortex. In each of these cases it is not very likely that the axon routes and synapses are entirely preprogrammed or *hardwired* by the genes, so one needs a mechanism of creating the appropriate topographic maps during development. Unsupervised learning is one approach (among several) to solving this problem by *softwiring*.

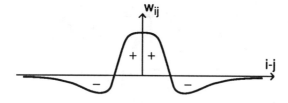

FIGURE 9.9 The "Mexican hat" form of lateral connection weights.

There are a number of ways to design an unsupervised learning network that self-organizes into a feature map:

1. For the second case just described, limited receptive fields for the output layer can be put in by hand. Unsupervised learning can then refine the mapping, as we saw in Linsker's model of orientational selectivity (page 211). However this begs the question for the softwiring problem itself.

2. We can use ordinary competitive learning with the addition of lateral connections within the output layer. These need to have the Mexican hat form, excitatory between nearby units and inhibitory at longer range, with strength falling off with distance, as in Fig. 9.9. Thus neighboring units gain neighboring responses, while units further away find select different patterns.

3. We can use ordinary competitive learning, but update weights going to the *neighbors* of the winning unit as well as those of the winning unit itself. This is **Kohonen's algorithm**.

We briefly discuss two examples of (2), and then focus on (3).

Willshaw and von der Malsburg's Model

An early model using lateral connections was proposed by Willshaw and von der Malsburg [1976] for the retinotopic map problem. They used the architecture of Fig. 9.8(b), but added lateral connections of Mexican hat form within the output layer. Their output units were not strictly winner-take-all, though a threshold ensured that only a few fired at a time. For learning they employed a plain Hebbian rule followed by renormalization of the weights.

To train their network Willshaw and von der Malsburg used **dipole stimuli** with two adjacent inputs turned on. Learning then converged on an ordered topographic map, such as that shown in Fig. 9.10. Note that this is a diagram of the *output weights plotted in the input space*. Specifically each output unit O_i is represented by the intersection between lines that connect it to its neighbors, and is plotted at the point $\sum_k w_{ik}\mathbf{r}_k$, where $\mathbf{r}_k$ is the location of input ξ_k in the input space. This gives the center of mass of the effective receptive field of unit O_i.

The mapping could converge on any of eight possible orientations; corners ABCD of the input array might map, for instance, onto BADC of the output. In order to select a particular orientation (as found biologically), Willshaw and von der Malsburg chose four of the inputs to be **polarity markers**, with initial strong

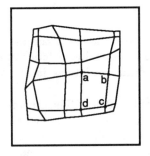

FIGURE 9.10 A mapping between two 6 × 6 arrays produced by Willshaw and von der Malsburg's network. The outer square represents the input array and the locations of the line intersections represent the output weights. The polarity markers are labelled a–d.

connections to the correct places in the output array. The polarity markers acted as nucleation regions to break the initial symmetry. Another approach would be to pin certain units at the boundary of the desired mapping [Cottrell and Fort, 1986].

Amari [1980] and Takeuchi and Amari [1979] analyzed a modified version of Willshaw and von der Malsburg's model using continuous one-dimensional lines for both input and output. They found that a correct feature map will form provided the width of the Mexican hat function is large enough compared to the width of typical input patterns. If this condition is not satisfied they find that the mapping develops discontinuous jumps.

It is not now believed that this unsupervised learning mechanism is solely responsible for the retinotopic maps found biologically. In particular it cannot reproduce the experimental phenomena of regeneration after damage. Current theories involve some degree of **chemoaffinity**, in which growing axons carry chemical markers that help to define appropriate target sites. However a degree of softwiring by unsupervised learning is often invoked to refine the map.

von der Malsburg's Model for Orientation Column Formation

Somewhat earlier, von der Malsburg [1973] used a very similar scheme to model the formation of the orientation columns in the visual cortex described in Section 8.4. There is no longer a simple mapping between position in the input space and position in the output layer. Instead, a mapping develops from *the angle of orientation* of an input object to position in the output layer. The preferred orientation of the output units changes smoothly (with occasional breaks) across the array, as found in nature.

It is important to notice that the network itself is not essentially different from the one which makes the position-to-position map; the difference lies in the set of input images it is trained on. Whereas the Willshaw-von der Malsburg model used spatially localized dipole patterns, the von der Malsburg orientation column model used elongated bar-like patterns, always centered at the middle of the image and presented in various orientations. The very different results illustrate dramatically the influence of experience on the properties of neural networks.

Kohonen's Algorithm

Kohonen's algorithm takes a computational shortcut to achieve the effect accomplished by the Mexican hat lateral interactions. There are no lateral connections, but the weight update rule is modified to involve neighborhood relations in the output array.

The algorithm is normally applied to the architecture of Fig. 9.8(a). Thus there are N continuous-valued inputs, ξ_1 to ξ_N, defining a point $\boldsymbol{\xi}$ in an N-dimensional real space. The output units O_i are arranged in an array (generally one- or two-dimensional), and are fully connected via w_{ij} to the inputs. A competitive learning rule is used, choosing the winner i^* as the output unit with weight vector closest to the current input $\boldsymbol{\xi}$:

$$|\mathbf{w}_{i^*} - \boldsymbol{\xi}| \le |\mathbf{w}_i - \boldsymbol{\xi}| \qquad \text{(for all } i\text{)}. \qquad (9.20)$$

As remarked before, this cannot be done by a linear network unless the weights are normalized, but the algorithm is generally used computationally rather than with an actual network.

Thus far we have exactly the same design as for e.g., vector quantization. The difference is in the learning rule, which is [Kohonen, 1982, 1989]

$$\Delta w_{ij} = \eta \Lambda(i, i^*)(\xi_j - w_{ij}) \qquad (9.21)$$

for *all* i and j. The **neighborhood function** $\Lambda(i, i^*)$ is 1 for $i = i^*$ and falls off with distance $|\mathbf{r}_i - \mathbf{r}_{i^*}|$ between units i and i^* in the output array. Thus units close to the winner, as well as the winner i^* itself, have their weights changed appreciably, while those further away, where $\Lambda(i, i^*)$ is small, experience little effect. It is here that the topological information is supplied; nearby units receive similar updates and thus end up responding to nearby input patterns.

The rule (9.21) drags the weight vector $\mathbf{w}_{i^*}$ belonging to the winner towards $\boldsymbol{\xi}$. But it also drags the $\mathbf{w}_i$'s of the closest units along with it. Therefore we can think of a sort of **elastic net** in input space that wants to come as close as possible to the inputs; the net has the topology of the output array (i.e., a line or a plane) and the points of the net have the weights as coordinates. This picture is particularly useful to keep in mind when interpreting Figs. 9.10–9.13.

To construct a practical algorithm we have to specify $\Lambda(i, i^*)$ and η. It turns out to be useful (and often essential for convergence) to change these dynamically during learning. We start with a wide range for $\Lambda(i, i^*)$ and a large η, and then reduce both the range of $\Lambda(i, i^*)$ and the value of η gradually as learning proceeds. This allows the network to organize its elastic net rapidly, and then refine it slowly with respect to the input patterns. Making η go to 0 eventually stops the learning process.

A typical choice for $\Lambda(i, i^*)$ is

$$\Lambda(i, i^*) = \exp(-|\mathbf{r}_i - \mathbf{r}_{i^*}|^2 / 2\sigma^2) \qquad (9.22)$$

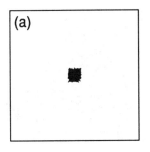
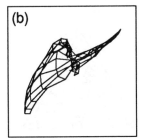
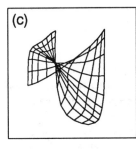
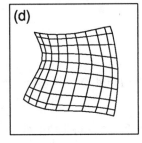

FIGURE 9.11 Kohonen feature mapping from a square region of the plane onto a 10 × 10 array of output units.

where σ is a width parameter that is gradually decreased. The time dependence of $\eta(t)$ and $\sigma(t)$ can take various forms, including $1/t$ and $a - bt$, though there is some theoretical reason to prefer $\eta \propto t^{-\alpha}$ with $0 < \alpha \leq 1$.

Note that (9.21) is Kohonen's choice of learning rule, but the same idea of updating the neighbors of the winner could be incorporated into other rules too.

Examples and Applications

Figure 9.11 shows an example of mapping 2 inputs ξ_1 and ξ_2 onto a 10 × 10 planar array. Input patterns were chosen randomly (with uniform weight) from the square $\{0 \leq \xi_1 < 1, 0 \leq \xi_2 < 1\}$, which is shown as the outer boundary in each diagram. The weights (w_{i1}, w_{i2}) for each output unit are shown in this input space by an intersection in the grid of lines, which connects all nearest neighbor pairs.

The weights were started from random values near to $(0.5, 0.5)$ as shown in Fig. 9.11(a). As learning progressed, indicated by the snapshots in (b)–(d), the weights were pulled apart by the input patterns and organized themselves into a square grid. The fold seen in (c) is not uncommon and takes a long time to eliminate. Eventually the grid would become even more regular than in (d), filling most[4] of the unit square.

Note that we used a uniform probability distribution for the random input patterns. If the distribution had been non-uniform we would have found more grid

---

[4]But not all: there is always a border which is not filled, with a width inversely proportional to the linear size of the output array [Kohonen, 1989].

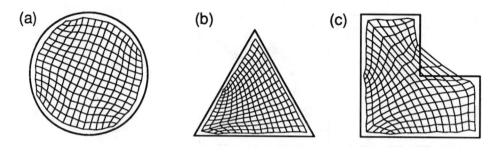

FIGURE 9.12 Kohonen feature mapping from various regions in the plane onto a 15 × 15 array of output units.

points of the weight network where the probability was higher. Ideally we might hope that the local density of grid points would be proportional to the input probability there, but this is not quite true. Nevertheless a higher probability generally leads to a higher density of points.

Figure 9.12 shows three further examples of mapping from two inputs to a square array. Now the input probability distribution has been chosen uniformly from a circle, a triangle, or an L-shape. We show only the final result of each map. The square shape of the output array has a clear effect on the result, making the point density somewhat non-uniform in (b) and (c), and leaving noticeable corners in all cases. In (c) there are four grid-points in a region with zero probability, but it is hard to find a better solution without clustering a lot of points near the inside corner. The outlying points are pulled this way and that as their neighbors in the arms of the L win in turn, but they have no stable place to go.

We can also try maps from two dimensions onto one dimension, despite the impossibility of preserving all the topology. Figure 9.13 shows the development of a map from a two-dimensional L shape to a one-dimensional *line* of 50 output units. The initial weights were random in the unit square, as shown in (a). They evolved rather quickly to a regular curve in (b), and then gradually developed a finer structure to cover the space as best they could. With a larger number of output units we get a good approximation to a space-filling (Peano) curve. Note that such a map is *not* entirely topology preserving; a small step in the input space can trigger a very different winner when we jump from one loop of the curve to another; this occurs most drastically near the point x in our example.

Maps from one dimension to one dimension are also possible, and much of the theoretical analysis has been for this case. Figure 9.14 shows an example, with the weights w_i plotted directly against i. The single input ξ_1 was taken uniformly from the interval $0 \leq \xi_i < 1$, so we expect the output weights to become linearly ordered, as seen in (c). Note that there are two possible solutions, with w_i increasing or decreasing. Indeed, at the intermediate time shown in (b) part of the network has adopted one solution and part has adopted the other. The way that this smooths out to produce (c) will be discussed in the next section.

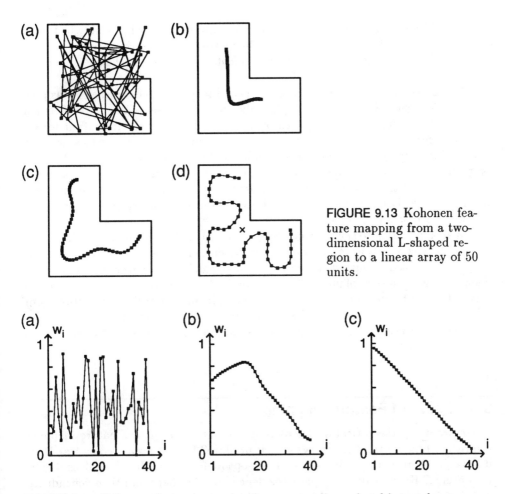

FIGURE 9.13 Kohonen feature mapping from a two-dimensional L-shaped region to a linear array of 50 units.

FIGURE 9.14 Kohonen feature mapping from a one-dimensional interval onto a linear array of 40 units.

Applications of feature mapping have been made in many areas, including sensory mapping, motor control, speech recognition, vector quantization, and combinatorial optimization. Kohonen [1989] and Ritter and Schulten [1988c] provide partial reviews. We discuss combinatorial optimization separately in Section 9.6, and mention just a few other applications here.

In a simple example of robot sensing, a robot arm has a number of angles specifying the state of each of its joints, and a given set of these angles corresponds to a point in space for the end effector. A feature mapping network can be given the angles as inputs, with samples taken randomly from the arm's workspace. It will then develop a map of that space, generally avoiding obstacles [Kohonen, 1989; Graf and LaLonde, 1988; Ritter and Schulten, 1988a]. Here a three-dimensional output array would clearly be appropriate. In a real application it would probably be better to construct an initial map theoretically, but unsupervised learning might well be useful for making small subsequent adjustments as conditions change or sensors degrade.

An interesting example of a one-dimensional map onto one dimension was provided by Kohonen [1982]. He fed single audio tones into a set of 20 *randomly tuned* bandpass filters (or resonators), each of which produced maximum output for its random center frequency. The 20 values were then used as the inputs to a network with a linear array of 10 units. After training, the output units had ranked the original input frequencies into increasing or decreasing order—a correct tonotopic map. Note that this is technically a map from 20 dimensions onto one dimension, but the data only occupy a one-dimensional subspace of that space. Alternatively it can be seen as an example of the layer-to-layer architecture of Fig. 9.8(b).

In an application to speech processing, Kohonen et al. [1984] used a similar set of 15 frequency channels as inputs with a *two*-dimensional array for the output. Then they took sampled Finnish speech as the input signal. The map evolved to have well-defined units or clusters of units for each Finnish phoneme, with "similar" phonemes generally close to one another. Such a "phonotopic map" is truly a projection from a high-dimensional space (15 in this case) onto a two-dimensional one, and may be very useful for visualizing similarities and structures in the original space.

9.5 Theory of Feature Mapping

There are many theoretical questions to ask about the Kohonen feature mapping algorithm. Exactly what mappings does it produce? Does it always converge? Does it have problems with getting stuck in non-optimum states? How long does it take to converge? How do all these questions depend on the shape and time dependence of $\Lambda(i, i^*)$ and on the value and time dependence of η? Are there optimum values for these parameters? Some of these questions are rather hard to answer in general, and much of the analysis has been done only in the case of one-dimensional mappings.

Once more there is a cost function, based on an extension of the competitive learning one (9.8) [Ritter and Schulten, 1988c]:

$$E\{w_{ij}\} = \frac{1}{2}\sum_{ijk\mu} M_k^\mu \Lambda(i,k)(\xi_j^\mu - w_{ij})^2 = \frac{1}{2}\sum_{i\mu} \Lambda(i,i^*)|\xi^\mu - \mathbf{w}_i|^2. \qquad (9.23)$$

Here M_i^μ is again the cluster membership matrix (9.9), equal to 1 if $i = i^*(\mu)$ and 0 otherwise. It is important to keep in mind that the winner i^* depends on μ.

Gradient descent on this cost function yields

$$\langle \Delta w_{ij} \rangle = -\eta \frac{\partial E}{\partial w_{ij}} = \eta \sum_{k\mu} M_k^\mu \Lambda(i,k)(\xi_j^\mu - w_{ij}) = \eta \sum_\mu \Lambda(i,i^*)(\xi_j^\mu - w_{ij}) \quad (9.24)$$

which is just the sum of the Kohonen rule (9.21) over all the patterns μ. Thus on average (for small enough η) the Kohonen rule decreases the cost (9.23) until we reach a local minimum. However, as we discussed for the competitive learning case (page 222), this result is a little deceptive: there may be local minima, and we have averaged over μ.

To investigate the equilibrium states we can set $\langle \Delta w_{ij} \rangle = 0$ as usual, giving

$$0 = \sum_{\mu} \Lambda(i, i^*)(\xi_j^{\mu} - w_{ij}) \tag{9.25}$$

for all i and j. This equation is not, however, easy to solve. In the case of a *uniform* distribution of $\boldsymbol{\xi}^{\mu}$ vectors we can convince ourselves that it is satisfied by a uniform distribution of $\mathbf{w}_i$'s in the same space *if* we ignore the boundaries. For further progress in the general case it is best to go to a **continuum approximation**, replacing (9.25) by

$$0 = \int \Lambda(\mathbf{r} - \mathbf{r}^*(\boldsymbol{\xi}))[\boldsymbol{\xi} - \mathbf{w}(\mathbf{r})]P(\boldsymbol{\xi}) \, d\boldsymbol{\xi} \,. \tag{9.26}$$

Here the sum over μ became $\int P(\boldsymbol{\xi}) \, d\boldsymbol{\xi}$, allowing for an arbitrary distribution of $\boldsymbol{\xi}$ vectors, and the index i for the output array became a position vector $\mathbf{r}$. We also wrote $\Lambda(\mathbf{r} - \mathbf{r}^*(\boldsymbol{\xi}))$ instead of $\Lambda(\mathbf{r}, \mathbf{r}^*(\boldsymbol{\xi}))$; we would normally choose a neighborhood function that only depends on the distance between $\mathbf{r}$ and the winner $\mathbf{r}^*(\boldsymbol{\xi})$.

It is possible to derive an implicit partial differential equation for the weights $\mathbf{w}(\mathbf{r})$ from (9.26) [Ritter and Schulten, 1986]. However, an explicit solution for a given $P(\boldsymbol{\xi})$ is only known for the one-dimensional case, and for certain cases that factorize into a product of one-dimensional cases. The solution for the one-dimensional case shows that the weights become regularly increasing or decreasing, as we saw in Fig. 9.14, with a density of output units proportional to $P(\xi)^{2/3}$ around point ξ. An ideal representation would have density proportional to $P(\xi)$ instead of $P(\xi)^{2/3}$, so the Kohonen algorithm tends to undersample high probability regions and oversample low probability ones.

One-Dimensional Equilibrium $\star$

A derivation of the $P(\xi)^{2/3}$ result is interesting. We replace $\mathbf{r}$ by x in (9.26) and assume that the neighborhood function $\Lambda(x - x^*(\xi))$ is fairly sharply peaked. Then we can expand the integrand in powers of

$$\varepsilon = x^*(\xi) - x \tag{9.27}$$

and ignore terms beyond ε^2. The relevant expansions are a little tricky:

- $\Lambda(x - x^*(\xi))$ becomes just $\Lambda(-\varepsilon)$. We assume that Λ is an even function, so this is the same as $\Lambda(\varepsilon)$.

- ξ becomes $w(x^*)$ or $w(x + \varepsilon)$, and thus $\xi - w(x)$ expands into $\varepsilon w' + \frac{1}{2}\varepsilon^2 w''$. We leave it understood that w' and w'' are evaluated at point x.
- $P(\xi)$ becomes $P(w(x^*)) = P(w(x + \varepsilon))$ and expands into $P(w) + \varepsilon P'(w)w'$.
- $d\xi$ becomes $dw(x + \varepsilon)$ giving $w'(x + \varepsilon)\,d\varepsilon$, which expands into $(w' + \varepsilon w'')\,d\varepsilon$.

Putting these into (9.26) and collecting powers of ε up to ε^2 (dropping higher order terms) produces

$$
\begin{aligned}
0 &= \int \Lambda(\varepsilon)\left(\varepsilon w' + \tfrac{1}{2}\varepsilon^2 w''\right)\left(P(w) + \varepsilon P'(w)w'\right)\left(w' + \varepsilon w''\right)d\varepsilon \\
&= \int \Lambda(\varepsilon)\left(\varepsilon w'^2 P(w) + \varepsilon^2 w'\left[\tfrac{3}{2}w'' P(w) + w'^2 P'(w)\right]\right)d\varepsilon \\
&= w'\left[\tfrac{3}{2}w'' P(w) + w'^2 P'(w)\right]\int \Lambda(\varepsilon)\varepsilon^2\,d\varepsilon\,.
\end{aligned}
\tag{9.28}
$$

The term of order ε disappeared because $\Lambda(\varepsilon)$ is even. Assuming that $w'(x)$ is nowhere zero, we must clearly have

$$
\tfrac{3}{2}w'' P(w) + w'^2 P'(w) = 0
\tag{9.29}
$$

giving

$$
\frac{d}{dx}\log|w'| = \frac{w''}{w'} = -\frac{2}{3}\frac{P'(w)w'}{P(w)} = -\frac{2}{3}\frac{d}{dx}\log P(w)
\tag{9.30}
$$

which shows that

$$
|w'| \propto P(w)^{-2/3}\,.
\tag{9.31}
$$

This is the result desired, because the density of output units in w (or ξ) space is $|dx/dw|$ or $|1/w'|$. Note that there are two possible solutions, with w' always positive or always negative, as we have already realized.

The differential equation (9.29) can be solved for the actual weights $w(x)$ if we are given a particular $P(\xi)$. For example $P(\xi) \propto \xi^\alpha$ gives $w(x) \propto x^\beta$ with $\beta = 1/(1 + 2\alpha/3)$.

Convergence

The above analysis of the equilibrium states tells us nothing about how the algorithm convergences to them. We *must* eventually get to one of them however, because there is nowhere else to go as long as η and σ remain non-zero; the whole Kohonen algorithm is a Markov process with only two absorbing states. A detailed analysis leads to conditions on $\eta(t)$ needed to ensure convergence [Cottrell and Fort, 1986; Ritter and Schulten, 1988b].

The two-dimensional case is much harder than the one-dimensional one, but can be formulated in terms of a Fokker-Planck (probability diffusion) equation for the probability of finding any particular set of weights at time t [Ritter and Schulten, 1988b]. Such an equation can also provide information about the likely fluctuations

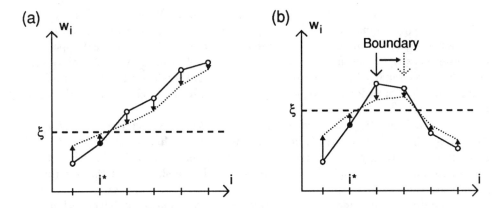

FIGURE 9.15 Kohonen maps in one dimension. (a) Any monotonic region of weights remains monotonic when the weights are updated. (b) The boundary between two monotonic regions can move one step at an update.

before the equilibrium is reached. It is found that wave-like fluctuations can occur in the output weights, as one would expect from interpreting the meshes in Figs. 9.11 and 9.12 as elastic nets.

In most cases the convergence process has two phases. The map first becomes untangled and fairly even, and only then adapts in detail to the input probability $P(\xi)$. The untangling phase can often dominate the total time required, because some types of tangle can take a long time to iron out. Two examples of tangles are evident in the our examples: in Fig. 9.11(c) there is a **twist**, and in Fig. 9.14 there is a **kink**.

Geszti [1990, chapter 10] studied the problem of kinks in the one-dimensional case. He observed (see also Kohonen [1989]) that a monotonically increasing or decreasing sequence of weights w_i remains monotonic at each update. We can see this by rewriting the learning rule

$$\Delta w_i = \eta \Lambda(i, i^*)(\xi - w_i) \tag{9.32}$$

in the form

$$[w_i - \xi]^{\text{new}} = (1 - \eta \Lambda(i, i^*))[w_i - \xi]^{\text{old}} \tag{9.33}$$

and looking at Fig. 9.15(a). The vertical distance $w_i - \xi$ of a point w_i from the input value ξ is multiplied by a factor $1 - \eta \Lambda(i, i^*)$, which becomes closer to 1 as we move further from the winner i^*. Thus the monotonic order of the weight values cannot change.

The interesting things happen at the **boundaries** between monotonically ordered regions. Such a boundary can jump one step to either side, as shown in Fig. 9.15(a). Boundaries can also disappear by annihilation when two meet, or by wandering off the edge. But no new boundaries can appear.

This tells us a lot about the time it takes to untangle the map. Consider for instance the single kink in Fig. 9.14(b); that kink has to diffuse to one end of the the chain in order to be eliminated. But it only moves at all when ξ happens to be chosen nearby (which happens with probability of order $1/N$ for N units), and even then it is equally likely to move left or right. It therefore executes a **random walk** and takes of order N^2 steps to get to one end of the chain. Thus the total time scales as N^3.

This conclusion is based on assuming that $\Lambda(i,j)$ is symmetric ($\Lambda(i,j) = \Lambda(j,i)$), so that left and right jumps are equally likely. Instead, Geszti suggested making $\Lambda(i,j)$ strongly asymmetric, to drive the boundaries in one direction. Then the time should scale as only N^2, and indeed simulation shows that the ordering is much faster.

In two dimensions there are many more types of "topological defects" than the simple boundaries found in one dimension, but it turns out that one of them—the twist shown in Fig. 9.11(c)—is dominant. Geszti found that an anisotropic Λ, with different ranges in the horizontal and vertical directions, speeds up learning considerably.

9.6 The Travelling Salesman Problem

We return here to the travelling salesman problem first introduced in Chapter 4 to see how it can be solved by feature mapping. We also discuss the **elastic ring method** which is closely related to the feature mapping approach, though originally derived from a different route..

Consider the case of N cities in a two-dimensional plane. A tour is line though all the points, and so can be regarded as a *mapping from the plane to a line*. Or actually a ring; the ends should be joined. We know how to make a Kohonen network produce that, using N units in a ring and two inputs (for the x and y coordinates). In the end we expect the weights to become equal to the coordinates of the cities, and yet have neighboring weights relatively close to one another. Thus the sequence of the cities defined by the units should be a good, if not optimal, tour for our salesman.

Some details must be filled in, and modifications made, to get this to work well in practice. Since more than one unit can be attracted to the same city it is actually best to have more units than cities. Or one can adapt the algorithm so that extra units are added or deleted as needed. With such modifications the method has been shown to perform fairly well compared to other algorithms [Angéniol et al., 1988; Hueter, 1988]. Figure 9.16 shows an example, starting from a small ring in the center of the cities.

One of the drawbacks of this approach is that there is no cost function associated with the process; it is built only on intuition. It works something like an elastic rubber ring that is gradually drawn towards the cities and otherwise stays as small

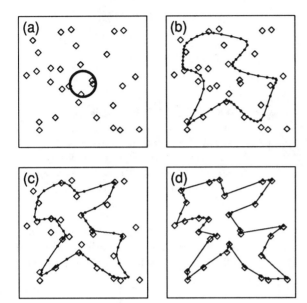

FIGURE 9.16 The gradual formation of a tour for the travelling salesman problem by a feature map, or by the Durbin-Willshaw method, for 30 cities and 70 units.

as possible. This physical picture were used explicitly by Durbin and Willshaw [1987] who used the updating rule

$$\Delta \mathbf{w}_i = \eta \left(\sum_\mu \Lambda^\mu(i)(\xi^\mu - \mathbf{w}_i) + \kappa(\mathbf{w}_{i+1} - 2\mathbf{w}_i + \mathbf{w}_{i-1}) \right). \tag{9.34}$$

where the weight $\mathbf{w}_i$ specifies a point on the rubber ring and ξ^μ is the position of city i. Durbin and Willshaw did not actually use the same terminology, but we have put their algorithm into our terms; on the mathematical level the methods are closely related. Thus (9.34) looks like the Kohonen algorithm (9.21) summed over all cities, except that $\Lambda(i, i^*)$ has been replaced by $\Lambda^\mu(i)$ and a second term with coefficient κ has been introduced. In Durbin and Willshaw's physical picture the first term is a force that drags each point $\mathbf{w}_i$ on the ring towards each city ξ^μ with influence coefficients $\Lambda^\mu(i)$, and the second term is an elastic force that drags each point towards neighboring points on the ring.

The coefficient $\Lambda^\mu(i)$ specifies how strongly $\mathbf{w}_i$ is pulled towards city ξ^μ. In the usual Kohonen algorithm it would be a Gaussian function of $|i - i^*|$, the number of units between $\mathbf{w}_i$ and the $\mathbf{w}$ closest to ξ^μ. The dependence on the number of units (i.e., neighborhoods in the *output* array) was needed to build in the topology of the array. But now the κ term takes care of keeping neighboring $\mathbf{w}_i$'s close to one another, and we can let $\Lambda^\mu(i)$ depend simply on the distance between $\mathbf{w}_i$ and ξ^μ. Durbin and Willshaw used a normalized Gaussian form

$$\Lambda^\mu(i) = \frac{\exp\left(-|\xi^\mu - \mathbf{w}_i|^2/2\sigma^2\right)}{\sum_j \exp\left(-|\xi^\mu - \mathbf{w}_j|^2/2\sigma^2\right)} \tag{9.35}$$

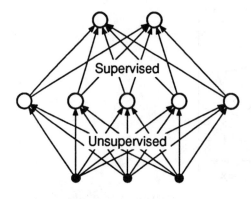

FIGURE 9.17 A hybrid learning network. Counterpropagation networks, hierarchical feature map classifiers, and radial basis function (locally tuned) networks all have this basic architecture.

with σ determining the effective range of the influence. The parameters η, σ, and κ are taken as relatively large initially, and then reduced gradually as the tour develops (Durbin and Willshaw took them all equal). Thus large-scale adjustments occur early on, and smaller refinements later.

Now there *is* a cost function

$$E\{\mathbf{w}_i\} = -\sigma^2 \sum_{\mu} \log\left[\sum_i \exp\left(-|\xi^{\mu} - \mathbf{w}_i|^2/2\sigma^2\right)\right] + \frac{\kappa}{2}\sum_i |\mathbf{w}_{i+1} - \mathbf{w}_i|^2 \quad (9.36)$$

for which (9.34) is a gradient descent algorithm. Thus (9.36) is minimized by the algorithm. In the limit $N \to \infty$ (where N is the number of units) and $\sigma \to 0$ the first term goes to zero and the second term is minimized by the shortest possible tour. Thus the algorithm should find the shortest tour *if* it manages to escape local minima.

In practice the Durbin-Willshaw approach seems to do fairly well, and is comparable to the Kohonen algorithm described above. However, a hybrid of the two, with additional modifications, appears to be much faster than either method [Burr, 1988]. It has also recently been shown [Simic, 1990] that there is a formal connection in statistical mechanics between the Durbin-Willshaw algorithm and the Hopfield-Tank network described in Section 4.2.

9.7 Hybrid Learning Schemes

In some problems it can be useful to combine supervised and unsupervised learning in the same network. The most common idea (see Fig. 9.17) is to have one layer that learns in an unsupervised way, followed by one (or more) layers trained by back-propagation. The problem being solved is hetero-association, and so could be approached with a purely supervised learning approach. But back-propagation is extremely slow, especially for networks with more than one hidden layer. So, if it does not greatly degrade our results, we can speed up learning considerably by training some layers in an unsupervised way.

This works well when the problem to be solved has the property that similar input vectors produce similar outputs. Then it is a sensible first step to categorize the inputs into clusters with competitive learning, and use only the category information for the supervised learning. On the other hand we would not expect success on something like the parity function (page 131), where changing a single input bit changes the target output.

The hybrid schemes are not optimal in the sense that back-propagation is, since the hidden layer responses are not optimized with respect to the output performance. Therefore we will probably need more hidden units to get comparable results. But the savings in training time may make up for this many times over.

Counterpropagation Networks

Networks with the architecture of Fig. 9.17 have been proposed by Hecht-Nielsen [1987b, 1988] and Huang and Lippmann [1988]. They have been called **counterpropagation networks** and **hierarchical feature map classifiers**. Counterpropagation networks [Hecht-Nielsen, 1987b, 1988] are actually more general, involving a common hidden layer between two separate pairs of input and output layers, one of which can be used for the inverse mapping of the first, but we describe just the "forward only" case.

The first (or hidden) layer uses the standard competitive learning rule, so the hidden units divide up the input space in a Voronoi tessellation. Then the output layer, which is *linear*, is trained with the usual delta rule (5.42)

$$\Delta w_{ij} = \eta(\zeta_i^\mu - O_i^\mu)V_j . \qquad (9.37)$$

Since only one of the hidden units V_j fires at a time, this may also be written

$$\Delta w_{ij} = \eta(\zeta_i^\mu - w_{ij})V_j \qquad (9.38)$$

(where V_j is 1 for the winner and 0 otherwise). In this form the delta rule is equivalent to one form of **outstar learning** [Grossberg, 1969; see also Caudill, 1989]. It leads to

$$\langle w_{ij} \rangle = \langle M_j^\mu \zeta_i^\mu \rangle \qquad (9.39)$$

where as usual M_j^μ is 1 if hidden unit j wins the competition for input pattern ξ^μ, and 0 otherwise. This says that in each of the Voronoi polyhedra the output is fixed at the average of all the output vectors ζ^μ whose corresponding input vector ξ^μ is in that polyhedron.

In effect the network is implementing a key-value **lookup table**. The keys are encoded in the input-to-hidden weights and the values are in the hidden-to-output weights. But the important difference from a conventional lookup table is that this one is self-programming; the key vectors are found by unsupervised learning, and should evolve a good statistical representation of the input space.

In some applications it may be desirable to interpolate the output between those for the different polyhedra. This can be done by relaxing the winner-take-all constraint (*after* the task has been learned), allowing several hidden units to fire at a time. If the total strength $\sum_j V_j$ of these multiple winners is normalized to 1, then the total output will interpolate linearly between the individual vectors. To select which units fire we could choose all those with weight vectors within a certain radius of the input ξ^μ, or just the K closest ones for some fixed K.

Many applications have been investigated, including pattern classification, data compression, speech recognition, function approximation, and statistical analysis [Hecht-Nielsen, 1988; Huang and Lippmann, 1988; Lippmann, 1989]. One example is vector quantization. In our earlier discussion (page 225) we were satisfied with finding a set of prototype vectors, the centroids of the polyhedra. But in a practical application for data compression we actually need to generate a binary code saying which vector to use. Clearly the counterpropagation network can be trained to generate appropriate codes in its output layer. We could for example use M output units and 2^M hidden units with such coding.

The greatest appeal of these networks is their speed. For a given problem they are typically a factor of 10–100 times faster to train than conventional back-propagation networks, with results that are often comparable.

Radial Basis Functions

Another example of a hybrid network, again with the architecture of Fig. 9.17, was examined by Moody and Darken [1988, 1989]. Similar schemes have also been suggested by Casdagli [1989], Specht [1990], Niranjan and Fallside [1990], Poggio and Girosi [1990], and many others.

The hidden units in the Moody-Darken network are neither linear winner-take-all units (as for competitive learning), nor sigmoidal units (as for back-propagation). Instead they have normalized **Gaussian activation functions**

$$g_j(\xi) = \frac{\exp[(\xi - \mu_j)^2/2\sigma_j^2]}{\sum_k \exp[(\xi - \mu_k)^2/2\sigma_k^2]} \tag{9.40}$$

where ξ is the input vector itself. By *normalized* we mean that $\sum_j g_j(\xi) = 1$ for any ξ. Thus unit j gives a maximum response to input vectors near μ_j. We say that each hidden unit has its own *receptive field* in the input space, a region centered on μ_j with size proportional to σ_j. The Gaussians are a particular example of **radial basis functions**.

The idea is now to pave the input space (or, better, the part of it where the input vectors lie) with these receptive fields. Then the problem is almost solved. Suppose a particular input vector ξ^μ lies in the middle of the receptive field for unit j, so $\xi^\mu = \mu_j$. If we ignore the overlaps between different receptive fields, then only hidden unit j will be activated, making it the only "winner". We could simply choose the output weights leading from that unit to be $w_{ij} = \zeta_i^\mu$ (for each i), which will produce the target pattern ζ_i^μ at the output assuming linear output units.

If another input lies, say, between two receptive field centers, then those two hidden units will be appreciably activated, and the output will be a weighted average of the corresponding targets. In this way the network makes a sensible smooth fit to the desired function.

The choice $w_{ij} = \zeta_i^\mu$ for the hidden-to-output weights is not optimal if the overlaps between receptive fields are taken into account. However, it is a simple matter to optimize them, just as in the counterpropagation network, using delta rule learning. The rule is simply

$$\Delta w_{ij} = \eta(\zeta_i^\mu - O_i^\mu)V_j \tag{9.41}$$

where $O_i^\mu = w_{ij}V_j$. When averaged over μ this minimizes the usual quadratic cost function

$$H\{w_{ij}\} = \tfrac{1}{2}\sum_{ij\mu}[\zeta_i^\mu - w_{ij}g_j(\xi^\mu)]^2. \tag{9.42}$$

The unsupervised part of the learning is the determination of the receptive field centers μ_j and widths σ_j. Appropriate μ's can be found by any vector quantization approach, including the usual competitive learning algorithm. The σ's are usually determined by an *ad hoc* choice, such as the mean distance to the first few nearest neighbor μ's. The performance of the network is not very sensitive to the precise values of the σ's.

Overall this scheme is very like the counterpropagation approach. The main difference is that the Moody-Darken approach uses locally tuned hidden units, whereas the counterpropagation network requires a winner-take-all search (or circuit). Interpolation between target patterns is also more natural in the present case. On the other hand normalized Gaussian activation functions do not so easily lend themselves to electronic implementation. Note that locally tuned receptive fields are widely found in biology, but usually emerge from a network and are not single-cell properties.

Moody and Darken tried their method out on the extrapolation problem for the Mackey-Glass equation described in Section 6.9 (pages 137–139). In this context it is interesting to note that the present method, with Gaussian receptive fields, allows one to fit an arbitrary function with just *one* hidden layer [Hartman et al., 1990]. With sigmoidal units, the arguments of Section 6.4 suggested that it takes two hidden layers to fit an arbitrary function. Indeed the "bump" construction described on page 142 essentially used two layers to construct hidden units with local receptive fields out of sigmoidal functions, so it is not surprising that we can save a layer by starting with units that already have bump-like responses.

The Lapedes-Farber network described in Section 6.9 had two hidden layers, so there were three layers of connections to train by back-propagation, requiring considerable supercomputer time. Moody and Darken were able to train their network several orders of magnitude faster than this, since they only had one layer of connections trained by supervised learning. They did not quite achieve the accuracy of Lapedes and Farber (or that obtained by Farmer and Sidorowich with their non-neural algorithm), but were not far behind.

The best results to date on the signal extrapolation problem have been obtained by Stokbro et al. [1990]. They generalize the Moody-Darken network so that each hidden unit j passes on to each output unit i a linear fit $w_{ij} + \mathbf{v}_{ij} \cdot \boldsymbol{\xi}$, instead of just a constant value w_{ij}. This allows for much better interpolation between the receptive field centers $\boldsymbol{\mu}_j$, but at the price of needing extra units and connections to represent the $\mathbf{v}_{ij} \cdot \boldsymbol{\xi}$ terms.

The receptive field centers $\boldsymbol{\mu}_j$ and widths σ_j are found as in the Moody-Darken case. Then delta rule supervised learning can be used to find the appropriate w_{ij}'s and $\mathbf{v}_{ij}$'s because there is a cost function

$$H\{w_{ij}, \mathbf{v}_{ij}\} = \tfrac{1}{2} \sum_{ij\mu} \left[\zeta_i^\mu - (w_{ij} + \mathbf{v}_{ij} \cdot \boldsymbol{\xi}^\mu) g_j(\boldsymbol{\xi}^\mu)\right]^2 \qquad (9.43)$$

on which we can perform gradient descent. The problem can be framed in network terms by associating an extra N hidden units with each of the original hidden units, where N is the input vector dimensionality. The kth one of these ($1 \leq k \leq N$) at the jth location computes $\xi_k g_j(\boldsymbol{\xi})$ and is connected to output unit i with weight $[\mathbf{v}_{ij}]_k$, the kth component of $\mathbf{v}_{ij}$.

Stokbro et al. find that it does pay to use $N+1$ times as many hidden units per receptive field in this way, even if the number of receptive fields is reduced by the same factor (i.e., with a fixed total number of hidden units). On the Mackey-Glass data, they were able to train a network as fast as Moody and Darken, achieving better prediction accuracy than either Lapedes and Farber or Farmer and Sidorowich.

TEN

Formal Statistical Mechanics of Neural Networks

As we have stressed in earlier chapters, neural networks are large interacting systems of simple units, like the physical systems we study in statistical mechanics. The formal methods and concepts of statistical physics are therefore natural tools to use for neural networks. In this chapter we illustrate the use of such methods in two different problems that we encountered earlier in the book: the recall of stored patterns in the Hopfield associative memory network, and the capacity of a simple perceptron.

This chapter is not for everyone. Up to this point, this book has been a fairly general introduction to the theory of neural networks, and has not required specialized knowledge of formal techniques. While what follows is also self-contained, it is much more formal mathematically, and readers without other exposure to techniques of this sort will probably find it hard going. We include these calculations here anyway because they illustrate how such theoretical methods can be brought to bear on problems in neural computation, with nontrivial results.

We do assume in this chapter some basic knowledge of statistical mechanics. The necessary ideas are reviewed briefly in the Appendix.

10.1 The Hopfield Model

In Chapter 2 we described the stochastic Hopfield model and obtained a number of its properties heuristically. The starting points were the Hebb rule (2.9) for the connection strengths, and the dynamics based on the stochastic evolution rule (2.40). We then calculated the average activations $\langle S_i \rangle$ in the heuristically motivated mean field scheme (2.50). Here we will take a more systematic approach, obtaining the quantities we want by first calculating the **partition function**

$$Z = \text{Tr}_S \exp(-\beta H\{S_i\}) \tag{10.1}$$

where the trace, Tr_S, means a sum over all possible states, $\{S_i = \pm 1\}$.[1] We can then take appropriate derivatives to obtain quantities of more direct interest, as outlined in the Appendix. Our treatment follows closely the classic article by Amit, Gutfreund, and Sompolinsky [1987a].

We start with the energy function

$$H_0 = -\frac{1}{2N} \sum_{\mu=1}^{p} \left(\sum_i S_i \xi_i^\mu\right)^2 + \frac{p}{2} \tag{10.2}$$

which is the same as (2.30) except for the second term (a constant), which cancels out the diagonal S_i^2 contributions from the first term. As we showed in (2.31), this energy function has the Hebbian connection strengths (2.9).

We again define $\alpha = p/N$, the ratio of the number of desired memories to the size of the system. We consider only the large N limit (the so-called **thermodynamic limit** $N \to \infty$), so when we discuss $\alpha \neq 0$ we mean that the number of patterns p scales proportionally with the number of units N. But first we discuss the simpler case $\alpha = 0$ with a fixed number p of patterns, independent of N.

Mean Field Theory for $\alpha = 0$

We start by adding to H_0 a set of "external fields" $h^\mu \xi_i^\mu$, one for each pattern ξ_i^μ:

$$H = H_0 - \sum_\mu h^\mu \sum_i \xi_i^\mu S_i . \tag{10.3}$$

We will set all the strengths h^μ to zero later, after these fields have served their purpose.

The partition function (10.1) is now

$$Z = e^{-\beta p/2} \text{Tr}_S \exp\left[\frac{\beta}{2N} \sum_\mu \left(\sum_i S_i \xi_i^\mu\right)^2 + \beta \sum_\mu h^\mu \sum_i \xi_i^\mu S_i\right]. \tag{10.4}$$

This would be easy to evaluate if both the terms in the exponent were linear in the S_i's. Then the trace would simply factor into a product of N independent terms, one for each i, every term being a simple sum over $S_i = +1$ and $S_i = -1$.

---

[1] Usually the trace of an operator (or matrix) means the sum of all the diagonal elements. This way of using it originates in quantum statistics.

Unfortunately the first term is quadratic. But we can use the "Gaussian integral trick" to make it linear, at the expense of some other complications. This trick exploits the identity

$$\int_{-\infty}^{\infty} dx\, e^{-ax^2 \pm bx} = \sqrt{\pi/a}\, e^{b^2/4a} \tag{10.5}$$

which can be used to turn an exponential in b^2 (on the right) into an exponential linear in b (on the left). The cost, of course, is the introduction of the auxiliary variable x, and the integral over it. Our price is actually p times higher, because (10.4) contains p quadratic terms, one for each μ. So we introduce p auxiliary variables m^μ, and take $a = \beta N/2$ and $b^\mu = \beta \sum_i S_i \xi_i^\mu$ to give

$$Z = e^{-\beta p/2} \left(\frac{\beta N}{2\pi}\right)^{p/2} i$$
$$\times \operatorname{Tr}_S \prod_\mu \int dm^\mu \exp\left(-\tfrac{1}{2}\beta N(m^\mu)^2 + \beta(m^\mu + h^\mu)\sum_i \xi_i^\mu S_i\right). \tag{10.6}$$

Now let us adopt a shorthand vector notation, taking $\mathbf{m}$, $\mathbf{h}$, and $\boldsymbol{\xi}_i$ to be p-component vectors with components m^μ, h^μ, and ξ_i^μ respectively. Then (10.6) becomes

$$Z = e^{-\beta p/2} \left(\frac{\beta N}{2\pi}\right)^{p/2} \int d\mathbf{m}\, e^{-\beta N \mathbf{m}^2/2} \prod_i \operatorname{Tr}_{S_i} e^{\beta(\mathbf{m}+\mathbf{h})\cdot \boldsymbol{\xi}_i S_i}. \tag{10.7}$$

The trace is now easy because the exponent is linear in S_i. Using $e^x + e^{-x} = 2\cosh x$, we obtain after a little reorganization

$$Z = \left(\frac{\beta N}{2\pi}\right)^{p/2} \int d\mathbf{m}\, e^{-\beta N f(\beta, \mathbf{m})} \tag{10.8}$$

with

$$f(\beta, \mathbf{m}) = \tfrac{1}{2}\alpha + \tfrac{1}{2}\mathbf{m}^2 - \frac{1}{\beta N}\sum_i \log\left(2\cosh[\beta(\mathbf{m}+\mathbf{h})\cdot \boldsymbol{\xi}_i]\right). \tag{10.9}$$

We still have a p-fold integral to do, but the fact that the exponent in (10.8) is proportional to N allows us to evaluate it in the limit of large N. The bigger N is, the more the integral is dominated by contributions from the region where f is smallest. So we can approximate it by finding the value of $\mathbf{m}$ which minimizes f, and expanding the integrand around there. This is called the **saddle-point method**, and is best understood through a simple example.

Suppose that we had a one-dimensional integral of the form

$$I = \sqrt{N} \int dx\, e^{-Ng(x)}. \tag{10.10}$$

Then expanding the exponent around the point x_0 where $g(x)$ is minimized we get

$$I = \sqrt{N} \int dx\, \exp\left(-N[g(x_0) + \tfrac{1}{2}g''(x_0)(x - x_0)^2 + \cdots]\right) \tag{10.11}$$

using $g'(x_0) = 0$. If we truncate the expansion at this point, the integral is just a Gaussian one, so

$$I = \sqrt{N}e^{-Ng(x_0)}\sqrt{\frac{2\pi}{Ng''(x_0)}} = e^{-Ng(x_0)}\sqrt{\frac{2\pi}{g''(x_0)}}. \tag{10.12}$$

For large N this result is dominated by the exponential factor, as can be clearly seen by putting it in the form

$$-\frac{1}{N}\log I = g(x_0) + \frac{1}{2N}(\log g''(x_0) - \log 2\pi)$$

$$\overset{N\to\infty}{\longrightarrow} g(x_0). \tag{10.13}$$

Thus all we need to do is to find x_0; this is often called the **saddle point**, from behavior in the complex x plane.

For (10.8) we use a p-dimensional version of the same idea, thereby obtaining

$$-\frac{1}{N}\log Z = \beta\min_{\mathbf{m}} f(\beta, \mathbf{m}) \tag{10.14}$$

in the $N\to\infty$ limit. Comparing this with (A.9) we see that

$$F/N = \min_{\mathbf{m}} f(\beta, \mathbf{m}) \tag{10.15}$$

where F is the free energy, so $\min_{\mathbf{m}} f(\beta, \mathbf{m})$ gives us the free energy per unit.

We now have to minimize $f(\beta, \mathbf{m})$, which requires

$$0 = \frac{\partial f}{\partial m^\mu} = m^\mu - \frac{1}{N}\sum_i \xi_i^\mu \tanh[\beta(\mathbf{m}+\mathbf{h})\cdot\boldsymbol{\xi}_i]. \tag{10.16}$$

Note that this is a set of p nonlinear simultaneous equations for the p unknowns m^μ. These equations appear to depend on the random patterns $\boldsymbol{\xi}_i$, but in fact the system is **self-averaging**; we can replace the average $N^{-1}\sum_i$ over *units* by an average over *patterns* at any one site, yielding

$$m^\mu = \langle\!\langle \xi^\mu \tanh[\beta(\mathbf{m}+\mathbf{h})\cdot\boldsymbol{\xi}]\rangle\!\rangle \tag{10.17}$$

where $\langle\!\langle \cdots \rangle\!\rangle$ indicates an average over the random distribution of $\boldsymbol{\xi}$ patterns. Similarly (10.9) becomes (with $\alpha\to 0$)

$$f = \tfrac{1}{2}\mathbf{m}^2 - \beta^{-1}\langle\!\langle\log(2\cosh[\beta(\mathbf{m}+\mathbf{h})\cdot\boldsymbol{\xi}])\rangle\!\rangle. \tag{10.18}$$

It is easy to see how the self-averaging property arises. As we go from unit to unit in the sum on i in (10.9) or (10.16), we are choosing N independent $\boldsymbol{\xi}_i$'s from the distribution $P(\boldsymbol{\xi})$, which we take to be uniform over the 2^p possibilities. So if N

is large compared to 2^p the average over sites is equivalent to an average over the distribution. This requires $p \ll \log N$, which *is* valid in our present $\alpha = 0$ case, but not for the $\alpha \neq 0$ case considered later.

The values of m^μ at the saddle point given by (10.17) admit a simple physical interpretation. To see this, we start from the free energy $F = -\beta^{-1} \log Z$ and differentiate with respect to h^μ. Using the original expression (10.4) for Z leads—as in (A.10)—to

$$\frac{\partial F}{\partial h^\mu} = -\beta^{-1} \frac{\partial \log Z}{\partial h^\mu} = -\sum_i \langle S_i \rangle \xi_i^\mu \qquad (10.19)$$

whereas (10.15), (10.17), and (10.18) give us

$$\frac{\partial F}{\partial h^\mu} = N \frac{\partial f}{\partial h^\mu} = -N \langle\!\langle \xi^\mu \tanh[\beta(\mathbf{m} + \mathbf{h}) \cdot \boldsymbol{\xi}] \rangle\!\rangle = -Nm^\mu . \qquad (10.20)$$

We can thus identify

$$m^\mu = \frac{1}{N} \sum_i \xi_i^\mu \langle S_i \rangle \qquad (10.21)$$

so the saddle-point value of m^μ is just the average overlap of the network configuration with pattern number μ.

It was to derive (10.21) that we needed the external field terms inserted in (10.3). Now they are no longer needed and we henceforth set $h^\mu = 0$. Thus the mean field equation (10.17) becomes simply

$$m^\mu = \langle\!\langle \xi^\mu \tanh(\beta \mathbf{m} \cdot \boldsymbol{\xi}) \rangle\!\rangle . \qquad (10.22)$$

There are many solutions of (10.22). The simplest and most important are the **memory states**, which have a finite correlation with just *one* of the patterns ξ_i^μ. So from (10.21) we expect the **m** vector for these solutions to have the form

$$\mathbf{m} = (m, 0, 0, \ldots) \qquad (10.23)$$

if we order the indices μ so that the "condensed" pattern is first. Then (10.22) reduces to

$$m^\mu = \langle\!\langle \xi^\mu \tanh \beta m \xi^1 \rangle\!\rangle = \langle\!\langle \xi^\mu \xi^1 \rangle\!\rangle \tanh \beta m = \delta_{\mu 1} \tanh \beta m . \qquad (10.24)$$

So (10.23) *does* give a solution—putting (10.23) into the right-hand side of (10.22) produces the same form on the left—provided the magnitude m of the average overlap with pattern 1 satisfies

$$m = \tanh \beta m . \qquad (10.25)$$

This is identical to the equation (2.54) that we found in our simpler analysis in Chapter 2. It implies stable memory states for $T < T_c$ with $T_c = 1$, and tells us what fraction of the bits will be correct at any such temperature; see Fig. 2.14.

TABLE 10.1
Critical Temperatures

n	T_n
1	1
3	0.46
5	0.39
7	0.35

There are also more complicated solutions of the mean field equations, corresponding to the **spurious states**. The simplest of these are the **symmetric mixture states** in which the **m** vector has the form

$$\mathbf{m} = (\underbrace{m, m, m, \ldots, m}_{n}, \underbrace{0, 0, \ldots, 0}_{p-n}) \tag{10.26}$$

with n nonzero entries of **m** equal to some value m. Note that there are $\binom{n}{p}$ ways we might have placed the nonzero elements, corresponding to many such spurious states. There is actually a further degeneracy factor of 2^n, because solutions like $(\pm m, \pm m, \pm m, 0, \ldots, 0)$ are all possible.

If we insert the form (10.26) into the mean field equations (10.22) we obtain

$$m^{\mu} = \left\langle\!\!\left\langle \xi^{\mu} \tanh\left(\beta m \sum_{\nu=1}^{n} \xi^{\nu}\right) \right\rangle\!\!\right\rangle \tag{10.27}$$

which vanishes if $\mu > n$ (because $\langle\!\langle \xi^{\mu}\xi^{\nu} \rangle\!\rangle = 0$ for $\mu \neq \nu$), and otherwise gives

$$m = \langle\!\langle z \tanh \beta m z \rangle\!\rangle / n \tag{10.28}$$

where z is the random variable

$$z = \sum_{\mu=1}^{n} \xi^{\mu} \tag{10.29}$$

which has a binomial distribution. Thus our symmetric combination pattern (10.26) solves the mean field equations if m satisfies (10.28). This has solutions at any n, as long as $T < 1$.

However, not all these solutions are stable. We want **m** to produce a *minimum* of $f(\beta, \mathbf{m})$, whereas our mean field equations only guarantee a stationary point, $\partial f/\partial m^{\mu} = 0$. So we also need the eigenvalues of the matrix

$$A_{\mu\nu} = \frac{\partial^2 f}{\partial m^{\mu} \partial m^{\nu}} \tag{10.30}$$

to be positive. This turns out to be satisfied only if n is odd, and then only if the temperature T is below a **critical temperature** T_n. The first few T_n's are shown in table 10.1.

There are also **asymmetric mixture states**, such as

$$\mathbf{m} = (\tfrac{1}{2}, \tfrac{1}{2}, \tfrac{1}{4}, \tfrac{1}{4}, \tfrac{1}{4}, 0, 0, 0, \ldots). \tag{10.31}$$

None of these is stable above T_3, however. This means that one can avoid *all* mixture states by going to temperatures above T_3. Of course raising the temperature from $T = 0$ degrades the memory states somewhat, but the amount is actually very small; $\langle N_{\text{correct}} \rangle$ falls only very slowly from N with increasing T, as seen in Fig. 2.14. At $T = 0.47$ we find from (10.25) that $\langle N_{\text{correct}} \rangle \approx 0.97N$, so only about 3% of the bits will be recalled incorrectly if we work just above T_3.

Mean Field Theory for $\alpha \neq 0$

As we observed already in Chapter 2, the crosstalk between different patterns on account of their random overlap begins to affect the recall of a given pattern when p becomes of the order of N. We now examine the statistical mechanics for this case. The self-averaging we used in the $\alpha = 0$ calculation breaks down, and we are forced to do the averaging over the distribution of patterns more systematically.

As always, the basic quantity we start from is $\log Z$. Now Z depends on the particular set of patterns used to compute the weights w_{ij} using the Hebb rule (2.9). What is of interest to us is the *average* $\langle\!\langle \log Z \rangle\!\rangle$ over the distribution of all random binary patterns; this gives us the average free energy whose derivatives give the average quantities we want to know, such as m^μ. Unfortunately this average is very hard to calculate directly, and is *not* the same thing as $\log\langle\!\langle Z \rangle\!\rangle$, which would be much easier. To get meaningful results we must average the relevant quantity, which is $\log Z$, not Z.

Luckily there is a technique, called the **replica method**, that lets us circumvent averaging $\log Z$. It is based on the identity

$$\log Z = \lim_{n \to 0} \frac{Z^n - 1}{n} \tag{10.32}$$

which allows us to compute $\langle\!\langle \log Z \rangle\!\rangle$ from knowledge of $\langle\!\langle Z^n \rangle\!\rangle$. Note that we need this for the parameter n close to 0, but we ignore that for a while and focus on $\langle\!\langle Z^n \rangle\!\rangle$ for *integer* n. In that case we can think of Z^n as the partition function of n copies, or **replicas**, of the original system, writing

$$Z^n = \text{Tr}_{S^1} \, \text{Tr}_{S^2} \ldots \text{Tr}_{S^n} \, e^{-\beta(E\{S_i^1\} + \cdots + E\{S_i^n\})}. \tag{10.33}$$

Each copy is labelled by a superscript **replica index** on its S_i's, running from 1 to n.

Proceeding as we did in the $\alpha = 0$ case for (10.6), using the Gaussian integral trick for each pattern and each replica, we now find

$$\langle\langle Z^n \rangle\rangle = e^{-\beta pn/2} \Big\langle\!\!\Big\langle \text{Tr}_S \prod_{\mu=1}^{p} \prod_{\rho=1}^{n} \int dm_\rho^\mu \left(\frac{\beta N}{2\pi}\right)^{1/2}$$

$$\times \exp\left(-\tfrac{1}{2}\beta N(m_\rho^\mu)^2 + \beta m_\rho^\mu \sum_i \xi_i^\mu S_i^\rho\right) \Big\rangle\!\!\Big\rangle \qquad (10.34)$$

where ρ labels the different replicas. Note that the pattern average $\langle\langle \cdots \rangle\rangle$ is still over the Np variables ξ_i^μ; there is no replica index on the patterns. We have omitted any external fields h^μ this time, although it would be easy to include them.

Henceforth we focus on states in which the configuration has appreciable overlap with only a *finite* number s of the p stored patterns, called the **condensed patterns**. Specifically we assume that the m_ρ^μ's[2] are only appreciable in size when $\mu \leq s$, with s independent of N. This will eventually allow us to use the self-averaging trick just on these μ's. For $\mu > s$ we assume $m_\rho^\mu \ll 1$.

Let us consider the contribution of the last term in the integrand of (10.34) for a particular $\mu > s$, one of the *small* m_ρ^μ's:

$$\Big\langle\!\!\Big\langle \prod_\rho \exp\left(\beta m_\rho^\mu \sum_i \xi_i^\mu S_i^\rho\right) \Big\rangle\!\!\Big\rangle = \prod_i \Big\langle\!\!\Big\langle \exp\left(\beta \xi_i^\mu \sum_\rho m_\rho^\mu S_i^\rho\right) \Big\rangle\!\!\Big\rangle$$

$$= \prod_i \cosh\left(\beta \sum_\rho m_\rho^\mu S_i^\rho\right)$$

$$= \exp\left[\sum_i \log\cosh\left(\beta \sum_\rho m_\rho^\mu S_i^\rho\right)\right]$$

$$\approx \exp\left[\sum_i \frac{1}{2}\left(\beta \sum_\rho m_\rho^\mu S_i^\rho\right)^2\right]$$

$$= \exp\left(\frac{\beta^2}{2} \sum_{i\rho\sigma} S_i^\rho S_i^\sigma m_\rho^\mu m_\sigma^\mu\right) \qquad (10.35)$$

where the approximation involved $\log\cosh x \approx x^2/2$ for small x. If we now define an $n \times n$ matrix $\tilde{\Lambda}_{\rho\sigma}$ by

$$\tilde{\Lambda}_{\rho\sigma} \equiv \delta_{\rho\sigma} - (\beta/N)\sum_i S_i^\rho S_i^\sigma \qquad (10.36)$$

we can write the whole exponential factor (for fixed $\mu > s$) in (10.34) as

$$E = \exp\left(-\frac{\beta N}{2}\sum_{\rho\sigma} \tilde{\Lambda}_{\rho\sigma} m_\rho^\mu m_\sigma^\mu\right). \qquad (10.37)$$

---

[2]Strictly speaking: the *saddle-point* values of the m_ρ^μ's. That is, we will again evaluate the multidimensional integral by the saddle-point method, and the values of the m_ρ^μ that will matter will be those at the saddle point.

This leaves us with an n-dimensional Gaussian integral, giving

$$\int \left(\prod_\rho dm_\rho^\mu \left(\frac{\beta N}{2\pi} \right)^{1/2} \right) E = \left(\frac{\beta N}{2\pi} \right)^{n/2} \sqrt{\frac{\pi^n}{\det(\frac{1}{2}\beta N \tilde{\Lambda})}} = (\det \tilde{\Lambda})^{-1/2}. \quad (10.38)$$

We get a contribution exactly like this for every value of μ greater than s (about p in all, since $p \gg s$), giving an overall factor

$$(\det \tilde{\Lambda})^{-p/2} = \exp\left(-\tfrac{1}{2}p \log \det \tilde{\Lambda}\right) = \exp\left(-\tfrac{1}{2}p \log \prod_\rho \tilde{\lambda}_\rho\right)$$

$$= \exp\left(-\tfrac{1}{2}p \sum_\rho \log \tilde{\lambda}_\rho\right) \quad (10.39)$$

where $\tilde{\lambda}_\rho$ are the eigenvalues of $\tilde{\Lambda}$.

The extra complications we encounter for $\alpha > 0$ all come from this factor (10.39), which, together with the other parts of (10.34), now has to be summed over all the S_i^ρ. Unfortunately the S-dependence is buried in the eigenvalues $\tilde{\lambda}_\rho$, and the trace is far from easy. So now we use some more auxiliary variable tricks. First let us define a generalized version of $\tilde{\Lambda}_{\rho\sigma}$:

$$\Lambda_{\rho\sigma} \equiv (1 - \beta)\delta_{\rho\sigma} - \beta q_{\rho\sigma}. \quad (10.40)$$

This is equal to $\tilde{\Lambda}_{\rho\sigma}$ if

$$q_{\rho\sigma} = \begin{cases} N^{-1} \sum_i S_i^\rho S_i^\sigma & \text{for } \rho \neq \sigma; \\ 0 & \text{otherwise.} \end{cases} \quad (10.41)$$

Thus we can write any function $G\{\tilde{\lambda}_\rho\}$ of the eigenvalues of $\tilde{\Lambda}$ in the form

$$G\{\tilde{\lambda}_\rho\} = \int \left[\prod_{(\rho\sigma)} dq_{\rho\sigma} \, \delta\left(q_{\rho\sigma} - \frac{1}{N}\sum_i S_i^\rho S_i^\sigma\right) \right] G\{\lambda_\rho\} \quad (10.42)$$

using a Dirac delta function, where the λ_ρ's are the eigenvalues of Λ, and are functions of the $q_{\rho\sigma}$'s. There are $n(n-1)/2$ integrals (the notation $(\rho\sigma)$ means all distinct pairs), and we leave it as understood that $q_{\rho\sigma} = q_{\sigma\rho}$ and $q_{\rho\rho} = 0$.

Now we introduce yet another set of auxiliary variables, this time for an integral representation of the delta-function:

$$\delta(x) = \int_{-i\infty}^{i\infty} \frac{d\tilde{r}}{2\pi i} \, e^{-\tilde{r}x}. \quad (10.43)$$

We need to use this $n(n-1)/2$ times, giving us

$$G\{\tilde{\lambda}_\rho\} \propto \int \left[\prod_{(\rho\sigma)} dq_{\rho\sigma} dr_{\rho\sigma} \exp\left(-N\alpha\beta^2 r_{\rho\sigma} q_{\rho\sigma} + \alpha\beta^2 r_{\rho\sigma} \sum_i S_i^\rho S_i^\sigma\right) \right] G\{\lambda_\rho\} \quad (10.44)$$

where we have left out unimportant prefactors and scaled the r variables by a factor of $N\alpha\beta^2$ for later convenience.

When we apply the transformation (10.44) to (10.39), we can write our full expression (10.34) for $\langle\!\langle Z^n \rangle\!\rangle$ as

$$
\langle\!\langle Z^n \rangle\!\rangle \propto e^{-\beta p n/2} \int \left(\prod_{\mu\rho} dm_\rho^\mu \left(\frac{\beta N}{2\pi} \right)^{1/2} \right) \left(\prod_{(\rho\sigma)} dq_{\rho\sigma} dr_{\rho\sigma} \right)
$$

$$
\times \ \exp\left(-\tfrac{1}{2}\beta N \sum_{\mu\rho} (m_\rho^\mu)^2 - \frac{\alpha N}{2} \sum_\rho \log \lambda_\rho - \tfrac{1}{2} N\alpha\beta^2 \sum_{\rho\sigma} r_{\rho\sigma} q_{\rho\sigma} \right)
$$

$$
\times \ \left\langle\!\!\left\langle \mathrm{Tr}_S \exp\left(\beta \sum_{\mu\rho} m_\rho^\mu \sum_i \xi_i^\mu S_i^\rho + \tfrac{1}{2}\alpha\beta^2 \sum_{i\rho\sigma} r_{\rho\sigma} S_i^\rho S_i^\sigma \right) \right\rangle\!\!\right\rangle \qquad (10.45)
$$

where the sums over μ now run only over the condensed patterns: $\mu = 1, 2, \ldots, s$. We have also written $\tfrac{1}{2}\sum_{\rho\sigma}$ instead of $\sum_{(\rho\sigma)}$ and again left it understood that diagonal $\rho\rho$ terms are zero.

Now at last we can get rid of the i indices through self-averaging. The last line of (10.45) is the pattern average of an expression with the form

$$
X \equiv \mathrm{Tr}_S \exp\left(\sum_i F\{\xi_i, S_i\} \right) \qquad (10.46)
$$

$$
= \prod_i \mathrm{Tr}_{S_i} \exp F\{\xi_i, S_i\} \qquad (10.47)
$$

$$
= \exp\left(\sum_i \log \mathrm{Tr}_{S_i} \exp F\{\xi_i, S_i\} \right). \qquad (10.48)
$$

The function F depends on ξ_i^1–ξ_i^s and S_i^1–S_i^n, but only one index i is needed at a time. The trace in (10.46) is over all the S_i^ρ's (all i's and all ρ's), but in (10.47) and (10.48) is over only the S_i^ρ's for a particular i. The result of the trace would be exactly the same for each i except for the dependence of F on the ξ_i^μ's, because i is otherwise a dummy index. But since N (the number of i's) is much larger as $N \to \infty$ than 2^s (the number of possible sets $\{\xi_i^\mu\}$ at fixed i), the sum over i is equivalent to an average over patterns. Thus

$$
X = \exp\left(N \langle\!\langle \log \mathrm{Tr}_S \exp F\{\xi_i, S_i\} \rangle\!\rangle \right) \qquad (10.49)
$$

where now all i's have disappeared and we have in effect a single unit with n different S^ρ's and p different ξ^μ's. Note that in the end we did not need the outer average $\langle\!\langle \cdots \rangle\!\rangle$ in (10.45), because the self-averaging of the inner i sum already performs all the pattern averaging. So we may drop the outer average.

Now we can write the whole expression for $\langle\!\langle Z^n \rangle\!\rangle$ as an integral of the exponential of something proportional to N:

$$
\langle\!\langle Z^n \rangle\!\rangle \propto e^{-\beta p n/2} \int \left(\prod_{\mu\rho} dm_\rho^\mu \left(\frac{\beta N}{2\pi} \right)^{1/2} \right) \left(\prod_{(\rho\sigma)} dq_{\rho\sigma} dr_{\rho\sigma} \right) e^{-N\beta f\{m, q, r\}} \qquad (10.50)
$$

where

$$f\{m, q, r\} = \tfrac{1}{2} \sum_{\mu\rho} (m_\rho^\mu)^2 + \frac{\alpha}{2\beta} \sum_\rho \log \lambda_\rho + \tfrac{1}{2}\alpha\beta \sum_{\rho\sigma} r_{\rho\sigma} q_{\rho\sigma}$$
$$- \frac{1}{\beta} \left\langle\!\!\!\left\langle \log \mathrm{Tr}_S \exp\left(\beta \sum_{\mu\rho} m_\rho^\mu \xi^\mu S^\rho + \tfrac{1}{2}\alpha\beta^2 \sum_{\rho\sigma} r_{\rho\sigma} S^\rho S^\sigma \right) \right\rangle\!\!\!\right\rangle. \quad (10.51)$$

The factor of N in the exponent allows us to use the saddle-point method again, minimizing this time with respect to the q's and r's as well as the m's. Thus we obtain the free energy per unit

$$F/N = -\frac{1}{\beta N}\langle\!\langle \log Z \rangle\!\rangle = -\frac{1}{\beta N} \lim_{n\to 0} \frac{1}{n}(\langle\!\langle Z^n \rangle\!\rangle - 1) \quad (10.52)$$

$$= -\frac{1}{\beta N} \lim_{n\to 0} \frac{1}{n} \log\langle\!\langle Z^n \rangle\!\rangle$$

$$= \frac{\alpha}{2} + \lim_{n\to 0} \frac{1}{n} \min f\{m, q, r\}. \quad (10.53)$$

In replacing $\langle\!\langle Z^n \rangle\!\rangle - 1$ by $\log\langle\!\langle Z^n \rangle\!\rangle$ we just assumed that $\langle\!\langle Z^n \rangle\!\rangle$ goes to 1 as $n \to 0$, as it must; that is why we didn't bother to keep all the prefactors earlier.

The location of the saddle point is determined by the equations

$$\frac{\partial f}{\partial m_\rho^\mu} = 0 \quad (10.54)$$

$$\frac{\partial f}{\partial q_{\rho\sigma}} = 0 \quad (10.55)$$

$$\frac{\partial f}{\partial r_{\rho\sigma}} = 0. \quad (10.56)$$

As in the simpler $\alpha = 0$ case, these equations lead to interpretations of the **order parameters** m_ρ^μ, $q_{\rho\sigma}$, and $r_{\rho\sigma}$ at the saddle point:

$$m_\rho^\mu = \frac{1}{N} \sum_i \xi_i^\mu \langle S_i^\rho \rangle \quad (10.57)$$

$$q_{\rho\sigma} = \left\langle\!\!\!\left\langle \frac{1}{N} \sum_i \langle S_i^\rho \rangle \langle S_i^\sigma \rangle \right\rangle\!\!\!\right\rangle \quad (10.58)$$

$$r_{\rho\sigma} = \frac{1}{\alpha} \sum_{\mu > s} \langle\!\langle m_\rho^\mu m_\sigma^\mu \rangle\!\rangle. \quad (10.59)$$

We omit the detailed derivations of these results, which require the inclusion of external field terms h^μ as in the $\alpha = 0$ case. Equation (10.59), which comes from $\partial f/\partial q_{\rho\sigma} = 0$, also involves rewriting the $\log \lambda_\rho$ term as a Gaussian integral. Note that (10.57) is just like the $\alpha = 0$ result (10.21) apart from the presence of the replica index ρ.

To proceed further we have to make an *ansatz* without *a priori* justification: that of **replica symmetry**. This means that we assume that the saddle-point values of the order parameters do not depend on their replica indices:

$$m_\rho^\mu = m^\mu \tag{10.60}$$

$$q_{\rho\sigma} = q \tag{10.61}$$

$$r_{\rho\sigma} = r. \tag{10.62}$$

The validity of this assumption can be tested afterwards, and one finds that it is exactly true except at very low temperatures, and that even there it is a good approximation.

With this simplification the meaning of the order parameters (10.57)–(10.59) is evident, and consistent with the heuristic treatment in Chapter 2: m^μ is (as before) the overlap between the network configuration and the μth pattern, q is the mean squared magnetization, and αr is the mean squared value of the overlap with the uncondensed patterns ($\mu > s$). Each m^μ for $\mu > s$ is of order $1/\sqrt{N}$, but r remains finite as $N \to \infty$ because there are of order N terms in the sum (10.59).

Using the replica symmetric *ansatz* the expression (10.51) for $f(m, q, r)$ simplifies to

$$f(\mathbf{m}, q, r) = \tfrac{1}{2} n m^2 + \frac{\alpha}{2\beta} \sum_\rho \log \lambda_\rho + \tfrac{1}{2} n(n-1)\alpha\beta r q + \tfrac{1}{2} n\alpha\beta r$$

$$- \frac{1}{\beta} \left\langle\!\left\langle \log \mathrm{Tr}_S \exp\left[\beta \mathbf{m} \cdot \boldsymbol{\xi} \sum_\rho S^\rho + \tfrac{1}{2}\alpha\beta^2 r \left(\sum_\rho S^\rho\right)^2\right]\right\rangle\!\right\rangle \tag{10.63}$$

where the last term on the first line is to cancel the diagonal part of the $(\sum_\rho S^\rho)^2$ term. We still have to evaluate the sum of the $\log \lambda_\rho$'s and compute the average of the Tr over the S^1–S^n but, thanks to the replica symmetry, these can now be done without too much trouble.

Let us first deal with the eigenvalue sum. The matrix $\Lambda_{\rho\sigma}$ now has the simple form

$$\Lambda_{\rho\sigma} = \begin{cases} 1 - \beta & \text{if } \rho = \sigma; \\ -\beta q & \text{otherwise.} \end{cases} \tag{10.64}$$

It is an elementary exercise to show that such a matrix has eigenvalues

$$\lambda_1 = 1 - \beta - (n-1)\beta q \tag{10.65}$$

with multiplicity 1 and

$$\lambda_2 = 1 - \beta(1 - q) \tag{10.66}$$

with multiplicity (i.e., number of eigenvectors with this eigenvalue) $n - 1$. Thus the sum over the logs of the eigenvalues becomes

$$\frac{1}{n} \sum_\rho \log \lambda_\rho = \frac{1}{n}\{\log[1 - \beta - (n-1)\beta q] + (n-1)\log[1 - \beta(1-q)]\}$$

$$\xrightarrow{n \to 0} \log[1 - \beta(1-q)] - \frac{\beta q}{1 - \beta(1-q)}. \tag{10.67}$$

To evaluate the Tr over the S's, we again use the Gaussian integral trick:

$$\exp\left[\tfrac{1}{2}\alpha\beta^2 r\left(\sum_\rho S^\rho\right)^2\right] = \int \frac{dz}{\sqrt{2\pi}} \exp\left(-\tfrac{1}{2}z^2 + \beta\sqrt{\alpha r}z \sum_\rho S^\rho\right) \qquad (10.68)$$

giving for the trace $X \equiv \text{Tr}\exp[\cdots]$ in (10.63):

$$
\begin{aligned}
X &= \text{Tr}_S \int \frac{dz}{\sqrt{2\pi}} \exp\left(-\tfrac{1}{2}z^2 + \beta(\sqrt{\alpha r}z + \mathbf{m}\cdot\boldsymbol{\xi})\sum_\rho S^\rho\right) \\
&= \int \frac{dz}{\sqrt{2\pi}} e^{-z^2/2}\left(2\cosh\beta(\sqrt{\alpha r}z + \mathbf{m}\cdot\boldsymbol{\xi})\right)^n \\
&= \int \frac{dz}{\sqrt{2\pi}} e^{-z^2/2} \exp\left(n\log[2\cosh\beta(\sqrt{\alpha r}z + \mathbf{m}\cdot\boldsymbol{\xi})]\right). \qquad (10.69)
\end{aligned}
$$

We actually want $1/n$ times the average of the log of this, in the $n \to 0$ limit. Expanding for small n gives

$$
\begin{aligned}
\frac{1}{n}\langle\!\langle\log X\rangle\!\rangle &= \frac{1}{n}\left\langle\!\!\left\langle \log\int\frac{dz}{\sqrt{2\pi}} e^{-z^2/2}\left(1 + n\log[2\cosh\beta(\sqrt{\alpha r}z + \mathbf{m}\cdot\boldsymbol{\xi})] + \cdots\right)\right\rangle\!\!\right\rangle \\
&= \frac{1}{n}\left\langle\!\!\left\langle n\int\frac{dz}{\sqrt{2\pi}} e^{-z^2/2}\log[2\cosh\beta(\sqrt{\alpha r}z + \mathbf{m}\cdot\boldsymbol{\xi})] + \cdots\right\rangle\!\!\right\rangle \\
&\xrightarrow{n\to 0} \langle\!\langle\log[2\cosh\beta(\sqrt{\alpha r}z + \mathbf{m}\cdot\boldsymbol{\xi})]\rangle\!\rangle \qquad (10.70)
\end{aligned}
$$

where now we take $\langle\!\langle\cdots\rangle\!\rangle$ to mean both the average over the condensed patterns $\mu \le s$ and the Gaussian average over z. Physically this means averaging over *all* the patterns, since the Gaussian random field z came from representing the effects of the uncondensed patterns $\mu > s$.

All we have left is to collect the terms from (10.53), (10.63), (10.67), and (10.70) to give the average free energy per site in the form

$$
\begin{aligned}
F/N &= \tfrac{1}{2}\alpha + \tfrac{1}{2}\mathbf{m}^2 + \frac{\alpha}{2\beta}\left(\log[1 - \beta(1-q)] - \frac{\beta q}{1 - \beta(1-q)}\right) \\
&\quad + \tfrac{1}{2}\alpha\beta r(1-q) - \frac{1}{\beta}\langle\!\langle\log[2\cosh\beta(\sqrt{\alpha r}z + \mathbf{m}\cdot\boldsymbol{\xi})]\rangle\!\rangle. \qquad (10.71)
\end{aligned}
$$

The saddle-point equations (10.54)–(10.56) are equivalent to setting the derivatives of F/N to zero, giving

$$m^\mu = \langle\!\langle\xi^\mu \tanh\beta(\sqrt{\alpha r}z + \mathbf{m}\cdot\boldsymbol{\xi})\rangle\!\rangle \qquad (10.72)$$

$$q = \langle\!\langle\tanh^2\beta(\sqrt{\alpha r}z + \mathbf{m}\cdot\boldsymbol{\xi})\rangle\!\rangle \qquad (10.73)$$

$$r = \frac{q}{[1 - \beta(1-q)]^2}. \qquad (10.74)$$

Only the second of these, which comes from $\partial F/\partial r = 0$, is a little tricky, needing the identity

$$\int \frac{dz}{\sqrt{2\pi}} e^{-z^2/2} z f(z) = \int \frac{dz}{\sqrt{2\pi}} e^{-z^2/2} f'(z) \qquad (10.75)$$

for any bounded function $f(z)$.

Equation (10.72) is just like (10.22) for the $\alpha = 0$ case, except for the addition of the effective Gaussian random field term, which represents the crosstalk from the uncondensed patterns. For $\alpha = 0$ it reduces directly to (10.22). Equation (10.73) is the obvious equation for the mean square magnetization. Equation (10.74) gives the (nontrivial) relation between q and the mean square value of the random field, and is identical to (2.67).

For **memory states**, i.e., **m**-vectors of the form $(m, 0, 0, \ldots)$, the saddle-point equations (10.72) and (10.73) become simply

$$m = \left\langle\!\left\langle \tanh \beta(\sqrt{\alpha r} z + m) \right\rangle\!\right\rangle_z \qquad (10.76)$$

$$q = \left\langle\!\left\langle \tanh^2 \beta(\sqrt{\alpha r} z + m) \right\rangle\!\right\rangle_z \qquad (10.77)$$

where the averaging is solely over the Gaussian random field. These are are identical to (2.65) and (2.68) that we found in the heuristic theory of Section 2.5. Their solution, and the consequent phase diagram of the model in $\alpha - T$ space, can be studied as we sketched there. Spurious states, such as the symmetric combinations (10.26), can also be analyzed at finite α using the full equations (10.72)–(10.74).

There are several subtle points in this replica method calculation:

- We started by calculating $\langle\!\langle Z^n \rangle\!\rangle$ for *integer* n but eventually interpreted n as a real number and took the $n \to 0$ limit. This is not the only possible continuation from the integers to the reals; we might for example have added a function like $\sin \pi n / n$.

- We treated the order of limits and averages in a cavalier fashion, and in particular reversed the order of $n \to 0$ and $N \to \infty$.

- We made the replica symmetry approximation (10.60)–(10.62) which was really only based on intuition.

Experience has shown that the replica method usually does work, but there are few rigorous mathematical results. It can be shown for the Sherrington-Kirkpatrick spin glass model, and probably for this one too, that the reversal of limits is justified, and that the replica symmetry assumption is correct *for integer* n [van Hemmen and Palmer, 1979]. But for some problems, including the spin glass, the method sometimes gives the wrong answer. This can be blamed on the integer-to-real continuation, and can be corrected by **replica symmetry breaking**, in which the replica symmetry assumption is replaced by a more complicated assumption. Then the natural continuation seems to give the right answer.

For the present problem Amit et al. showed that the replica symmetric approximation is valid except at very low temperatures where there is replica symmetry breaking. This seems to lead only to very small corrections in the results. However, the predicted change in the capacity—α_c becomes 0.144 instead of 0.138—can be detected in numerical simulations [Crisanti et al., 1986].

10.2 Gardner Theory of the Connections

The second classic statistical mechanical *tour de force* in neural networks is the computation by Gardner [1987, 1988] of the capacity of a simple perceptron. The calculation applies in the same form to a Hopfield-like recurrent network for auto-associative memory if the connections are allowed to be asymmetric.

This theory is very general; it is not specific to any particular algorithm for determining the connections. On the other hand, it does not provide us with a specific set of connections even when it has told us that such a set exists. As in Section 6.5, the basic idea is to consider the fraction of **weight space** that implements a particular input-output function; recall that weight space is the space of all possible connection weights $\mathbf{w} = \{w_{ij}\}$.

In Section 6.5 we used relatively simple methods to calculate weight space volumes. The present approach is more complicated, though often more powerful. We use many of the techniques introduced in the previous section, including replicas, auxiliary variables, and the saddle-point method.

We consider a simple perceptron with N binary inputs $\xi_j = \pm 1$ and M binary threshold units that compute the outputs

$$O_i = \text{sgn}\left(N^{-1/2} \sum_j w_{ij}\xi_j\right). \tag{10.78}$$

The $N^{-1/2}$ factor will be discussed shortly. Given a desired set of associations $\xi_j^\mu \to \zeta_i^\mu$ for $\mu = 1, 2, \ldots, p$, we want to know in what fraction of weight space the equations

$$\zeta_i^\mu = \text{sgn}\left(N^{-1/2} \sum_j w_{ij}\xi_j^\mu\right) \tag{10.79}$$

are satisfied (for all i and μ). Or equivalently, in what fraction of this space are the inequalities

$$\zeta_i^\mu N^{-1/2} \sum_j w_{ij}\xi_j^\mu > 0 \tag{10.80}$$

true?

It is also interesting to ask the corresponding question if the condition (10.80) is strengthened so there is a **margin size** $\kappa > 0$ as in (5.20):

$$\zeta_i^\mu N^{-1/2} \sum_j w_{ij}\xi_j^\mu > \kappa. \tag{10.81}$$

A nonzero κ guarantees correction of small errors in the input pattern.

Until (10.81) the factor $N^{-1/2}$ was irrelevant. We include it because it is convenient to work with w_{ij}'s of order unity, and a sum of N such terms of random sign gives a result of order $N^{1/2}$. Thus the explicit factor $N^{-1/2}$ makes the left-hand side of (10.81) of order unity, and it is appropriate to think about κ's that are independent of N. Of course this is only appropriate if the terms in the sum over j are really of random sign, but that turns out to be the case of most interest here. On the other hand, in Chapter 5 we were mainly dealing with a correlated sum, and so used a factor N instead of $N^{1/2}$.

For a recurrent autoassociative network, the same equations with $\zeta_i^\mu = \xi_i^\mu$ give the condition for the stability of the patterns, and a nonzero κ ensures finite basins of attraction.

The Capacity of a Simple Perceptron

The fundamental quantity that we want to calculate is the volume fraction of weight space in which (10.81) is satisfied. Adding an additional constraint

$$\sum_j w_{ij}^2 = N \tag{10.82}$$

for each unit i, so as to keep the weights within bounds, this fraction is

$$V = \frac{\int d\mathbf{w} \left(\prod_{i\mu} \Theta\left(\zeta_i^\mu N^{-1/2} \sum_j w_{ij} \xi_j^\mu - \kappa \right) \right) \prod_i \delta\left(\sum_j w_{ij}^2 - N \right)}{\int d\mathbf{w} \prod_i \delta\left(\sum_j w_{ij}^2 - N \right)}. \tag{10.83}$$

Here we enforce the constraint (10.82) with the delta functions, and restrict the numerator to regions satisfying (10.81) with the step functions $\Theta(x)$.

The expression (10.83) is rather like a statistical-mechanical partition function (10.1), but the conventional exponential weight is replaced by an all-or-nothing one given by the step functions. It is also important to recognize that here it is the *weights* w_{ij} that are the fundamental statistical-mechanical variables, not the activations of the units.

We observe immediately that (10.83) factors into a product of identical terms, one for each i. Therefore we can drop the index i altogether, reducing without loss of generality the calculation to the case of a single output unit. The corresponding step also works for the recurrent network if w_{ij} and w_{ji} are independent, but the calculation cannot be done this way if a symmetry constraint $w_{ij} = w_{ji}$ is imposed.

In the same way as for Z in the previous section, the statistically relevant quantity is the average over the pattern distribution, not of V itself, but *of its logarithm*. Therefore, we introduce replicas again and compute the average

$$\langle\!\langle V^n \rangle\!\rangle = \frac{\left\langle\!\!\!\left\langle \prod_{\alpha=1}^m \int d\mathbf{w}^\alpha \left(\prod_\mu \Theta\left(\zeta^\mu N^{-1/2} \sum_j w_j^\alpha \xi_j^\mu - \kappa \right) \right) \delta\left(\sum_j (w_j^\alpha)^2 - N \right) \right\rangle\!\!\!\right\rangle}{\prod_{\alpha=1}^m \int d\mathbf{w}^\alpha \, \delta\left(\sum_j (w_j^\alpha)^2 - N \right)}$$

$$\tag{10.84}$$

where the integrals are over all the w_j^α's and the average $\langle\langle \cdots \rangle\rangle$ is over the ξ_j^μ's and the ζ^μ's.

To proceed we use the same kinds of tricks as in the previous section. First we work on the step functions, using the integral representation

$$\Theta(z - \kappa) = \int_\kappa^\infty d\lambda \, \delta(\lambda - z) = \int_\kappa^\infty d\lambda \int \frac{dx}{2\pi} e^{ix(\lambda - z)} . \qquad (10.85)$$

We have step functions for each α and μ, so at this point we need auxiliary variables λ_α^μ and x_α^μ. Thus a particular step function becomes

$$\Theta\left(\zeta^\mu N^{-1/2} \sum_j w_j^\alpha \xi_j^\mu - \kappa\right) = \int_\kappa^\infty \frac{d\lambda_\alpha^\mu}{2\pi} \int dx_\alpha^\mu \, e^{ix_\alpha^\mu \lambda_\alpha^\mu} \, e^{-ix_\alpha^\mu z_\alpha^\mu} \qquad (10.86)$$

where

$$z_\alpha^\mu = \zeta^\mu N^{-1/2} \sum_j w_j^\alpha \xi_j^\mu . \qquad (10.87)$$

It is now easy to average over the patterns, which occur only in the last factor of (10.86). We consider the case of independent binary patterns, for which we have

$$
\begin{aligned}
\left\langle\!\!\!\left\langle \prod_{\mu\alpha} e^{-ix_\alpha^\mu z_\alpha^\mu} \right\rangle\!\!\!\right\rangle &= \prod_{j\mu} \left\langle\!\!\!\left\langle \exp\left(-i\zeta^\mu \xi_j^\mu N^{-1/2} \sum_\alpha x_\alpha^\mu w_j^\alpha\right) \right\rangle\!\!\!\right\rangle \\
&= \exp\left(\sum_{j\mu} \log \cos\left[N^{-1/2} \sum_\alpha x_\alpha^\mu w_j^\alpha\right]\right) \\
&\xrightarrow{N\to\infty} \exp\left(-\frac{1}{2N} \sum_{\mu\alpha\beta} x_\alpha^\mu x_\beta^\mu \sum_j w_j^\alpha w_j^\beta\right) .
\end{aligned}
\qquad (10.88)
$$

The resulting $\sum_j w_j^\alpha w_j^\beta$ term is not easy to deal with directly, so we replace it by a new variable $q_{\alpha\beta}$ defined by

$$q_{\alpha\beta} = \frac{1}{N} \sum_j w_j^\alpha w_j^\beta . \qquad (10.89)$$

This gives $q_{\alpha\alpha} = 1$ from (10.82), but we prefer to treat the $\alpha = \beta$ terms explicitly and use $q_{\alpha\beta}$ only for $\alpha \neq \beta$. Thus we rewrite (10.88) as

$$\left\langle\!\!\!\left\langle \prod_{\mu\alpha} e^{-ix_\alpha^\mu z_\alpha^\mu} \right\rangle\!\!\!\right\rangle = \prod_\mu \exp\left(-\tfrac{1}{2} \sum_\alpha (x_\alpha^\mu)^2 - \sum_{\alpha<\beta} q_{\alpha\beta} x_\alpha^\mu x_\beta^\mu\right) \qquad (10.90)$$

using $q_{\alpha\beta} = q_{\beta\alpha}$. The $q_{\alpha\beta}$'s play the same role in this problem that $q_{\alpha\beta}$ and $r_{\alpha\beta}$ did in the previous section.

When we insert (10.90) into (10.86) we see that we get an identical result for each μ, so we can drop all the μ's and write

$$\left\langle\!\!\!\left\langle \prod_{\mu\alpha} \Theta(z_\alpha^\mu - \kappa) \right\rangle\!\!\!\right\rangle = \left[\int_\kappa^\infty \left(\prod_\alpha \frac{d\lambda_\alpha}{2\pi}\right) \int \left(\prod_\alpha dx_\alpha\right) e^{K\{\lambda, x, q\}}\right]^p \qquad (10.91)$$

where

$$K\{\lambda, x, q\} = i \sum_\alpha x_\alpha \lambda_\alpha - \tfrac{1}{2} \sum_\alpha x_\alpha^2 - \sum_{\alpha < \beta} q_{\alpha\beta} x_\alpha x_\beta . \tag{10.92}$$

Now we turn to the delta functions. Using the basic integral representation

$$\delta(z) = \int \frac{dr}{2\pi i} \, e^{-rz} \tag{10.93}$$

we choose $r = E_\alpha/2$ for each α to write the delta functions in (10.84) as

$$\delta\left(\sum_j (w_j^\alpha)^2 - N\right) = \int \frac{dE_\alpha}{4\pi i} \, e^{NE_\alpha/2 - E_\alpha (w_j^\alpha)^2/2} . \tag{10.94}$$

In the same way we enforce the condition (10.89) for each pair $\alpha\beta$ (with $\alpha > \beta$) using $r = N F_{\alpha\beta}$:

$$\delta\left(q_{\alpha\beta} - N^{-1} \sum_j w_j^\alpha w_j^\beta\right) = N \int \frac{dF_{\alpha\beta}}{2\pi i} \, e^{-N F_{\alpha\beta} q_{\alpha\beta} + F_{\alpha\beta} \sum_j w_j^\alpha w_j^\beta} . \tag{10.95}$$

We also have to add an integral over each of the $q_{\alpha\beta}$'s, so that the delta function can pick out the desired value.

A factorization of the integrals over the w's is now possible. Taking everything not involving w_j^α outside, the numerator of (10.84) includes a factor

$$\int \left(\prod_\alpha dw_j^\alpha\right) e^{-\sum_\alpha E_\alpha (w_j^\alpha)^2/2 + \sum_{\alpha < \beta} F_{\alpha\beta} w_j^\alpha w_j^\beta} \tag{10.96}$$

for each j. These factors are all identical—w_j^α is a dummy variable, and j no longer appears elsewhere—so we can drop the j's and rewrite (10.96) as

$$\left[\int \left(\prod_\alpha dw_\alpha\right) e^{-\sum_\alpha E_\alpha w_\alpha^2/2 + \sum_{\alpha < \beta} F_{\alpha\beta} w_\alpha w_\beta}\right]^N . \tag{10.97}$$

The same transformation applies to the denominator of (10.84), except that there are no $F_{\alpha\beta}$ terms.

It is now time to collect together our factors from (10.92), (10.94), (10.95), and (10.97). Writing A^k as $\exp(k \log A)$, and omitting prefactors, (10.84) becomes

$$\langle\!\langle V^n \rangle\!\rangle = \frac{\int (\prod_\alpha dE_\alpha)(\prod_{\alpha < \beta} dq_{\alpha\beta} dF_{\alpha\beta}) \, e^{NG\{q, F, E\}}}{\int (\prod_\alpha dE_\alpha) e^{NH\{E\}}} \tag{10.98}$$

where

$$\begin{aligned}
G\{q, F, E\} &= \frac{p}{N} \log\left[\int_\kappa^\infty \left(\prod_\alpha \frac{d\lambda_\alpha}{2\pi}\right) \int \left(\prod_\alpha dx_\alpha\right) e^{K\{\lambda, x, q\}}\right] \\
&\quad + \log\left[\int \left(\prod_\alpha dw_\alpha\right) e^{-\sum_\alpha E_\alpha w_\alpha^2/2 + \sum_{\alpha < \beta} F_{\alpha\beta} w_\alpha w_\beta}\right] \\
&\quad - \sum_{\alpha < \beta} F_{\alpha\beta} q_{\alpha\beta} + \tfrac{1}{2} \sum_\alpha E_\alpha
\end{aligned} \tag{10.99}$$

and

$$H\{E\} = \log\left[\int\left(\prod_\alpha dw_\alpha\right)e^{-\sum_\alpha E_\alpha w_\alpha^2/2}\right] + \frac{1}{2}\sum_\alpha E_\alpha. \tag{10.100}$$

Since the exponents inside the integrals in (10.98) are proportional to N, we will be able to evaluate them exactly using the saddle-point method in the large-N limit.

As before, we make a replica-symmetric *ansatz*:

$$q_{\alpha\beta} = q \qquad F_{\alpha\beta} = F \qquad E_\alpha = E \tag{10.101}$$

(where the first two apply for $\alpha \neq \beta$ only). This allows us to evaluate each term of G.

For the first term we can rewrite K from (10.92) as

$$K\{\lambda, x, q\} = i\sum_\alpha x_\alpha \lambda_\alpha - \frac{1-q}{2}\sum_\alpha x_\alpha^2 - \frac{q}{2}\left(\sum_\alpha x_\alpha\right)^2 \tag{10.102}$$

and linearize the last term with the usual Gaussian integral trick

$$e^{-\frac{1}{2}q\left(\sum_\alpha x_\alpha\right)^2} = \int \frac{dt}{\sqrt{2\pi}} e^{-t^2/2 + it\sqrt{q}\sum_\alpha x_\alpha} \tag{10.103}$$

derived from (10.5). Then the x_α integrals can be done, leaving a product of identical integrals over the λ_α's. Upon replacing these by a single integral to the nth power we obtain for the whole first line of (10.99):

$$\alpha \log\left\{\int \frac{dt}{\sqrt{2\pi}} e^{-t^2/2}\left[\int_\kappa^\infty \frac{d\lambda}{\sqrt{2\pi(1-q)}} \exp\left(-\frac{(\lambda + t\sqrt{q})^2}{2(1-q)}\right)\right]^n\right\}$$

$$\xrightarrow{n\to 0} n\alpha \int \frac{dt}{\sqrt{2\pi}} e^{-t^2/2} \log\left[\int_\kappa^\infty \frac{d\lambda}{\sqrt{2\pi(1-q)}} \exp\left(-\frac{(\lambda + t\sqrt{q})^2}{2(1-q)}\right)\right] \tag{10.104}$$

where $\alpha \equiv p/N$.

The second term in G can be evaluated in the same way, linearizing the $(\sum_\alpha w_\alpha)^2$ term with a Gaussian integral trick, then performing in turn the w_α integrals and the Gaussian integral. The final result in the small n limit is

$$\frac{1}{2}n\left(\log(2\pi) - \log(E + F) + \frac{F}{E + F}\right). \tag{10.105}$$

Finally, the third term of G gives simply (again for small n)

$$\frac{1}{2}n(E + qF). \tag{10.106}$$

Now we are in a position to find the saddle point of G with respect to q, F, and E. The most important order parameter is q. Its value at the saddle point is the most probable value of the overlap (10.89) between a pair of solutions. If, as at small α, there is a large region of $\mathbf{w}$-space that solves (10.80), then different solutions can be quite uncorrelated and q will be small. As we increase α, it becomes harder and harder to find solutions, and the typical overlap between a pair of them increases. Finally, when there is just a single solution, q becomes equal to 1. This point defines the **optimal perceptron**: the one with the largest capacity for a given stability parameter κ, or equivalently the one with highest stability for a given α. We focus on this case henceforth, taking $q \to 1$ shortly.

The saddle-point equations $\partial G/\partial E = 0$ and $\partial G/\partial F = 0$ can readily be solved to express E and F in terms of q:

$$F = \frac{q}{(1-q)^2}$$

$$E = \frac{1-2q}{(1-q)^2}.$$
(10.107)

Substituting these into the expression for G (and making a change of variable in the $d\lambda$ integral), we get

$$\frac{1}{n}G(q) = \alpha \int \frac{dt}{\sqrt{2\pi}} e^{-t^2/2} \log\left[\int_{\frac{t\sqrt{q}+\kappa}{\sqrt{1-q}}}^{\infty} \frac{dz}{\sqrt{2\pi}} e^{-z^2/2}\right]$$

$$+ \tfrac{1}{2}\log(2\pi) + \tfrac{1}{2}\log(1-q) + \frac{q}{2(1-q)} + \tfrac{1}{2}.$$
(10.108)

Setting $\partial G/\partial q = 0$ to find the saddle point gives

$$\alpha \int \frac{dt}{\sqrt{2\pi}} e^{-t^2/2} \left[\int_u^\infty dz\, e^{-z^2/2}\right]^{-1} e^{-u^2/2} \frac{t + \kappa\sqrt{q}}{2\sqrt{q}(1-q)^{3/2}} = \frac{q}{2(1-q)^2}$$
(10.109)

where $u = (\kappa + t\sqrt{q})/\sqrt{1-q}$. Taking the limit $q \to 1$ is a little tricky, but can be done using L'Hospital's rule, yielding the final result

$$\alpha_c(\kappa) = \left[\int_{-\kappa}^\infty \frac{dt}{\sqrt{2\pi}} e^{-t^2/2}(t+\kappa)^2\right]^{-1}.$$
(10.110)

Equation (10.110) gives the capacity for fixed κ. Alternatively we can use it to find the appropriate κ for the optimal perceptron to store $N\alpha$ patterns. In the limit $\kappa = 0$ it gives

$$\alpha_c(0) = 2$$
(10.111)

in agreement with the result found geometrically by Cover that was outlined in Chapter 5.

One can also perform the corresponding calculation for biased patterns with a distribution

$$P(\xi_i^\mu) = \tfrac{1}{2}(1+m)\delta(\xi_i^\mu - 1) + \tfrac{1}{2}(1-m)\delta(\xi_i^\mu + 1)$$
(10.112)

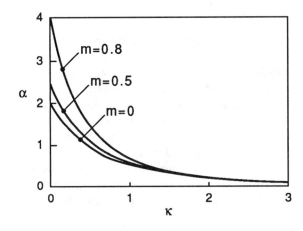

FIGURE 10.1 Capacity α_c as a function of κ for three values of m (from Gardner [1988]).

so that $\langle\langle\xi_i^\mu\rangle\rangle = m$. The calculation is just a little bit more complicated, with an extra set of variables $M^\alpha = N^{-1/2}\sum_j w_j^\alpha$ with respect to which G has to be maximized. The results for the storage capacity as a function of m and κ are shown in Fig. 10.1.

An interesting limit is that of $m \to 1$ (sparse patterns). Then the result for $\kappa = 0$ is

$$\alpha_c = \frac{1}{(1-m)\log(\frac{1}{1-m})} \tag{10.113}$$

which shows that one can store a great many sparse patterns. But there is nothing very surprising about this, because very sparse patterns have very small information content. Indeed, if we work out the total information capacity—the maximum information we can store, in bits—given by

$$I = -\frac{N^2\alpha_c}{\log 2}\left[\tfrac{1}{2}(1-m)\log\left(\frac{1-m}{2}\right) + \tfrac{1}{2}(1+m)\log\left(\frac{1+m}{2}\right)\right], \tag{10.114}$$

then we obtain

$$I = \frac{N^2}{2\log 2} \tag{10.115}$$

in the limit $m \to 1$. This is *less* than the result for the unbiased case ($m = 0$, $\alpha_c = 2$), which is $I = 2N^2$. In fact the total information capacity is always of the order N^2, depending only slightly on m.

It is interesting to note that a capacity of the order of the optimal one (10.113) is obtained for a Hopfield network from a simple Hebb-like rule [Willshaw et al., 1969; Tsodyks and Feigel'man, 1988], as we mentioned in Chapter 2.

A number of extensions of this work have been made, notably to patterns with a finite fraction of errors, binary weights, diluted connections, and (in the recurrent network) connections with differing degrees of correlation between w_{ij} and w_{ji} [Gardner and Derrida, 1988; Gardner et al., 1989].

Generalization Ability

A particularly interesting application is to the calculation of the generalization ability of a simple perceptron. Recall from Section 6.5 that the generalization ability of a network was defined as the probability of its giving the correct output for the mapping it is trained to implement when tested on a random example of the mapping, not restricted to the training set. This can be calculated analytically by Gardner's methods [Györgyi and Tishby, 1990; Györgyi, 1990; Opper et al., 1990].

The basic idea, first used by Gardner and Derrida [1989], is to perform a calculation of the weight-space volume like the one just described, but, instead of considering *random* input-target pairs (ξ_j^μ, ζ^μ), using pairs which are examples of a particular function $f(\xi) = \text{sgn}(\mathbf{v} \cdot \xi)$ that the perceptron could learn. That is, we think of our perceptron as learning to imitate a **teacher perceptron** whose weights are v_i.

Under learning, the pupil perceptron's weight vector $\mathbf{w}$ will come to line up with that of its teacher. Its generalization ability will depend on one parameter, the dot product of the two vectors:

$$R = \frac{1}{N}\mathbf{w} \cdot \mathbf{v}. \tag{10.116}$$

Here both $\mathbf{w}$ and $\mathbf{v}$ are normalized as in (10.82). R is introduced into the calculation in the same way that q was earlier, by inserting a delta function and integrating over it. Ultimately one obtains saddle-point equations for both q and R.

To find the generalization ability from R, consider the two variables

$$x = N^{-1/2} \sum_j w_j \xi_j \quad \text{and} \quad y = N^{-1/2} \sum_j v_j \xi_j \tag{10.117}$$

which are the net inputs to the pupil and the teacher respectively. For large N, x and y are Gaussian variables, each of zero mean and unit variance, with covariance $\langle xy \rangle = R$. Thus their joint distribution is

$$P(x, y) = \frac{1}{2\pi\sqrt{1-R^2}} \exp\left(-\frac{x^2 - 2Rxy + y^2}{2(1-R^2)}\right). \tag{10.118}$$

Having averaged over all inputs, the generalization ability $g(f)$ no longer depends on the specific mapping of the teacher (parametrized by $\mathbf{v}$), but only on the number of examples. We therefore write it as $g(\alpha)$. Clearly $g(\alpha)$ is the probability that x and y have the same sign. Simple geometry then leads to

$$g(\alpha) = 1 - \frac{1}{\pi}\cos^{-1} R \tag{10.119}$$

where R is obtained from the saddle-point condition as described above.

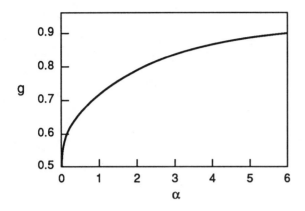

FIGURE 10.2 The generalization ability, $g(\alpha)$, as a function of relative training-set size, α. Adapted from Opper et al. [1990].

Figure 10.2 shows the resulting $g(\alpha)$. The necessary number of examples for good generalization is clearly of order N, in agreement with the estimate (6.81). In the limit of many training examples perfect generalization is approached:

$$1 - g(\alpha) = \frac{1}{\alpha}. \tag{10.120}$$

This form of approach means that the *a priori* generalization ability distribution $\rho_0(g)$ discussed in Section 6.5 has *no gap* around $g = 1$.

This example shows how one can actually do an explicit calculation (for the simple perceptron) which fits into the theoretical framework for generalization introduced in Section 6.5. We hope it will guide us in future calculations for less trivial architectures.

All the preceding has been algorithm-independent—it is about the existence of connection weights that implement the desired association, not about how they are found. It is also possible to apply statistical mechanics methods to particular algorithms [Kinzel and Opper, 1990; Hertz et al., 1989; Hertz, 1990], including their dynamics, but these calculations lie outside the scope of the framework we have presented here.

APPENDIX

Statistical Mechanics

In this appendix we describe some of the basic results from statistical mechanics. We consider only discrete systems, but most of the results are easily generalized to continuous systems by replacing the sums by integrals.

A.1 The Boltzmann-Gibbs Distribution

The starting point of statistical mechanics is an energy function. We consider a physical system with a set of states α, each of which has an energy H_α. Note that a particular α specifies the state or configuration of the entire system, not just of one unit or one spin, and H_α is the *total* energy of the system in that state.

If the system is at an absolute temperature $T > 0$, its state α will vary with time, and quantities such as H_α that depend on the state will themselves fluctuate. Although there must be some driving mechanism for these fluctuations, part of the idea of temperature involves ignoring this and treating them as random; thermal energy is by definition disordered. When a system is first prepared, or after a change of parameters, the fluctuations may have on average a definite direction, such as that of decreasing energy H_α. But after a while any such trend ceases and the system just fluctuates around constant average values. Then we say that it is in **thermal equilibrium**.

A fundamental result from physics tells us that in thermal equilibrium each of the possible states α occurs with probability

$$P_\alpha = \frac{1}{Z} e^{-H_\alpha/k_B T} \tag{A.1}$$

where the normalizing factor

$$Z = \sum_\alpha e^{-H_\alpha/k_B T} \tag{A.2}$$

is called the **partition function**. k_B is **Boltzmann's constant**, with value 1.38×10^{-16} erg/K.

Equation (A.1) is called the Boltzmann-Gibbs distribution. We will not attempt to justify it here. It is usually derived from very general assumptions about microscopic dynamics, but can also be interpreted from the viewpoint of information theory. It usually works well, but can fail when a system does not explore all of its possible states.

In applications to neural networks, the "temperature" of a stochastic network is of course *not* related to the physical temperature, but is simply a parameter controlling the update rule. So its scale is irrelevant, and we can choose to measure it in units such that $k_B = 1$. We have implicitly done this in all the statistical mechanics calculations in this book, simply omitting the k_B factor, as we do in the rest of this Appendix.

If we know the energy function H_α for a system we can in principle use (A.1) to calculate the probability of finding it in each of its states α. Then we can compute the average value $\langle A \rangle$ of any quantity A that has a particular value A_α in each state, using

$$\langle A \rangle = \sum_\alpha A_\alpha P_\alpha . \tag{A.3}$$

This is called a **thermal average**, and for a large system usually corresponds closely to the measured value of A. This is how we can compute useful quantities for prediction or comparison with experiment.

In practice the hardest part is calculating the partition function Z. On the other hand, once we have done so, we can usually compute the desired averages $\langle A \rangle$ from Z itself, and do not have to evaluate summations like (A.3) explicitly.

Let us see how this works for the example of a Hopfield network, with the energy function

$$H\{S\} = -\frac{1}{2} \sum_{ij} w_{ij} S_i S_j - \sum_i b_i S_i . \tag{A.4}$$

Here we have added explicit bias terms b_i for the purposes of the present calculation; we can set $b_i = 0$ later if we so desire. Note that H is a function of all the activations S_i; the set of all S_i's plays the role of the generic state label α used above.

The partition function is a sum over all the possible states, all the 2^N combinations of $S_i = \pm 1$:

$$Z = \sum_{S_1 = \pm 1} \cdots \sum_{S_N = \pm 1} \exp\left(\frac{\beta}{2} \sum_{ij} w_{ij} S_i S_j + \beta \sum_i b_i S_i\right) \tag{A.5}$$

where $\beta = 1/T$. The sum over all states is called a **trace** and instead of writing out all the summations it is common to write (A.5) as

$$Z = \mathrm{Tr}_S \exp\left(\frac{\beta}{2}\sum_{ij} w_{ij} S_i S_j + \beta \sum_i b_i S_i\right). \tag{A.6}$$

The average activation $\langle S_i \rangle$ of unit i is given by

$$\langle S_i \rangle = \frac{1}{Z}\mathrm{Tr}_S\left[S_i \exp\left(\frac{\beta}{2}\sum_{ij} w_{ij} S_i S_j + \beta \sum_i b_i S_i\right)\right] \tag{A.7}$$

and it looks at first as though we have to do *two* traces to evaluate the average, one in (A.6) and one in (A.7). But if we know the Z as a function of the b_i, then $\langle S_i \rangle$ can be obtained by differentiation:

$$\langle S_i \rangle = \frac{1}{\beta Z}\frac{\partial Z}{\partial b_i} = T\frac{\partial}{\partial b_i}\log Z. \tag{A.8}$$

Of course we still haven't evaluated Z (see Section 10.1), but at least we have reduced two problems to one.

A.2 Free Energy and Entropy

Many averages can be calculated in this same sort of way, and that is one reason why it is a good idea to define the **free energy**

$$F = -T\log Z. \tag{A.9}$$

In terms of the free energy, (A.8) becomes simply

$$\langle S_i \rangle = -\frac{\partial F}{\partial b_i}. \tag{A.10}$$

Similarly, the correlation function is

$$\langle S_i S_j \rangle = -\frac{\partial F}{\partial w_{ij}} \tag{A.11}$$

as we showed on page 171 and used in deriving the Boltzmann learning algorithm. Almost every quantity of interest can be written as a suitable derivative of F, so computing F or $\log Z$ is a central goal of the statistical mechanics approach to a problem.

It is interesting to calculate the difference between the average energy $\langle H \rangle$ and the free energy, using $\sum_\alpha P_\alpha = 1$:

$$
\begin{aligned}
\langle H \rangle - F &= \sum_\alpha P_\alpha H_\alpha - T \log Z \\
&= T \sum_\alpha P_\alpha (\beta H_\alpha - \log Z) \\
&= -T \sum_\alpha P_\alpha \log \frac{e^{-\beta H_\alpha}}{Z} \\
&= -T \sum_\alpha P_\alpha \log P_\alpha .
\end{aligned}
\tag{A.12}
$$

Apart from a factor of T, the result is called the **entropy**:

$$
S = - \sum_\alpha P_\alpha \log P_\alpha
\tag{A.13}
$$

and the free energy can be written as

$$
F = \langle H \rangle - TS .
\tag{A.14}
$$

The entropy S has several interpretations. One is as the width of the probability distribution P_α; the more states α that have appreciable probability, the larger S. An easy way to see this is to consider a simple case where K states have the same probability $1/K$ and the rest have zero probability. Then

$$
S = - \sum_\alpha P_\alpha \log P_\alpha = \log K .
\tag{A.15}
$$

It is sometimes useful to invert this and use $\exp(S)$ as a measure of the number of appreciably weighted states, even when the probabilities are not all equal. This number can also be thought of as the volume of **state space**—the space of all possible α's—in which the system resides.

Another interpretation of entropy comes from information theory. If we have a set of possible states α with probabilities P_α, then

$$
I = - \sum_\alpha P_\alpha \log_2 P_\alpha
\tag{A.16}
$$

can be shown to be the average amount of additional information required to specify *one* of the states. The larger I, the more uncertain the actual state α. The thermodynamic entropy S and the information entropy I are the same except for a factor of $\log 2$; the information theoretic version is normally written with a base 2 logarithm so as to give a result in **bits**. For the case of K equally likely states we find $I = \log_2 K$, which is indeed the number of bits needed to specify one choice out of K alternatives.

The free energy F can be interpreted further by writing (A.2) and (A.9) as

$$e^{-F/T} = Z = \sum_\alpha e^{-H_\alpha/T} . \tag{A.17}$$

This shows that F is something like an exponentially weighted sum of energies. More importantly, dividing out a factor of Z we obtain

$$\frac{e^{-F/T}}{Z} = \sum_\alpha \frac{e^{-H_\alpha/T}}{Z} = \sum_\alpha P_\alpha . \tag{A.18}$$

Thus $e^{-F/T}/Z$ (which is just 1) can be thought of as the sum of the probabilities of the individual states. The significance of this is that it still applies if the sum over *all* α's is replaced by a restricted sum over a subset of them; then $e^{-F'/T}/Z$ gives the probability of finding the system in that subset, where F' is the restricted free energy. This is the approach used in Section 7.1.

A.3 Stochastic Dynamics

In some cases we have detailed knowledge of the stochastic sequence of states α_1, α_2, α_3, etc. For example in the Hopfield network we used the update rule (2.48)

$$\text{Prob}(S_i = \pm 1) = f_\beta(\pm h_i) = \frac{1}{1 + \exp(\mp 2\beta h_i)} \tag{A.19}$$

where h_i was the net input $\sum_j w_{ij} S_j + b_i$ to unit i. This can be rewritten as a **transition probability** to flip unit i from S_i to $-S_i$

$$W(S_i \to -S_i) = \frac{1}{1 + \exp(\beta \Delta H_i)} \tag{A.20}$$

where

$$\Delta H_i = H(S_1 \ldots - S_i \ldots S_N) - H(S_1 \ldots S_i \ldots S_N) = 2h_i S_i \tag{A.21}$$

is the energy change resulting from such a flip.

The transition probabilities (A.20) give us a complete description of the stochastic sequence of states in the Hopfield model. In the general case we might know the transition probabilities $W(\alpha \to \alpha')$ for all pairs of states α and α'. Typically many of these would be zero; in the Hopfield model we choose zero transition probability for any change involving two or more units at a time.

A set of transition probabilities $W(\alpha \rightarrow \alpha')$ describes a very general kind of model which is not necessarily describable by equilibrium statistical mechanics; an arbitrary set of $W(\alpha \rightarrow \alpha')$'s can produce a limit cycle or chaotic behavior instead of leading to thermal equilibrium. Usually however, we are chiefly interested in equilibrium, in part because of the availability of powerful methods of analysis. It is therefore important to seek a condition on $W(\alpha \rightarrow \alpha')$ to guarantee equilibrium.

In equilibrium the probability P_α of finding the system in a given state α is independent of time and is given by the Boltzmann-Gibbs distribution (A.1). A sufficient condition for maintaining this state of affairs is that the average numbers of transitions from α to α' and from α' to α be equal:

$$P_\alpha W(\alpha \rightarrow \alpha') = P_{\alpha'} W(\alpha' \rightarrow \alpha) \tag{A.22}$$

or

$$\frac{W(\alpha \rightarrow \alpha')}{W(\alpha' \rightarrow \alpha)} = \frac{P_{\alpha'}}{P_\alpha} = e^{-\beta(H_{\alpha'} - H_\alpha)} = e^{-\beta \Delta H} \tag{A.23}$$

where $\Delta H = H_{\alpha'} - H_\alpha$. It is easy to check that the rule (A.20) does indeed satisfy this condition, so we can expect to find thermal equilibrium in the Hopfield model. Indeed it is the condition (A.23) that motivates the choice (A.20) for the update rule.

Other evolution rules satisfying (A.23) are also possible, and also lead to the Boltzmann-Gibbs distribution in equilibrium. In simulating physical systems a very common choice is the Metropolis algorithm [Metropolis et al., 1953] where

$$W(\alpha \rightarrow \alpha') = \begin{cases} 1 & \text{if } \Delta H < 0; \\ e^{-\beta \Delta H} & \text{otherwise.} \end{cases} \tag{A.24}$$

This has the advantage that it makes more flips and therefore reaches equilibrium more rapidly.

Note that we haven't actually shown that the system relaxes to thermal equilibrium from an arbitrary starting state, but this can be demonstrated with more powerful methods.

Bibliography

Abu-Mostafa, Y.S. (1989). The Vapnik-Chervonenkis Dimension: Information versus Complexity in Learning. *Neural Computation* **1**, 312–317.

Abu-Mostafa, Y.S. and D. Psaltis (1987). Optical Neural Computers. *Scientific American* **256**, March 1987, 88–95.

Ackley, D.H., G.E. Hinton, and T.J. Sejnowski (1985). A Learning Algorithm for Boltzmann Machines. *Cognitive Science* **9**, 147–169. Reprinted in Anderson and Rosenberg [1988].

Ackley, D.H. and M.S. Littman (1990). Generalization and Scaling in Reinforcement Learning. In *Advances in Neural Information Processing Systems II* (Denver 1989), ed. D.S. Touretzky, 550–557. San Mateo: Morgan Kaufmann.

Ahalt, S.C., A.K. Krishnamurthy, P. Chen, and D.E. Melton (1990). Competitive Learning Algorithms for Vector Quantization. *Neural Networks* **3**, 277–290.

Almeida, L.B. (1987). A Learning Rule for Asynchronous Perceptrons with Feedback in a Combinatorial Environment. In *IEEE First International Conference on Neural Networks* (San Diego 1987), eds. M. Caudill and C. Butler, vol. II, 609–618. New York: IEEE.

Almeida, L.B. (1988). Backpropagation in Perceptrons with Feedback. In *Neural Computers* (Neuss 1987), eds. R. Eckmiller and Ch. von der Malsburg, 199–208. Berlin: Springer-Verlag.

Alspector, J. and R.B. Allen (1987). A Neuromorphic VLSI Learning System. In *Advanced Research in VLSI: Proceedings of the 1987 Stanford Conference*, ed. P. Losleben, 313–349. Cambridge: MIT Press.

Alspector, J., R.B. Allen, V. Hu, and S. Satyanarayana (1988). Stochastic Learning Networks and Their Electronic Implementation. In *Neural Information Processing Systems* (Denver 1987), ed. D.Z. Anderson, 9–21. New York: American Institute of Physics.

Amari, S.A. (1977). Dynamics of Pattern Formation in Lateral-Inhibition Type Neural Fields. *Biological Cybernetics* **27**, 77—87.

Amari, S.A. (1980). Topographic Organization of Nerve Fields. *Bulletin of Mathematical Biology* **42**, 339–364.

Amit, D. (1988). Neural Networks for Counting Chimes. *Proceedings of the National Academy of Sciences, USA* **85**, 2141–2145.

Amit, D. (1989). *Modelling Brain Function.* Cambridge: Cambridge University Press.

Amit, D., H. Gutfreund, and H. Sompolinsky (1985a). Spin-Glass Models of Neural Networks. *Physical Review A* **32**, 1007–1018.

Amit, D., H. Gutfreund, and H. Sompolinsky (1985b). Storing Infinite Numbers of Patterns in a Spin-Glass Model of Neural Networks. *Physical Review Letters* **55**, 1530–1533.

Amit, D., H. Gutfreund, and H. Sompolinsky (1987a). Statistical Mechanics of Neural Networks Near Saturation. *Annals of Physics* **173**, 30–67.

Amit, D., H. Gutfreund, and H. Sompolinsky (1987b). Information Storage in Neural Networks with Low Levels of Activity. *Physical Review A* **35**, 2293–2303.

Anderson, C.H. and D.C. Van Essen (1987). Shifter Circuits: A Computational Strategy for Dynamic Aspects of Visual Processing. *Proceedings of the National Academy of Sciences, USA* **84**, 6297–6301.

Anderson, D.Z. (1986). Coherent Optical Eigenstate Memory. *Optics Letters* **11**, 56–58.

Anderson, J.A. (1968). A Memory Model Using Spatial Correlation Functions. *Kybernetik* **5**, 113–119.

Anderson, J.A. (1970). Two Models for Memory Organization. *Mathematical Biosciences* **8**, 137–160.

Anderson, J.A. and M.C. Mozer (1981). Categorization and Selective Neurons. In *Parallel Models of Associative Memory*, eds. G.E. Hinton and J.A. Anderson, 213–236. Hillsdale: Erlbaum.

Anderson, J.A. and E. Rosenfeld, eds. (1988). *Neurocomputing: Foundations of Research.* Cambridge: MIT Press.

Anderson, S., J.W.L. Merrill, and R. Port (1989). Dynamic Speech Categorization with Recurrent Networks. In *Proceedings of the 1988 Connectionist Models Summer School* (Pittsburg 1988), eds. D. Touretzky, G. Hinton, and T. Sejnowski, 398–406. San Mateo: Morgan Kaufmann.

Angéniol, B., G. de La Croix Vaubois, and J.-Y. Le Texier (1988). Self-Organizing Feature Maps and the Travelling Salesman Problem. *Neural Networks* **1**, 289–293.

Arbib, M.A. (1987). *Brains, Machines, and Mathematics.* Berlin: Springer-Verlag.

Bachmann, C.M., L.N. Cooper, A. Dembo, and O. Zeitouni (1987). A Relaxation Model for Memory with High Storage Density. *Proceedings of the National Academy of Sciences, USA* **84**, 7529–7531.

Bachrach, J. (1988). Learning to Represent State. Master's Thesis, University of Massachusetts, Amherst.

Baldi, P. and K. Hornik (1989). Neural Networks and Principal Component Analysis: Learning from Examples Without Local Minima. *Neural Networks* **2**, 53–58.

Barhen, J., S. Gulati, and M. Zak (1989). Neural Learning of Constrained Nonlinear Transformations. *Computer*, June 1989, 67–76.

Barlow, H.B. (1989). Unsupervised Learning. *Neural Computation* **1**, 295–311.

Barto, A.G. and P. Anandan (1985). Pattern Recognizing Stochastic Learning Automata. *IEEE Transactions on Systems, Man, and Cybernetics* **15**, 360–375.

Barto, A.G. and M.I. Jordan (1987). Gradient Following Without Back-Propagation in Layered Networks. In *IEEE First International Conference on Neural Networks* (San Diego 1987), eds. M. Caudill and C. Butler, vol. II, 629–636. New York: IEEE.

Barto, A.G., R.S. Sutton, and C.W. Anderson (1983). Neuronlike Adaptive Elements That Can Solve Difficult Learning Control Problems. *IEEE Transactions on Systems, Man, and Cybernetics* **13**, 834–846. Reprinted in Anderson and Rosenberg [1988].

Barto, A.G., R.S. Sutton, and C.J.C.H. Watkins (1991). Learning and Sequential Decision Making. In *Learning and Computational Neuroscience*, eds. M. Gabriel and J.W. Moore. Cambridge: MIT Press.

Baum, E.B. (1986). Towards Practical "Neural" Computation for Combinatorial Optimization Problems. In *Neural Networks for Computing* (Snowbird 1986), ed. J.S. Denker, 53–58. New York: American Institute of Physics.

Baum, E.B. and D. Haussler (1989). What Size Net Gives Valid Generalization? *Neural Computation* **1**, 151–160.

Baum, E.B. and F. Wilczek (1988). Supervised Learning of Probability Distributions by Neural Networks. In *Neural Information Processing Systems* (Denver 1987), ed. D.Z. Anderson, 52–61. New York: American Institute of Physics.

Becker, S. and Y. Le Cun (1989). Improving the Convergence of Back-Propagation Learning with Second Order Methods. In *Proceedings of the 1988 Connectionist Models Summer School* (Pittsburg 1988), eds. D. Touretzky, G. Hinton, and T. Sejnowski, 29–37. San Mateo: Morgan Kaufmann.

Bengio, Y., R. Cardin, and R. De Mori (1990). Speaker Independent Speech Recognition with Neural Networks and Speech Knowledge. In *Advances in Neural Information Processing Systems II* (Denver 1989), ed. D.S. Touretzky, 218–225. San Mateo: Morgan Kaufmann.

Beurle, R.L. (1956). Properties of a Mass of Cells Capable of Regenerating Pulses. *Philosophical Transactions of the Royal Society of London B* **240**, 55–94.

Bienenstock, E.L., L.N. Cooper, and P.W. Munro (1982). Theory for the Development of Neuron Selectivity: Orientation Specificity and Binocular Interaction in Visual Cortex. *Journal of Neuroscience* **2**, 32–48. Reprinted in Anderson and Rosenberg [1988].

Bilbro, G., R. Mann, T.K. Miller, W.E. Snyder, D.E. Van den Bout, and M. White (1989). Optimization by Mean Field Annealing. In *Advances in Neural Information Processing Systems I* (Denver 1988), ed. D.S. Touretzky, 91–98. San Mateo: Morgan Kaufmann.

Bilbro, G.L. and W. Snyder (1989). Range Image Restoration Using Mean Field Annealing. In *Advances in Neural Information Processing Systems I* (Denver 1988), ed. D.S. Touretzky, 594–601. San Mateo: Morgan Kaufmann.

Binder, K. and D.W. Heerman (1988). *Monte Carlo Simulation in Statistical Mechanics.* Berlin: Springer-Verlag.

Binder, K. and A.P. Young (1986). Spin Glasses: Experimental Facts, Theoretical Concepts, and Open Questions. *Reviews of Modern Physics* **58**, 801–976.

Blakemore, C. and G.F. Cooper (1970). Development of the Brain Depends on the Visual Environment. *Nature* **228**, 477–478.

Block, H.D. (1962). The Perceptron: A Model for Brain Functioning. *Reviews of Modern Physics* **34**, 123–135. Reprinted in Anderson and Rosenberg [1988].

Blumer, A., A. Ehrenfeucht, D. Haussler, and M. Warmuth (1986). Classifying Learnable Geometric Concepts with the Vapnik-Chervonenkis Dimension. In *Proceedings of the Eighteenth Annual ACM Symposium on Theory of Computing* (Berkeley 1986), 273–282. Salem: ACM.

Bohr, H., J. Bohr, S. Brunak, R.M.J. Cotterill, B. Lautrup, L. Nørskov, O.H. Olsen, and S.B. Petersen (1988). Protein Secondary Structure and Homology by Neural Networks: The α-Helices in Rhodopsin. *FEBS Letters* **241**, 223–228.

Bourland, H. and Y. Kamp (1988). Auto-Association by Multilayer Perceptrons and Singular Value Decomposition. *Biological Cybernetics* **59**, 291–294.

Brandt, R.D., Y. Wang, A.J. Laub, and S.K. Mitra (1988). Alternative Networks for Solving the Travelling Salesman Problem and the List-Matching Problem. In *IEEE International Conference on Neural Networks* (San Diego 1988), vol. II, 333–340. New York: IEEE.

Brunak, S. and B. Lautrup (1989). Liniedeling med et Neuralt Netværk. *Skrifter for Anvendt Matematik og Lingvistik* **14**, 55–74.

Brunak, S. and B. Lautrup (1990). *Neural Networks: Computers with Intuition.* Singapore: World Scientific.

Bryson, A.E. and Y.-C. Ho (1969). *Applied Optimal Control.* New York: Blaisdell.

Buhmann, J. and K. Schulten (1987). Noise-Driven Temporal Association in Neural Networks. *Europhysics Letters* **4**, 1205–1209.

Buhmann, J. and K. Schulten (1988). Storing Sequences of Biased Patterns in Neural Networks with Stochastic Dynamics. In *Neural Computers* (Neuss 1987), eds. R. Eckmiller and Ch. von der Malsburg, 231–242. Berlin: Springer-Verlag.

Burr, D.J. (1988). An Improved Elastic Net Method for the Travelling Salesman Problem. In *IEEE International Conference on Neural Networks* (San Diego 1988), vol. I, 69–76. New York: IEEE.

Caianiello, E.R. (1961). Outline of a Theory of Thought and Thinking Machines. *Journal of Theoretical Biology* **1**, 204–235.

Canning, A. and E. Gardner (1988). Partially Connected Models of Neural Networks. *Journal of Physics A* **21**, 3275–3284.

Carpenter, G.A. and S. Grossberg (1987a). A Massively Parallel Architecture for a Self-Organizing Neural Pattern Recognition Machine. *Computer Vision, Graphics, and Image Processing* **37**, 54–115.

Carpenter, G.A. and S. Grossberg (1987b). ART2: Self-Organization of Stable Category Recognition Codes for Analog Input Patterns. *Applied Optics* **26**, 4919–4930.

Carpenter, G.A. and S. Grossberg (1988). The ART of Adaptive Pattern Recognition by a Self-Organizing Neural Network. *Computer*, March 1988, 77–88.

Casdagli, M. (1989). Nonlinear Prediction of Chaotic Time Series. *Physica* **35D**, 335–356.

Cater, J.P. (1987). Successfully Using Peak Learning Rates of 10 (and Greater) in Back-Propagation Networks with the Heuristic Learning Algorithm. In *IEEE First International Conference on Neural Networks* (San Diego 1987), eds. M. Caudill and C. Butler, vol. II, 645–651. New York: IEEE.

Caudill, M. (1989). *Neural Networks Primer*. San Francisco: Miller Freeman.

Chauvin, Y. (1989). A Back-Propagation Algorithm with Optimal Use of Hidden Units. In *Advances in Neural Information Processing Systems I* (Denver 1988), ed. D.S. Touretzky, 519–526. San Mateo: Morgan Kaufmann.

Cleeremans, A., D. Servan-Schreiber, and J.L. McClelland (1989). Finite State Automata and Simple Recurrent Networks. *Neural Computation* **1**, 372–381.

Cohen, M.A. and S. Grossberg (1983). Absolute Stability of Global Pattern Formation and Parallel Memory Storage by Competitive Neural Networks. *IEEE Transactions on Systems, Man, and Cybernetics* **13**, 815–826.

Cohen, M.S. (1986). Design of a New Medium for Volume Holographic Information Processing. *Applied Optics* **25**, 2228–2294.

Coolen, A.C.C. and C.C.A.M. Gielen (1988). Delays in Neural Networks. *Europhysics Letters* **7**, 281–285.

Cooper, L.N. (1973). A Possible Organization of Animal Memory and Learning. In *Collective Properties of Physical Systems* (24th Nobel Symposium), eds. B. Lundqvist and S. Lundqvist, 252–264. New York: Academic Press. Reprinted in Anderson and Rosenberg [1988].

Cortes, C. and J.A. Hertz (1989). A Network System for Image Segmentation. In *International Joint Conference on Neural Networks* (Washington 1989), vol. I, 121–127. New York: IEEE.

Cortes, C., A. Krogh, and J.A. Hertz (1987). Hierarchical Associative Networks. *Journal of Physics A* **20**, 4449–4455.

Cottrell, G.W., P. Munro, and D. Zipser (1987). Learning Internal Representations from Gray-Scale Images: An Example of Extensional Programming. In *Ninth Annual Conference of the Cognitive Science Society* (Seattle 1987), 462–473. Hillsdale: Erlbaum.

Cottrell, M. and J.C. Fort (1986). A Stochastic Model of Retinotopy: A Self Organizing Process. *Biological Cybernetics* **53**, 405–411.

Cover, T.M. (1965). Geometrical and Statistical Properties of Systems of Linear Inequalities with Applications in Pattern Recognition. *IEEE Transactions on Electronic Computers* **14**, 326–334.

Cowan, J.D. and D.H. Sharp (1988a). Neural Nets and Artificial Intelligence. *Dædalus, Proceedings of the American Academy of Arts and Sciences* **117**, 85–121.

Cowan, J.D. and D.H. Sharp (1988b). Neural Nets. *Quarterly Reviews of Biophysics* **21**, 365–427.

Cragg, B.G. and H.N.V. Temperley (1954). The Organization of Neurones: A Cooperative Analogy. *EEG and Clinical Neurophysiology* **6**, 85–92.

Cragg, B.G. and H.N.V. Temperley (1955). Memory: The Analogy with Ferromagnetic Hysteresis. *Brain* **78**, II,304–316.

Crick, F. (1989). The Recent Excitement About Neural Networks. *Nature* **337**, 129–132.

Crisanti, A. and H. Sompolinsky (1987). Dynamics of Spin Systems with Randomly Asymmetric Bonds: Langevin Dynamics and a Spherical Model. *Physical Review A* **36**, 4922–4939.

Crisanti, A., D.J. Amit, and H. Gutfreund (1986). Saturation Level of the Hopfield Model for Neural Network. *Europhysics Letters* **2**, 337–341.

Cybenko, G. (1988). Continuous Valued Neural Networks with Two Hidden Layers Are Sufficient. Technical Report, Department of Computer Science, Tufts University, Medford, MA.

Cybenko, G. (1989). Approximation by Superpositions of a Sigmoidal Function. *Mathematics of Control, Signals, and Systems* **2**, 303–314.

Dehaene, S., J.-P. Changeux, and J.-P. Nadal (1987). Neural Networks That Learn Temporal Sequences by Selection. *Proceedings of the National Academy of Sciences, USA* **84**, 2727–2731.

Denker, J. (1986). Neural Network Refinements and Extensions. In *Neural Networks for Computing* (Snowbird 1986), ed. J.S. Denker, 121–128. New York: American Institute of Physics.

Denker, J., D. Schwartz, B. Wittner, S. Solla, R. Howard, L. Jackel, and J. Hopfield (1987). Large Automatic Learning, Rule Extraction, and Generalization. *Complex Systems* **1**, 877–922.

Derrida, B., E. Gardner, and A. Zippelius (1987). An Exactly Soluble Asymmetric Neural Network Model. *Europhysics Letters* **4**, 167–173.

Derthick, M. (1984). Variations on the Boltzmann Machine. Technical Report CMU–CS–84–120, Department of Computer Science, Carnegie Mellon University, Pittsburgh, PA.

DeSieno, D. (1988). Adding a Conscience to Competitive Learning. In *IEEE International Conference on Neural Networks* (San Diego 1988), vol. I, 117–124. New York: IEEE.

Devijver, P.A. and J. Kittler (1982). *Pattern Recognition: A Statistical Approach*. London: Prentice-Hall.

Diederich, S. and M. Opper (1987). Learning of Correlated Patterns in Spin-Glass Networks by Local Learning Rules. *Physical Review Letters* **58**, 949–952.

Duda, R.O. and P.E. Hart (1973). *Pattern Classification and Scene Analysis*. New York: Wiley.

Durbin, R. and D. Willshaw (1987). An Analogue Approach to the Travelling Salesman Problem Using an Elastic Net Method. *Nature* **326**, 689–691.

Elman, J.L. (1990). Finding Structure in Time. *Cognitive Science* **14**, 179–211.

Elman, J.L. and D. Zipser (1988). Learning the Hidden Structure of Speech. *Journal of the Acoustical Society of America* **83**, 1615–1626.

Fahlman, S.E. (1989). Fast-Learning Variations on Back-Propagation: An Empirical Study. In *Proceedings of the 1988 Connectionist Models Summer School* (Pittsburg 1988), eds. D. Touretzky, G. Hinton, and T. Sejnowski, 38–51. San Mateo: Morgan Kaufmann.

Fahlman, S.E. and C. Lebiere (1990). The Cascade-Correlation Learning Architecture. In *Advances in Neural Information Processing Systems II* (Denver 1989), ed. D.S. Touretzky, 524–532. San Mateo: Morgan Kaufmann.

Farhat, N.H. (1987). Optoelectronic Analogs of Self-Programming Neural Nets: Architectures and Methods for Implementing Fast Stochastic Learning by Simulated Annealing. *Applied Optics* **26**, 5093–5103.

Farhat, N.H., D. Psaltis, A. Prata, and E. Paek (1985). Optical Implementation of the Hopfield Model. *Applied Optics* **24**, 1469–1475. Reprinted in Anderson and Rosenberg [1988].

Farmer, D. and J. Sidorowich (1987). Predicting Chaotic Time Series. *Physical Review Letters* **59**, 845–848.

Farmer, D. and J. Sidorowich (1988). Exploiting Chaos to Predict the Future and Reduce Noise. In *Evolution, Learning, and Cognition*, ed. W.C. Lee, 277–330. Singapore: World Scientific.

Feldman, J.A. and D.H. Ballard (1982). Connectionist Models and Their Properties. *Cognitive Science* **6**, 205–254. Reprinted in Anderson and Rosenberg [1988].

Feller, W. (1968). *An Introduction to Probability Theory and Its Applications*, vol. I. New York: Wiley.

Fischer, K.H. and J.A. Hertz (1990). *Spin Glasses*. Cambridge: Cambridge University Press.

Földiák, P. (1989). Adaptive Network for Optimal Linear Feature Extraction. In *International Joint Conference on Neural Networks* (Washington 1989), vol. I, 401–405. New York: IEEE.

Franzini, M.A. (1987). Speech Recognition with Back Propagation. In *Proceedings of the Ninth Annual Conference of the IEEE Engineering in Medicine and Biology Society* (Boston 1987), 1702–1703. New York: IEEE.

Frean, M (1990). The Upstart Algorithm: A Method for Constructing and Training Feedforward Neural Networks. *Neural Computation* **2**, 198–209.

Fu, Y. and P.W. Anderson (1986). Application of Statistical Mechanics to NP-Complete Problems in Combinatorial Optimization. *Journal of Physics A* **19**, 1605–1620.

Fukushima, K. (1975). Cognitron: A Self-Organizing Multilayered Neural Network. *Biological Cybernetics* **20**, 121–136.

Fukushima, K. (1980). Neocognitron: A Self-Organizing Neural Network Model for a Mechanism of Pattern Recognition Unaffected by Shift in Position. *Biological Cybernetics* **36**, 193–202.

Fukushima, K., S. Miyake, and T. Ito (1983). Neocognitron: A Neural Network Model for a Mechanism of Visual Pattern Recognition. *IEEE Transactions on Systems, Man, and Cybernetics* **13**, 826–834. Reprinted in Anderson and Rosenberg [1988].

Gallant, S.I. (1986). Optimal Linear Discriminants. In *Eighth International Conference on Pattern Recognition* (Paris 1986), 849–852. New York: IEEE.

Gardner, E. (1987). Maximum Storage Capacity in Neural Networks. *Europhysics Letters* **4**, 481–485.

Gardner, E. (1988). The Space of Interactions in Neural Network Models. *Journal of Physics A* **21**, 257–270.

Gardner, E. and B. Derrida (1988). Optimal Storage Properties of Neural Network Models. *Journal of Physics A* **21**, 271–284.

Gardner, E. and B. Derrida (1989). Three Unfinished Works on the Optimal Storage Capacity of Networks. *Journal of Physics A* **22**, 1983–1994.

Gardner, E., H. Gutfreund, and I. Yekutieli (1989). The Phase Space of Interactions in Neural Networks with Definite Symmetry. *Journal of Physics A* **22**, 1995–2008.

Garey, M.R. and D.S. Johnson (1979). *Computers and Intractability: A Guide to the Theory of NP-Completeness.* New York: Freeman.

Geman, S and D. Geman (1984). Stochastic Relaxation, Gibbs Distributions, and the Bayesian Restoration of Images. *IEEE Transactions on Pattern Analysis and Machine Intelligence* **6**, 721–741. Reprinted in Anderson and Rosenberg [1988].

Geszti, T. (1990). *Physical Models of Neural Networks.* Singapore: World Scientific.

Geszti, T. and F. Pázmándi (1987). Learning Within Bounds and Dream Sleep. *Journal of Physics A* **20**, L1299–L1303.

Glauber, R.J. (1963). Time-Dependent Statistics of the Ising Model. *Journal of Mathematical Physics* **4**, 294–307.

Goldberg, D.E. (1989). *Genetic Algorithms in Search, Optimization, and Machine Learning.* Reading: Addison-Wesley.

Golomb, D., N. Rubin, and H. Sompolinsky (1990). Willshaw Model: Associative Memory with Sparse Coding and Low Firing Rates. *Physical Review A* **41**, 1843–1854.

Gorman, R.P. and T.J. Sejnowski (1988a). Analysis of Hidden Units in a Layered Network Trained to Classify Sonar Targets. *Neural Networks* **1**, 75–89.

Gorman, R.P. and T.J. Sejnowski (1988b). Learned Classification of Sonar Targets Using a Massively-Parallel Network. *IEEE Transactions on Acoustics, Speech, and Signal Processing* **36**, 1135–1140.

Graf, H.P., L.D. Jackel, R.E. Howard, B. Straughn, J.S. Denker, W. Hubbard, D.M. Tennant, and D. Schwartz (1986). VLSI Implementation of a Neural Network Memory with Several Hundreds of Neurons. In *Neural Networks for Computing* (Snowbird 1986), ed. J.S. Denker, 182–187. New York: American Institute of Physics.

Graf, D.H. and W.R. LaLonde (1988). A Neural Controller for Collision-Free Movement of General Robot Manipulators. In *IEEE International Conference on Neural Networks* (San Diego 1988), vol. I, 77–84. New York: IEEE.

Gray, R.M. (1984). Vector Quantization. *IEEE ASSP Magazine*, April 1984, 4–29.

Grossberg, S. (1967). Nonlinear Difference-Differential Equations in Prediction and Learning Theory. *Proceedings of the National Academy of Sciences, USA* **58**, 1329–1334.

Grossberg, S. (1968a). Some Nonlinear Networks Capable of Learning a Spatial Pattern of Arbitrary Complexity. *Proceedings of the National Academy of Sciences, USA* **59**, 368–372.

Grossberg, S. (1968b). Some Physiological and Biochemical Consequences of Psychological Postulates. *Proceedings of the National Academy of Sciences, USA* **60**, 758–765.

Grossberg, S. (1969). Embedding Fields: A Theory of Learning with Physiological Implications. *Journal of Mathematical Psychology* **6**, 209–239.

Grossberg, S. (1972). Neural Expectation: Cerebellar and Retinal Analogs of Cells Fired by Learnable or Unlearned Pattern Classes. *Kybernetik* **10**, 49–57.

Grossberg, S. (1976a). Adaptive Pattern Classification and Universal Recoding: I. Parallel Development and Coding of Neural Feature Detectors. *Biological Cybernetics* **23**, 121–134. Reprinted in Anderson and Rosenberg [1988].

Grossberg, S. (1976b). Adaptive Pattern Classification and Universal Recoding: II. Feedback, Expectation, Olfaction, Illusions. *Biological Cybernetics* **23**, 187–202.

Grossberg, S. (1980). How Does the Brain Build a Cognitive Code? *Psychological Review* **87**, 1–51. Reprinted in Anderson and Rosenberg [1988].

Grossberg, S. (1987a). *The Adaptive Brain*, 2 vols. Amsterdam: Elsevier.

Grossberg, S. (1987b). Competitive Learning: From Interactive Activation to Adaptive Resonance. *Cognitive Science* **11**, 23–63.

Grossman, T. (1990). The CHIR Algorithm for Feed Forward Networks with Binary Weights. In *Advances in Neural Information Processing Systems II* (Denver 1989), ed. D.S. Touretzky, 516–523. San Mateo: Morgan Kaufmann.

Grossman, T., R. Meir, and E. Domany (1989). Learning by Choice of Internal Representations. *Complex Systems* **2**, 555–575.

Gutfreund, H. (1988). Neural Networks with Hierarchically Correlated Patterns. *Physical Review A* **37**, 570–577.

Gutfreund, H. and M. Mézard (1988). Processing of Temporal Sequences in Neural Networks. *Physical Review Letters* **61**, 235–238.

Gutzmann, K. (1987). Combinatorial Optimization Using a Continuous State Boltzmann Machine. In *IEEE First International Conference on Neural Networks* (San Diego 1987), eds. M. Caudill and C. Butler, vol. III, 721–734. New York: IEEE.

Györgyi, G. (1990). Inference of a Rule by a Neural Network with Thermal Noise. *Physical Review Letters* **64**, 2957–2960.

Györgyi, G. and N. Tishby (1990). Statistical Theory of Learning a Rule. In *Neural Networks and Spin Glasses*, eds. W.K. Theumann and R. Koeberle. Singapore: World Scientific.

Hanson, S.J. and L. Pratt (1989). A Comparison of Different Biases for Minimal Network Construction with Back-Propagation. In *Advances in Neural Information Processing Systems I* (Denver 1988), ed. D.S. Touretzky, 177–185. San Mateo: Morgan Kaufmann.

Harp, S.A., T. Samad, and A. Guha (1990). Designing Application-Specific Neural Networks Using the Genetic Algorithm. In *Advances in Neural Information Processing Systems II* (Denver 1989), ed. D.S. Touretzky, 447–454. San Mateo: Morgan Kaufmann.

Hartman, E.J., J.D. Keeler, and J.M. Kowalski (1990). Layered Neural Networks with Gaussian Hidden Units As Universal Approximations. *Neural Computation* **2**, 210–215.

Hebb, D.O. (1949). *The Organization of Behavior*. New York: Wiley. Partially reprinted in Anderson and Rosenberg [1988].

Hecht-Nielsen, R. (1987a). Combinatorial Hypercompression. In *IEEE First International Conference on Neural Networks* (San Diego 1987), eds. M. Caudill and C. Butler, vol. II, 455–461. New York: IEEE.

Hecht-Nielsen, R. (1987b). Counterpropagation Networks. *Applied Optics* **26**, 4979–4984.

Hecht-Nielsen, R. (1988). Applications of Counterpropagation Networks. *Neural Networks* **1**, 131–139.

Hecht-Nielsen, R. (1989). Theory of the Backpropagation Neural Network. In *International Joint Conference on Neural Networks* (Washington 1989), vol. I, 593–605. New York: IEEE.

Hegde, S.U., J.L. Sweet, and W.B. Levy (1988). Determination of Parameters in a Hopfield/Tank Computational Network. In *IEEE International Conference on Neural Networks* (San Diego 1988), vol. II, 291–298. New York: IEEE.

van Hemmen, J.L., L.B. Ioffe, R. Kühn, and M. Vaas (1990). Increasing the Efficiency of a Neural Network through Unlearning. *Physica* **163A**, 386–392.

van Hemmen, J.L. and R. Kühn (1986). Nonlinear Neural Networks. *Physical Review Letters* **57**, 913–916.

van Hemmen, J.L. and R.G. Palmer (1979). The Replica Method and a Solvable Spin Glass Model. *Journal of Physics A* **12**, 563–580.

Hertz, J.A. (1989). A Gauge Theory in Computational Vision: A Model for Outline Extraction. *Physica Scripta* **39**, 161–167.

Hertz, J.A. (1990). Statistical Dynamics of Learning. Preprint 90/34 S, Nordita, Copenhagen, Denmark.

Hertz, J.A., G. Grinstein, and S. Solla (1986). Memory Networks with Asymmetric Bonds. In *Neural Networks for Computing* (Snowbird 1986), ed. J.S. Denker, 212–218. New York: American Institute of Physics.

Hertz, J.A., G. Grinstein, and S. Solla (1987). Irreversible Spin Glasses and Neural Networks. In *Heidelberg Colloquium on Glassy Dynamics* (Heidelberg 1986), eds. J.L. van Hemmen and I. Morgenstern, 538–546. Berlin: Springer-Verlag.

Hertz, J.A., A. Krogh, and G.I. Thorbergsson (1989). Phase Transitions in Simple Learning. *Journal of Physics A* **22**, 2133–2150.

Herz, A., B. Sulzer, R. Kühn, and J.L. van Hemmen (1989). Hebbian Learning Reconsidered: Representation of Static and Dynamic Objects in Associative Neural Nets. *Biological Cybernetics* **60**, 457–467.

Hinton, G.E. (1986). Learning Distributed Representations of Concepts. In *Proceedings of the Eighth Annual Conference of the Cognitive Science Society* (Amherst 1986), 1–12. Hillsdale: Erlbaum.

Hinton, G.E. (1989). Deterministic Boltzmann Learning Performs Steepest Descent in Weight Space. *Neural Computation* **1**, 143–150.

Hinton, G.E. and T.J. Sejnowski (1983). Optimal Perceptual Inference. In *Proceedings of the IEEE Conference on Computer Vision and Pattern Recognition* (Washington 1983), 448–453. New York: IEEE.

Hinton, G.E. and T.J. Sejnowski (1986). Learning and Relearning in Boltzmann Machines. In *Parallel Distributed Processing*, vol. 1, chap. 7. *See* Rumelhart, McClelland, et al. [1986].

Hopfield, J.J. (1982). Neural Networks and Physical Systems with Emergent Collective Computational Abilities. *Proceedings of the National Academy of Sciences, USA* **79**, 2554–2558. Reprinted in Anderson and Rosenberg [1988].

Hopfield, J.J. (1984). Neurons with Graded Responses Have Collective Computational Properties Like Those of Two-State Neurons. *Proceedings of the National Academy of Sciences, USA* **81**, 3088–3092. Reprinted in Anderson and Rosenberg [1988].

Hopfield, J.J. (1987). Learning Algorithms and Probability Distributions in Feed-Forward and Feed-Back Networks. *Proceedings of the National Academy of Sciences, USA* **84**, 8429–8433.

Hopfield, J.J., D.I. Feinstein, and R.G. Palmer (1983). "Unlearning" Has a Stabilizing Effect in Collective Memories. *Nature* **304**, 158–159.

Hopfield, J.J. and D.W. Tank (1985). "Neural" Computation of Decisions in Optimization Problems. *Biological Cybernetics* **52**, 141–152.

Hopfield, J.J. and D.W. Tank (1986). Computing with Neural Circuits: A Model. *Science* **233**, 625–633.

Hopfield, J.J. and D.W. Tank (1989). Neural Architecture and Biophysics for Sequence Recognition. In *Neural Models of Plasticity*, eds. J.H. Byrne and W.O. Berry, 363–377. San Diego: Academic Press.

Hornik, K., M. Stinchcombe, and H. White (1989). Multilayer Feedforward Networks Are Universal Approximators. *Neural Networks* **2**, 359–366.

Hsu, K, D. Brady, and D. Psaltis (1988). Experimental Demonstration of Optical Neural Computers. In *Neural Information Processing Systems* (Denver 1987), ed. D.Z. Anderson, 377–386. New York: American Institute of Physics.

Huang, K. (1987). *Statistical Mechanics*. New York: Wiley.

Huang, W.Y. and R.P. Lippmann (1988). Neural Net and Traditional Classifiers. In *Neural Information Processing Systems* (Denver 1987), ed. D.Z. Anderson, 387–396. New York: American Institute of Physics.

Hubel, D.H. and T.N. Wiesel (1962). Receptive Fields, Binocular Interaction, and Functional Architecture in the Cat's Visual Cortex. *Journal of Physiology (London)* **160**, 106–154.

Hueter, G.J. (1988). Solution of the Travelling Salesman Problem with an Adaptive Ring. In *IEEE International Conference on Neural Networks* (San Diego 1988), vol. I, 85–92. New York: IEEE.

Hush, D.R. and J.M. Salas (1988). Improving the Learning Rate of Back-Propagation with the Gradient Reuse Algorithm. In *IEEE International Conference on Neural Networks* (San Diego 1988), vol. I, 441–447. New York: IEEE.

Jacobs, R.A. (1988). Increased Rates of Convergence Through Learning Rate Adaptation. *Neural Networks* **1**, 295–307.

Johnson, D.S., C.R. Aragon, L.A. McGeoch, and C. Schevon (1989). Optimization by Simulated Annealing: An Experimental Evaluation; Part I, Graph Partitioning. *Operations Research* **37**, 865–891. Parts II and III are expected to appear in 1990.

Jolliffe, I.T. (1986). *Principal Component Analysis*. New York: Springer-Verlag.

Jordan, M.I. (1986). Attractor Dynamics and Parallelism in a Connectionist Sequential Machine. In *Proceedings of the Eighth Annual Conference of the Cognitive Science Society* (Amherst 1986), 531–546. Hillsdale: Erlbaum.

Jordan, M.I. (1989). Serial Order: A Parallel, Distributed Processing Approach. In *Advances in Connectionist Theory: Speech*, eds. J.L. Elman and D.E. Rumelhart. Hillsdale: Erlbaum.

Kahng, A.B. (1989). Travelling Salesman Heuristics and Embedding Dimension in the Hopfield Model. In *International Joint Conference on Neural Networks* (Washington 1989), vol. I, 513–520. New York: IEEE.

Kammen, D.M. and A.L. Yuille (1988). Spontaneous Symmetry-Breaking Energy Functions and the Emergence of Orientation Selective Cortical Cells. *Biological Cybernetics* **59**, 23–31.

Kandel, E.R. and J.H. Schwartz (1985). *Principles of Neural Science* (2nd ed.). New York: Elsevier.

Kanter, I. and H. Sompolinsky (1987). Associative Recall of Memory Without Errors. *Physical Review A* **35**, 380–392.

Kerszberg, M. and A. Zippelius (1990). Synchronization in Neural Assemblies. *Physica Scripta* **T33**, 54–64.

Kinzel, W. and M. Opper (1990). Dynamics of Learning. In *Physics of Neural Networks*, eds. E. Domany, J.L. van Hemmen, and K. Schulten, vol. I. Berlin: Springer-Verlag.

Kirkpatrick, S., C.D. Gelatt Jr., and M.P. Vecchi (1983). Optimization by Simulated Annealing. *Science* **220**, 671–680. Reprinted in Anderson and Rosenberg [1988].

Kirkpatrick, S. and G. Toulouse (1985). Configuration Space Analysis of Travelling Salesman Problems. *Journal de Physique (Paris)* **46**, 1277–1292.

Kleinfeld, D. (1986). Sequential State Generation by Model Neural Networks. *Proceedings of the National Academy of Sciences, USA* **83**, 9469–9473.

Kleinfeld, D. and H. Sompolinsky (1989). Associative Network Models for Central Pattern Generators. In *Methods in Neuronal Modeling: From Synapses to Networks*, eds. C. Koch and I. Segev, 195–246. Cambridge: MIT Press.

Klopf, A.H. (1982). *The Hedonistic Neuron: A Theory of Memory, Learning, and Intelligence*. Washington: Hemisphere.

Koch, C., J. Luo, C. Mead, and J. Hutchinson (1988). Computing Motion Using Resistive Networks. In *Neural Information Processing Systems* (Denver 1987), ed. D.Z. Anderson, 422–431. New York: American Institute of Physics.

Koch, C., J. Marroquin, and A. Yuille (1986). Analog "Neuronal" Networks in Early Vision. *Proceedings of the National Academy of Sciences, USA* **83**, 4263–4267.

Kohonen, T. (1974). An Adaptive Associative Memory Principle. *IEEE Transactions on Computers* **C-23**, 444–445.

Kohonen, T. (1982). Self-Organized Formation of Topologically Correct Feature Maps. *Biological Cybernetics* **43**, 59–69. Reprinted in Anderson and Rosenberg [1988].

Kohonen, T. (1989). *Self-Organization and Associative Memory* (3rd ed.). Berlin: Springer-Verlag.

Kohonen, T., G. Barna, and R. Chrisley (1988). Statistical Pattern Recognition with Neural Networks: Benchmarking Studies. In *IEEE International Conference on Neural Networks* (San Diego 1988), vol. I, 61–68. New York: IEEE.

Kohonen, T., K. Mäkisara, and T. Saramäki (1984). Phonotopic Maps—Insightful Representation of Phonological Features for Speech Recognition. In *Proceedings of the Seventh International Conference on Pattern Recognition* (Montreal 1984), 182–185. New York: IEEE.

Kramer, A.H. and A. Sangiovanni-Vincentelli (1989). Efficient Parallel Learning Algorithms for Neural Networks. In *Advances in Neural Information Processing Systems I* (Denver 1988), ed. D.S. Touretzky, 40–48. San Mateo: Morgan Kaufmann.

Krauth, W. and M. Mézard (1989). The Cavity Method and the Travelling-Salesman Problem. *Europhysics Letters* **8**, 213–218.

Krishnaiah, P.R. and L.N. Kanal, eds. (1982). *Classification, Pattern Recognition, and Reduction of Dimensionality*. Handbook of Statistics, vol. 2. Amsterdam: North Holland.

Krogh, A. and J.A. Hertz (1988). Mean Field Analysis of Hierarchical Associative Networks with Magnetization. *Journal of Physics A* **21**, 2211–2224.

Krogh, A. and J.A. Hertz (1990). Hebbian Learning of Principal Components. In *Parallel Processing in Neural Systems and Computers* (Düsseldorf 1990), eds. R. Eckmiller, G. Hartmann, and G. Hauske, 183–186. Amsterdam: Elsevier.

Krogh, A., G.I. Thorbergsson, and J.A. Hertz (1990). A Cost Function for Internal Representations. In *Advances in Neural Information Processing Systems II* (Denver 1989), ed. D.S. Touretzky, 733–740. San Mateo: Morgan Kaufmann.

Kullback, S. (1959). *Information Theory and Statistics*. New York: Wiley.

Kushner, H.J. and D.S. Clark (1978). *Stochastic Approximation Methods for Constrained and Unconstrained Systems*. New York: Springer-Verlag.

Lapedes, A. and R. Farber (1986a). A Self-Optimizing, Nonsymmetrical Neural Net for Content Addressable Memory and Pattern Recognition. *Physica* **22D**, 247–259.

Lapedes, A. and R. Farber (1986b). Programming a Massively Parallel, Computation Universal System: Static Behavior. In *Neural Networks for Computing* (Snowbird 1986), ed. J.S. Denker, 283–298. New York: American Institute of Physics.

Lapedes, A. and R. Farber (1987). Nonlinear Signal Processing Using Neural Networks: Prediction and System Modelling. Technical Report LA–UR–87–2662, Los Alamos National Laboratory, Los Alamos, NM.

Lapedes, A. and R. Farber (1988). How Neural Nets Work. In *Neural Information Processing Systems* (Denver 1987), ed. D.Z. Anderson, 442–456. New York: American Institute of Physics.

Lawler, E.L. (1976). *Combinatorial Optimization: Networks and Matroids*. New York: Holt-Rinehart-Winston.

Lawler, E.L., J.K. Lenstra, A.H.G. Rinnooy Kan, and D.B. Shmoys, eds. (1985). *The Travelling Salesman Problem*. Chichester: Wiley.

Le Cun, Y. (1985). Une Procédure d'Apprentissage pour Réseau à Seuil Assymétrique. In *Cognitiva 85: A la Frontière de l'Intelligence Artificielle des Sciences de la Connaissance des Neurosciences* (Paris 1985), 599–604. Paris: CESTA.

Le Cun, Y., B. Boser, J.S. Denker, D. Henderson, R.E. Howard, W. Hubbard, and L.D. Jackel (1989). Backpropagation Applied to Handwritten Zip Code Recognition. *Neural Computation* **1**, 541–551.

Le Cun, Y., B. Boser, J.S. Denker, D. Henderson, R.E. Howard, W. Hubbard, and L.D. Jackel (1990). Handwritten Digit Recognition with a Back-Propagation Network. In *Advances in Neural Information Processing Systems II* (Denver 1989), ed. D.S. Touretzky, 396–404. San Mateo: Morgan Kaufmann.

Le Cun, Y., J.S. Denker, and S.A. Solla (1990). Optimal Brain Damage. In *Advances in Neural Information Processing Systems II* (Denver 1989), ed. D.S. Touretzky, 598–605. San Mateo: Morgan Kaufmann.

von Lehman, A., E.G. Paek, P.F. Liao, A. Marrakchi, and J.S. Patel (1988). Factors Influencing Learning by Back-Propagation. In *IEEE International Conference on Neural Networks* (San Diego 1988), vol. I, 335–341. New York: IEEE.

Liang, F.M. (1983). Ph.D. Thesis. Stanford University. See Appendix H of *The TEXBook*, D.E. Knuth (Reading: Addison Wesley 1986).

Lin, S. and B.W. Kernighan (1973). An Effective Heuristic Algorithm for the Travelling Salesman Problem. *Operations Research* **21**, 498–516.

Linsker, R. (1986). From Basic Network Principles to Neural Architecture. *Proceedings of the National Academy of Sciences, USA* **83**, 7508–7512, 8390–8394, 8779–8783.

Linsker, R. (1988). Self-Organization in a Perceptual Network. *Computer*, March 1988, 105–117.

Lippmann, R.P. (1987). An Introduction to Computing with Neural Nets. *IEEE ASSP Magazine*, April 1987, 4–22.

Lippmann, R.P. (1989). Review of Neural Networks for Speech Recognition. *Neural Computation* **1**, 1–38.

Lister, R. (1990). Segment Reversal and the TSP. In *International Joint Conference on Neural Networks* (Washington 1990), vol. 1, 424–427. Hillsdale: Erlbaum.

Little, W.A. (1974). The Existence of Persistent States in the Brain. *Mathematical Biosciences* **19**, 101–120.

Little, W.A. and G.L. Shaw (1975). A Statistical Theory of Short and Long Term Memory. *Behavioral Biology* **14**, 115–133. Reprinted in Anderson and Rosenberg [1988].

Little, W.A. and G.L. Shaw (1978). Analytic Study of the Memory Storage Capacity of a Neural Network. *Mathematical Biosciences* **39**, 281–290.

Luenberger, D.G. (1986). *Linear and Nonlinear Programming*. Reading: Addison-Wesley.

Ma, S.-K. (1985). *Statistical Mechanics*. Philadelphia: World Scientific.

MacKay, D.J.C. and K.D. Miller (1990). Analysis of Linsker's Simulation of Hebbian Rules. *Neural Computation* **2**, 173–187.

Mackey, M.C. and L. Glass (1977). Oscillation and Chaos in Physiological Control Systems. *Science* **197**, 287.

Makram-Ebeid, S., J.-A. Sirat, and J.-R. Viala (1989). A Rationalized Back-Propagation Learning Algorithm. In *International Joint Conference on Neural Networks* (Washington 1989), vol. II, 373–380. New York: IEEE.

von der Malsburg, Ch. (1973). Self-Organization of Orientation Sensitive Cells in the Striate Cortex. *Kybernetik* **14**, 85–100. Reprinted in Anderson and Rosenberg [1988].

von der Malsburg, Ch. and J.D. Cowan (1982). Outline of a Theory for the Ontogenesis of Iso-Orientation Domains in Visual Cortex. *Biological Cybernetics* **45**, 49–56.

Mandelbrot, B.B. (1982). *The Fractal Geometry of Nature*. San Francisco: Freeman.

Marchand, M., M. Golea, and P. Ruján (1990). A Convergence Theorem for Sequential Learning in Two-Layer Perceptrons. *Europhysics Letters* **11**, 487–492.

Marr, D. (1969). A Theory of Cerebellar Cortex. *Journal of Physiology (London)* **202**, 437–470.

Marr, D. (1970). A Theory for Cerebral Neocortex. *Proceedings of the Royal Society of London B* **176**, 161–234.

Marr, D. (1971). Simple Memory: A Theory for Archicortex. *Philosophical Transactions of the Royal Society of London B* **262**, 23–81.

Marr, D. (1982). *Vision*. San Francisco: Freeman. Partially reprinted in Anderson and Rosenberg [1988].

Marr, D. and T. Poggio (1976). Cooperative Computation of Stereo Disparity. *Science* **194**, 283–287. Reprinted in Anderson and Rosenberg [1988].

Mattis, D. (1976). Solvable Spin Systems with Random Interactions. *Physics Letters* **56A**, 421–422.

Mazaika, P.K. (1987). A Mathematical Model of the Boltzmann Machine. In *IEEE First International Conference on Neural Networks* (San Diego 1987), eds. M. Caudill and C. Butler, vol. III, 157–163. New York: IEEE.

McClelland, J.L. and J.L. Elman (1986). Interactive Processes in Speech Perception: The TRACE Model. In *Parallel Distributed Processing*, vol. 2, chap. 15. *See* Rumelhart, McClelland, et al. [1986].

McClelland, J.L. and D.E. Rumelhart (1988). *Explorations in Parallel Distributed Processing*. Cambridge: MIT Press.

McCulloch, W.S. and W. Pitts (1943). A Logical Calculus of Ideas Immanent in Nervous Activity. *Bulletin of Mathematical Biophysics* **5**, 115–133. Reprinted in Anderson and Rosenberg [1988].

McEliece, R.J., E.C. Posner, E.R. Rodemich, and S.S. Venkatesh (1987). The Capacity of the Hopfield Associative Memory. *IEEE Transactions on Information Theory* **33**, 461–482.

McInerny, J.M., K.G. Haines, S. Biafore, and R. Hecht-Nielsen (1989). Back Propagation Error Surfaces Can Have Local Minima. In *International Joint Conference on Neural Networks* (Washington 1989), vol. II, 627. New York: IEEE.

Mead, C. (1989). *Analog VLSI and Neural Systems*. Reading: Addison Wesley.

Metropolis, N., A.W. Rosenbluth, M.N. Rosenbluth, A.H. Teller, and E. Teller (1953). Equation of State Calculations for Fast Computing Machines. *Journal of Chemical Physics* **21**, 1087–1092.

Mézard, M. and J.-P. Nadal (1989). Learning in Feedforward Layered Networks: The Tiling Algorithm. *Journal of Physics A* **22**, 2191–2204.

Mézard, M. and G. Parisi (1985). Replicas and Optimization. *Journal de Physique Lettres (Paris)* **46**, 771–778.

Mézard, M. and G. Parisi (1986). A Replica Analysis of the Travelling Salesman Problem. *Journal de Physique (Paris)* **47**, 1285–1296.

Mézard, M. and G. Parisi (1988). The Euclidean Matching Problem. *Journal de Physique (Paris)* **49**, 2019–2025.

Mézard, M., G. Parisi, and M.A. Virasoro (1987). *Spin Glass Theory and Beyond*. Singapore: World Scientific.

Miller, G.F., P.M. Todd, and S.U. Hegde (1989). Designing Neural Networks Using Genetic Algorithms. In *Proceedings of the Third International Conference on Genetic Algorithms* (Arlington 1989), ed. J.D. Schaffer, 379–384. San Mateo: Morgan Kaufmann.

Minsky, M.L. (1967). *Computation: Finite and Infinite Machines.* Englewood Cliffs: Prentice-Hall.

Minsky, M.L. and S.A. Papert (1969). *Perceptrons.* Cambridge: MIT Press. Partially reprinted in Anderson and Rosenberg [1988].

Mitchison, G.J. and R.M. Durbin (1989). Bounds on the Learning Capacity of Some Multi-Layer Networks. *Biological Cybernetics* **60**, 345–356.

Montana, D.J. and L. Davis (1989). Training Feedforward Networks Using Genetic Algorithms. In *Eleventh International Joint Conference on Artificial Intelligence* (Detroit 1989), ed. N.S. Sridharan, 762–767. San Mateo: Morgan Kaufmann.

Moody, J. and C. Darken (1988). Learning with Localized Receptive Fields. In *Proceedings of the 1988 Connectionist Models Summer School* (Pittsburg 1988), eds. D. Touretzky, G. Hinton, and T. Sejnowski, 133–143. San Mateo: Morgan Kaufmann.

Moody, J. and C. Darken (1989). Fast Learning in Networks of Locally-Tuned Processing Units. *Neural Computation* **1**, 281–294.

Moore, B. (1988). ART1 and Pattern Clustering. In *Proceedings of the 1988 Connectionist Models Summer School* (Pittsburg 1988), eds. D. Touretzky, G. Hinton, and T. Sejnowski, 174–185. San Mateo: Morgan Kaufmann.

Mozer, M.C. (1989). A Focused Back-Propagation Algorithm for Temporal Pattern Recognition. *Complex Systems* **3**, 349–381.

Munro, P. (1987). A Dual Back-Propagation Scheme for Scalar Reward Learning. In *The Ninth Annual Conference of the Cognitive Science Society* (Seattle 1987), 165–176. Hillsdale: Erlbaum.

Nadal, J.-P., G. Toulouse, J.-P. Changeux, and S. Dehaene (1986). Networks of Formal Neurons and Memory Palimpsests. *Europhysics Letters* **1**, 535–542.

Narendra, K. and M.A.L. Thathachar (1989). *Learning Automata: An Introduction.* Englewood Cliffs: Prentice-Hall.

Nasrabadi, N.M. and Y. Feng (1988). Vector Quantization of Images Based upon the Kohonen Self-Organizing Feature Maps. In *IEEE International Conference on Neural Networks* (San Diego 1988), vol. I, 101–108. New York: IEEE.

Nasrabadi, N.M. and R.A. King (1988). Image Coding Using Vector Quantization: A Review. *IEEE Transactions on Communications* **36**, 957–971.

Nass, M.M. and L.N. Cooper (1975). A Theory for the Development of Feature Detecting Cells in Visual Cortex. *Biological Cybernetics* **19**, 1–18.

Naylor, J. and K.P. Li (1988). Analysis of a Neural Network Algorithm for Vector Quantization of Speech Parameters. *Neural Networks Supplement* **1**, 310.

von Neumann, J. (1956). Probabilistic Logics and the Synthesis of Reliable Organisms from Unreliable Components. In *Automata Studies*, eds. C.E. Shannon and J. McCarthy, 43–98. Princeton: Princeton University Press.

Niranjan, M. and F. Fallside (1990). Neural Networks and Radial Basis Functions in Classifying Static Speech Patterns. *Computer Speech and Language* **4**, 275–289.

Nishimori, H., T. Nakamura, and M. Shiino (1990). Retrieval of Spatio-Temporal Sequence in Asynchronous Neural Network. *Physical Review A* **41**, 3346–3354.

Nowlan, S.J. (1988). Gain Variation in Recurrent Error Propagation Networks. *Complex Systems* **2**, 305–320.

Oja, E. (1982). A Simplified Neuron Model As a Principal Component Analyzer. *Journal of Mathematical Biology* **15**, 267–273.

Oja, E. (1989). Neural Networks, Principal Components, and Subspaces. *International Journal of Neural Systems* **1**, 61–68.

Oja, E. and J. Karhunen (1985). On Stochastic Approximation of the Eigenvectors and Eigenvalues of the Expectation of a Random Matrix. *Journal of Mathematical Analysis and Applications* **106**, 69–84.

Opper, M., W. Kinzel, J. Kleinz, and R. Nehl (1990). On the Ability of the Optimal Perceptron to Generalize. *Journal of Physics A* **23**, L581–L586.

Orland, H. (1985). Mean-Field Theory for Optimization Problems. *Journal de Physique Lettres (Paris)* **46**, 763–770.

Owens, A.J. and D.L. Filkin (1989). Efficient Training of the Back Propagation Network by Solving a System of Stiff Ordinary Differential Equations. In *International Joint Conference on Neural Networks* (Washington 1989), vol. II, 381–386. New York: IEEE.

Packard, N.H., J.P Crutchfield, J.D. Farmer, and R.S. Shaw (1980). Geometry from a Time Series. *Physical Review Letters* **45**, 712–716.

Palmer, R.G. (1988). Statistical Mechanics Approaches to Complex Optimization Problems. In *The Economy As an Evolving Complex System*, eds. P.W. Anderson, K.J. Arrow, and D. Pines. SFI Studies in the Sciences of Complexity, proc. vol. 5, 177–193. Redwood City: Addison-Wesley.

Palmer, R.G. (1989). Neural Nets. In *Lectures in the Sciences of Complexity*, ed. D.L. Stein. SFI Studies in the Sciences of Complexity, lect. vol. 1, 439–461. Redwood City: Addison-Wesley.

Papadimitriou, C.H. and K. Steiglitz (1982). *Combinatorial Optimization: Algorithms and Complexity*. Englewood Cliffs: Prentice-Hall.

Parga, N. and M.A. Virasoro (1986). The Ultrametric Organization of Memories in a Neural Network. *Journal de Physique (Paris)* **47**, 1857–1864.

Parisi, G. (1986). Asymmetric Neural Networks and the Process of Learning. *Journal of Physics A* **19**, L675–L680.

Parker, D.B. (1985). Learning Logic. Technical Report TR–47, Center for Computational Research in Economics and Management Science, Massachusetts Institute of Technology, Cambridge, MA.

Parker, D.B. (1987). Optimal Algorithms for Adaptive Networks: Second Order Back Propagation, Second Order Direct Propagation, and Second Order Hebbian Learning. In *IEEE First International Conference on Neural Networks* (San Diego 1987), eds. M. Caudill and C. Butler, vol. II, 593–600. New York: IEEE.

Parks, M. (1987). Characterization of the Boltzmann Machine Learning Rate. In *IEEE First International Conference on Neural Networks* (San Diego 1987), eds. M. Caudill and C. Butler, vol. III, 715–719. New York: IEEE.

Pearlmutter, B.A. (1989a). Learning State Space Trajectories in Recurrent Neural Networks. In *International Joint Conference on Neural Networks* (Washington 1989), vol. II, 365–372. New York: IEEE.

Pearlmutter, B.A. (1989b). Learning State Space Trajectories in Recurrent Neural Networks. *Neural Computation* 1, 263–269.

Pearlmutter, B.A. and G.E. Hinton (1986). G-Maximization: An Unsupervised Learning Procedure for Discovering Regularities. In *Neural Networks for Computing* (Snowbird 1986), ed. J.S. Denker, 333–338. New York: American Institute of Physics.

Peretto, P. (1984). Collective Properties of Neural Networks: A Statistical Physics Approach. *Biological Cybernetics* 50, 51–62.

Peretto, P. (1988). On Learning Rules and Memory Storage Abilities of Asymmetrical Neural Networks. *Journal de Physique (Paris)* 49, 711–726.

Peretto, P. and J.J. Niez (1986). Collective Properties of Neural Networks. In *Disordered Systems and Biological Organization* (Les Houches 1985), eds. E. Bienenstock, F. Fogelman-Soulié, and G. Weisbuch, 171–185. Berlin: Springer-Verlag.

Pérez, R., L. Glass, and R. Shlaer (1975). Development of Specificity in the Cat Visual Cortex. *Journal of Mathematical Biology* 1, 275–288.

Personnaz, L., I. Guyon, and G. Dreyfus (1985). Information Storage and Retrieval in Spin-Glass-Like Neural Networks. *Journal de Physique Lettres (Paris)* 46, 359–365.

Personnaz, L., I. Guyon, and G. Dreyfus (1986). Collective Computational Properties of Neural Networks: New Learning Mechanisms. *Physical Review A* 34, 4217–4228.

Peterson, C. and J.R. Anderson (1987). A Mean Field Theory Learning Algorithm for Neural Networks. *Complex Systems* 1, 995–1019.

Peterson, C., S. Redfield, J.D. Keeler, and E. Hartman (1990). An Optoelectronic Architecture for Multilayer Learning in a Single Photorefractive Crystal. *Neural Computation* 2, 25–34.

Peterson, C. and B. Söderberg (1989). A New Method for Mapping Optimization Problems onto Neural Networks. *International Journal of Neural Systems* 1, 3–22.

Pineda, F.J. (1987). Generalization of Back-Propagation to Recurrent Neural Networks. *Physical Review Letters* 59, 2229–2232.

Pineda, F.J. (1988). Dynamics and Architecture for Neural Computation. *Journal of Complexity* 4, 216–245.

Pineda, F.J. (1989). Recurrent Back-Propagation and the Dynamical Approach to Adaptive Neural Computation. *Neural Computation* 1, 161–172.

Plaut, D., S. Nowlan, and G. Hinton (1986). Experiments on Learning by Back Propagation. Technical Report CMU–CS–86–126, Department of Computer Science, Carnegie Mellon University, Pittsburgh, PA.

Poggio, T. and F. Girosi (1990). Regularization Algorithms for Learning That Are Equivalent to Multilayer Networks. *Science* **247**, 978–982.

Poggio, T., V. Torre, and C. Koch (1985). Computational Vision and Regularization Theory. *Nature* **317**, 314–319.

Pomerleau, D.A. (1989). ALVINN: An Autonomous Land Vehicle in a Neural Network. In *Advances in Neural Information Processing Systems I* (Denver 1988), ed. D.S. Touretzky, 305–313. San Mateo: Morgan Kaufmann.

Press, W.H., B.P. Flannery, S.A. Teukolsky, and W.T. Vetterling (1986). *Numerical Recipes*. Cambridge: Cambridge University Press.

Qian, N. and T.J. Sejnowski (1988a). Predicting the Secondary Structure of Globular Proteins Using Neural Network Models. *Journal of Molecular Biology* **202**, 865–884.

Qian, N. and T.J. Sejnowski (1988b). Learning to Solve Random-Dot Stereograms of Dense Transparent Surfaces with Recurrent Back-Propagation. In *Proceedings of the 1988 Connectionist Models Summer School* (Pittsburg 1988), eds. D. Touretzky, G. Hinton, and T. Sejnowski, 435–443. San Mateo: Morgan Kaufmann.

Rao, C.R. and S.K. Mitra (1971). *Generalized Inverse of Matrices and Its Applications*. New York: Wiley.

Ramanujam, J. and P. Sadayappan (1988). Optimization by Neural Networks. In *IEEE International Conference on Neural Networks* (San Diego 1988), vol. II, 325–332. New York: IEEE.

Rashevsky, N. (1938). *Mathematical Biophysics*. Chicago: University of Chicago Press.

Rescorla, R.A. and A.R. Wagner (1972). A Theory of Pavlovian Conditioning: The Effectiveness of Reinforcement and Nonreinforcement. In *Classical Conditioning II: Current Research and Theory*, eds. A.H. Black and W.F. Prokasy, 64–69. New York: Appleton-Century-Crofts.

Ricotti, L.P., S. Ragazzini, and G. Martinelli (1988). Learning of Word Stress in a Sub-Optimal Second Order Back-Propagation Neural Network. In *IEEE International Conference on Neural Networks* (San Diego 1988), vol. I, 355–361. New York: IEEE.

Riedel, U., R. Kühn, and J.L. van Hemmen (1988). Temporal Sequences and Chaos in Neural Nets. *Physical Review A* **38**, 1105–1108.

Ritter, H. and K. Schulten (1986). On the Stationary State of Kohonen's Self-Organizing Sensory Mapping. *Biological Cybernetics* **54**, 99–106.

Ritter, H. and K. Schulten (1988a). Extending Kohonen's Self-Organizing Mapping Algorithm to Learn Ballistic Movements. In *Neural Computers* (Neuss 1987), eds. R. Eckmiller and Ch. von der Malsburg, 393–406. Berlin: Springer-Verlag.

Ritter, H. and K. Schulten (1988b). Convergence Properties of Kohonen's Topology Conserving Maps: Fluctuations, Stability, and Dimension Selection. *Biological Cybernetics* **60**, 59–71.

Ritter, H. and K. Schulten (1988c). Kohonen's Self-Organizing Maps: Exploring Their Computational Capabilities. In *IEEE International Conference on Neural Networks* (San Diego 1988), vol. I, 109–116. New York: IEEE.

Robinson, A.J. and F. Fallside (1988). Static and Dynamic Error Propagation Networks with Application to Speech Coding. In *Neural Information Processing Systems* (Denver 1987), ed. D.Z. Anderson, 632–641. New York: American Institute of Physics.

Rohwer, R. (1990). The "Moving Targets" Training Algorithm. In *Advances in Neural Information Processing Systems II* (Denver 1989), ed. D.S. Touretzky, 558–565. San Mateo: Morgan Kaufmann.

Rohwer, R. and B. Forrest (1987). Training Time-Dependence in Neural Networks. In *IEEE First International Conference on Neural Networks* (San Diego 1987), eds. M. Caudill and C. Butler, vol. II, 701–708. New York: IEEE.

Romeo, F.I. (1989). Simulated Annealing: Theory and Applications to Layout Problems. Ph.D. Thesis, University of California at Berkeley Memorandum UCB/ERL–M89/29.

Rose, D. and V.G. Dobson, eds. (1985). *Models of the Visual Cortex*. Chichester: Wiley.

Rosenblatt, F. (1962). *Principles of Neurodynamics*. New York: Spartan.

Rubner, J. and K. Schulten (1990). Development of Feature Detectors by Self-Organization. *Biological Cybernetics* **62**, 193–199.

Rubner, J. and P. Tavan (1989). A Self-Organizing Network for Principal-Component Analysis. *Europhysics Letters* **10**, 693–698.

Rumelhart, D.E., G.E. Hinton, and R.J. Williams (1986a). Learning Representations by Back-Propagating Errors. *Nature* **323**, 533–536. Reprinted in Anderson and Rosenberg [1988].

Rumelhart, D.E., G.E. Hinton, and R.J. Williams (1986b). Learning Internal Representations by Error Propagation. In *Parallel Distributed Processing*, vol. 1, chap. 8. *See* Rumelhart, McClelland, et al. [1986]. Reprinted in Anderson and Rosenberg [1988].

Rumelhart, D.E., J.L. McClelland, and the PDP Research Group (1986). *Parallel Distributed Processing: Explorations in the Microstructure of Cognition*, 2 vols. Cambridge: MIT Press.

Rumelhart, D.E. and D. Zipser (1985). Feature Discovery by Competitive Learning. *Cognitive Science* **9**, 75–112. Also reprinted in *Parallel Distributed Processing*, vol. 1, chap. 5. *See* Rumelhart, McClelland, et al. [1986].

Saad, D. and E. Marom (1990a). Learning by Choice of Internal Representations —An Energy Minimization Approach. Preprint, Faculty of Engineering, Tel Aviv University, Ramat-Aviv, Israel.

Saad, D. and E. Marom (1990b). Training Feed Forward Nets with Binary Weights via a Modified CHIR Algorithm. Preprint, Faculty of Engineering, Tel Aviv University, Ramat-Aviv, Israel.

Salamon, P., J.D. Nulton, J. Robinson, J. Petersen, G. Ruppeiner, and L. Liao (1988). Simulated Annealing with Constant Thermodynamic Speed. *Computer Physics Communications* **49**, 423–428.

Sanger, T.D. (1989a). Optimal Unsupervised Learning in a Single-Layer Linear Feedforward Neural Network. *Neural Networks* **2**, 459–473.

Sanger, T.D. (1989b). An Optimality Principle for Unsupervised Learning. In *Advances in Neural Information Processing Systems I* (Denver 1988), ed. D.S. Touretzky, 11–19. San Mateo: Morgan Kaufmann.

Sato, M. (1990). A Real Time Learning Algorithm for Recurrent Analog Neural Networks. *Biological Cybernetics* **62**, 237–241.

Scalettar, R. and A. Zee (1988). Emergence of Grandmother Memory in Feed Forward Networks: Learning with Noise and Forgetfulness. In *Connectionist Models and Their Implications: Readings from Cognitive Science*, eds. D. Waltz and J.A. Feldman, 309–332. Norwood: Ablex.

Schwartz, D.B., V.K. Samalam, S.A. Solla, and J.S. Denker (1990). Exhaustive Learning. *Neural Computation* **2**, 371–382.

Scofield, C.L. (1988). Learning Internal Representations in the Coulomb Energy Network. In *IEEE International Conference on Neural Networks* (San Diego 1988), vol. I, 271–276. New York: IEEE.

Sejnowski, T.J., P.K. Kienker, and G. Hinton (1986). Learning Symmetry Groups with Hidden Units: Beyond the Perceptron. *Physica* **22D**, 260–275.

Sejnowski, T.J. and C.R. Rosenberg (1987). Parallel Networks that Learn to Pronounce English Text. *Complex Systems* **1**, 145–168.

Sherrington, D. and S. Kirkpatrick (1975). Solvable Model of a Spin Glass. *Physical Review Letters* **35**, 1792–1796.

Shimohara, K., T. Uchiyama, and Y. Tokunaga (1988). Back-Propagation Networks for Event-Driven Temporal Sequence Processing. In *IEEE International Conference on Neural Networks* (San Diego 1988), vol. I, 665–672. New York: IEEE.

Sietsma, J. and R.J.F. Dow (1988). Neural Net Pruning—Why and How. In *IEEE International Conference on Neural Networks* (San Diego 1988), vol. I, 325–333. New York: IEEE.

Silverman, R.H. and A.S. Noetzel (1988). Time-Sequential Self-Organization of Hierarchical Neural Networks. In *Neural Information Processing Systems* (Denver 1987), ed. D.Z. Anderson, 709–714. New York: American Institute of Physics.

Simard, P.Y., M.B. Ottaway, and D.H. Ballard (1989). Analysis of Recurrent Backpropagation. In *Proceedings of the 1988 Connectionist Models Summer School* (Pittsburg 1988), eds. D. Touretzky, G. Hinton, and T. Sejnowski, 103–112. San Mateo: Morgan Kaufmann.

Simic, P.D. (1990). Statistical Mechanics As the Underlying Theory of "Elastic" and "Neural" Optimizations. *Network* **1**, 89–103.

Sirat, J.-A. and J.-P. Nadal (1990). Neural Trees: A New Tool for Classification. Preprint, Laboratoires d'Electronique Philips, Limeil-Brévannes, France.

Sivilotti, M.A., M.A. Mahowald, and C.A. Mead (1987). Real-Time Visual Computations Using Analog CMOS Processing Arrays. In *Advanced Research in VLSI: Proceedings of the 1987 Stanford Conference*, ed. P. Losleben, 295–312. Cambridge: MIT Press. Reprinted in Anderson and Rosenberg [1988].

Smolensky, P. (1986). Information Processing in Dynamical Systems: Foundations of Harmony Theory. In *Parallel Distributed Processing*, vol. 1, chap. 6. *See* Rumelhart, McClelland, et al. [1986].

Soffer, B.H., G.J. Dunning, Y. Owechko, and E. Marom (1986). Associative Holographic Memory with Feedback Using Phase-Conjugate Mirrors. *Optics Letters* **11**, 118–120.

Solla, S.A. (1989). Learning and Generalization in Layered Neural Networks: The Contiguity Problem. In *Neural Networks from Models to Applications* (Paris 1988), eds. L. Personnaz and G. Dreyfus, 168–177. Paris: I.D.S.E.T.

Solla, S.A., E. Levin, and M. Fleisher (1988). Accelerated Learning in Layered Neural Networks. *Complex Systems* **2**, 625–639.

Sompolinsky, H. (1987). The Theory of Neural Networks: The Hebb Rules and Beyond. In *Heidelberg Colloquium on Glassy Dynamics* (Heidelberg 1986), eds. J.L. van Hemmen and I. Morgenstern, 485–527. Berlin: Springer-Verlag.

Sompolinsky, H., A. Crisanti, and H.J. Sommers (1988). Chaos in Random Neural Networks. *Physical Review Letters* **61**, 259–262.

Sompolinsky, H. and I. Kanter (1986). Temporal Association in Asymmetric Neural Networks. *Physical Review Letters* **57**, 2861–2864.

Soukoulis, C.M., K. Levin, and G.S. Grest (1983). Irreversibility and Metastability in Spin-Glasses. I. Ising Model. *Physical Review B* **28**, 1495–1509.

Specht, D.F. (1990). Probabilistic Neural Networks. *Neural Networks* **3**, 109–118.

Steinbuch, K. (1961). Die Lernmatrix. *Kybernetik* **1**, 36–45.

Stokbro, K., D.K. Umberger, and J.A. Hertz (1990). Exploiting Neurons with Localized Receptive Fields to Learn Chaos. Preprint 90/28 S, Nordita, Copenhagen, Denmark.

Stornetta, W.S., T. Hogg, and B.A. Huberman (1988). A Dynamical Approach to Temporal Pattern Processing. In *Neural Information Processing Systems* (Denver 1987), ed. D.Z. Anderson, 750–759. New York: American Institute of Physics.

Sutton, R.S. (1984). Temporal Credit Assignment in Reinforcement Learning. Ph. D. Thesis, University of Massachusetts, Amherst.

Sutton, R.S. (1988). Learning to Predict by the Methods of Temporal Differences. *Machine Learning* **3**, 9–44.

Sutton, R.S. and A.G. Barto (1991). Time Derivative Models of Pavlovian Reinforcement. In *Learning and Computational Neuroscience*, eds. M. Gabriel and J.W. Moore. Cambridge: MIT Press.

Szu, H. (1986). Fast Simulated Annealing. In *Neural Networks for Computing* (Snowbird 1986), ed. J.S. Denker, 420–425. New York: American Institute of Physics.

Takens, F. (1981). Detecting Strange Attractors In Turbulence. In *Dynamical Systems and Turbulence*. Lecture Notes in Mathematics, vol. 898 (Warwick 1980), eds. D.A. Rand and L.-S. Young, 366–381. Berlin: Springer-Verlag.

Takeuchi, A. and S. Amari (1979). Formation of Topographic Maps and Columnar Microstructures in Nerve Fields. *Biological Cybernetics* **35**, 63–72.

Tank, D.W. and J.J. Hopfield (1986). Simple "Neural" Optimization Networks: An A/D Converter, Signal Decision Circuit, and a Linear Programming Circuit. *IEEE Transactions on Circuits and Systems* **33**, 533–541.

Tank, D.W. and J.J. Hopfield (1987a). Neural Computation by Time Compression. *Proceedings of the National Academy of Sciences, USA* **84**, 1896–1900.

Tank, D.W. and J.J. Hopfield (1987b). Concentrating Information in Time: Analog Neural Networks with Applications to Speech Recognition Problems. In *IEEE First International Conference on Neural Networks* (San Diego 1987), eds. M. Caudill and C. Butler, vol. IV, 455–468. New York: IEEE.

Taylor, W.K. (1956). Electrical Simulation of Some Nervous System Functional Activities. In *Information Theory* (London 1985), ed. C. Cherry, 314–328. London: Butterworths.

Tesauro, G. (1986). Simple Neural Models of Classical Conditioning. *Biological Cybernetics* **55**, 187–200.

Tesauro, G. (1990). Neurogammon Wins Computer Olympiad. *Neural Computation* **1**, 321–323.

Tesauro, G. and B. Janssens (1988). Scaling Relationships in Back-Propagation Learning. *Complex Systems* **2**, 39–44.

Tesauro, G. and T.J. Sejnowski (1988). A "Neural" Network That Learns to Play Backgammon. In *Neural Information Processing Systems* (Denver 1987), ed. D.Z. Anderson, 442–456. New York: American Institute of Physics.

Thakoor, A.P., A. Moopenn, J. Lambe, and S.K. Khanna (1987). Electronic Hardware Implementations of Neural Networks. *Applied Optics* **26**, 5085–5092.

Ticknor, A.J. and H. Barrett (1987). Optical Implementations of Boltzmann Machines. *Optical Engineering* **26**, 16–21.

Tishby, N., E. Levin, and S.A. Solla (1989). Consistent Inference of Probabilities in Layered Networks: Predictions and Generalization. In *International Joint Conference on Neural Networks* (Washington 1989), vol. II, 403–410. New York: IEEE.

Toulouse, G., S. Dehaene, and J.-P. Changeux (1986). Spin Glass Model of Learning by Selection. *Proceedings of the National Academy of Sciences, USA* **83**, 1695–1698.

Touretzky, D.S. and D.A. Pomerleau (1989). What's Hidden in the Hidden Layers? *BYTE*, August 1989, 227–233.

Tsodyks, M.V. and M.V. Feigel'man (1988). The Enhanced Storage Capacity in Neural Networks with Low Activity Level. *Europhysics Letters* **6**, 101–105.

Van den Bout, D.E. and T.K. Miller (1988). A Travelling Salesman Objective Function That Works. In *IEEE International Conference on Neural Networks* (San Diego 1988), vol. II, 299–303. New York: IEEE.

Van den Bout, D.E. and T.K. Miller (1989). Improving the Performance of the Hopfield-Tank Neural Network Through Normalization and Annealing. *Biological Cybernetics* **62**, 129–139.

Vapnik, V.N. (1982). *Estimation of Dependences Based on Empirical Data.* Berlin: Springer-Verlag.

Vapnik, V.N. and A.Y. Chervonenkis (1971). On the Uniform Convergence of Relative Frequencies of Events to Their Probabilities. *Theory of Probability and Its Applications* **16**, 264–280.

Vogl, T.P., J.K. Mangis, A.K. Rigler, W.T. Zink, and D.L. Alkon (1988). Accelerating the Convergence of the Back-Propagation Method. *Biological Cybernetics* **59**, 257–263.

Wagner, K. and D. Psaltis (1987). Multilayer Optical Learning Networks. *Applied Optics* **26**, 5061–5076.

Waibel, A. (1989). Modular Construction of Time-Delay Neural Networks for Speech Recognition. *Neural Computation* **1**, 39–46.

Waibel, A., T. Hanazawa, G. Hinton, K. Shikano, and K. Lang (1989). Phoneme Recognition Using Time-Delay Neural Networks. *IEEE Transactions on Acoustics, Speech, and Signal Processing* **37**, 328–339.

Watrous, R.L. (1987). Learning Algorithms for Connectionist Networks: Applied Gradient Methods of Nonlinear Optimization. In *IEEE First International Conference on Neural Networks* (San Diego 1987), eds. M. Caudill and C. Butler, vol. II, 619–627. New York: IEEE.

Webster (1988). *Webster's Ninth New Collegiate Dictionary.* Springfield: Merriam-Webster.

Weisbuch, G. and F. Fogelman-Soulié (1985). Scaling Laws for the Attractors of Hopfield Networks. *Journal de Physique Lettres (Paris)* **46**, 623–630.

Werbos, P. (1974). Beyond Regression: New Tools for Prediction and Analysis in the Behavioral Sciences. Ph.D. Thesis, Harvard University.

Werbos, P.J. (1987). Building and Understanding Adaptive Systems: A Statistical/Numerical Approach to Factory Automation and Brain Research. *IEEE Transactions on Systems, Man, and Cybernetics* **17**, 7–20.

Werbos, P.J. (1988). Generalization of Backpropagation with Application to a Recurrent Gas Market Model. *Neural Networks* **1**, 339–356.

Whitley, D. and T. Hanson (1989). Optimizing Neural Networks Using Faster, More Accurate Genetic Search. In *Proceedings of the Third International Conference on Genetic Algorithms* (Arlington 1989), ed. J.D. Schaffer, 391–396. San Mateo: Morgan Kaufmann.

Widrow, B. (1962). Generalization and Information Storage in Networks of Adaline "Neurons". In *Self-Organizing Systems 1962* (Chicago 1962), eds. M.C. Yovits, G.T. Jacobi, and G.D. Goldstein, 435–461. Washington: Spartan.

Widrow, B., N.K. Gupta, and S. Maitra (1973). Punish/Reward: Learning with a Critic in Adaptive Threshold Systems. *IEEE Transactions on Systems, Man, and Cybernetics* **3**, 455–465.

Widrow, B. and M.E. Hoff (1960). Adaptive Switching Circuits. In *1960 IRE WESCON Convention Record*, part 4, 96–104. New York: IRE. Reprinted in Anderson and Rosenberg [1988].

Wiener, N. (1948). *Cybernetics, or Control and Communication in the Animal and the Machine.* New York: Wiley.

Williams, R.J. (1987). A Class of Gradient-Estimating Algorithms for Reinforcement Learning in Neural Networks. In *IEEE First International Conference on Neural Networks* (San Diego 1987), eds. M. Caudill and C. Butler, vol. II, 601–608. New York: IEEE.

Williams, R.J. (1988a). On the Use of Back-Propagation in Associative Reinforcement Learning. In *IEEE International Conference on Neural Networks* (San Diego 1988), vol. I, 263–270. New York: IEEE.

Williams, R.J. (1988b). Towards a Theory of Reinforcement-Learning Connectionist Systems. Technical Report NU–CCS–88–3, College of Computer Science, Northeastern University, Boston, MA.

Williams, R.J. and J. Peng (1989). Reinforcement Learning Algorithms As Function Optimizers. In *International Joint Conference on Neural Networks* (Washington 1989), vol. II, 89–95. New York: IEEE.

Williams, R.J. and D. Zipser (1989a). A Learning Algorithm for Continually Running Fully Recurrent Neural Networks. *Neural Computation* **1**, 270–280.

Williams, R.J. and D. Zipser (1989b). Experimental Analysis of the Real-Time Recurrent Learning Algorithm. *Connection Science* **1**, 87–111.

Willshaw, D.J., O.P. Buneman, and H.C. Longuet-Higgins (1969). Non-Holographic Associative Memory. *Nature* **222**, 960–962. Reprinted in Anderson and Rosenberg [1988].

Willshaw, D.J. and C. von der Malsburg (1976). How Patterned Neural Connections Can Be Set Up by Self-Organization. *Proceedings of the Royal Society of London B* **194**, 431–445.

Wilson, G.V. and G.S. Pawley (1988). On the Stability of the Travelling Salesman Problem Algorithm of Hopfield and Tank. *Biological Cybernetics* **58**, 63–70.

Wilson, H.R. and J.D. Cowan (1973). A Mathematical Theory of the Functional Dynamics of Cortical and Thalamic Nervous Tissue. *Kybernetik* **13**, 55–80.

Winograd, S. and J.D. Cowan (1963). *Reliable Computation in the Presence of Noise*. Cambridge: MIT Press.

Winters, J.H. and C. Rose (1989). Minimum Distance Automata in Parallel Networks for Optimum Classification. *Neural Networks* **2**, 127–132.

Wittgenstein, L. (1958). *Philosophical Investigations*. 2nd ed., translated by G.E.M. Anscombe. Oxford: Blackwell.

Wittner, B.S. and J.S. Denker (1988). Strategies for Teaching Layered Networks Classification Tasks. In *Neural Information Processing Systems* (Denver 1987), ed. D.Z. Anderson, 850–859. New York: American Institute of Physics.

Yuille, A.L., D.M. Kammen, and D.S. Cohen (1989). Quadrature and the Development of Orientation Selective Cortical Cells by Hebb Rules. *Biological Cybernetics* **61**, 183–194.

Zak, M. (1988). Terminal Attractors for Addressable Memory in Neural Networks. *Physics Letters* **133A**, 18–22.

Zak, M. (1989). Terminal Attractors in Neural Networks. *Neural Networks* **2**, 259–274.

Subject Index

M

N

Author Index

Page numbers given in *italics* are within the bibliography.

The Addison-Wesley **Advanced Book Program** and the SANTA FE INSTITUTE would like to offer you the opportunity to learn about our new "Studies In the Sciences of Complexity" titles and workshops in advance. To be placed on our mailing list and receive pre-publication notices and special offers, just **fill out this card completely** and return to us.

Title, Author, and Code # of this book: **Date purchased:**

Name _____

Title _____

School/Company _____

Department _____

Street Address _____

City _____ State _____ Zip _____

Telephone ( )

Where did you buy this book?

- [ ] Bookstore
- [ ] Mail Order
- [ ] School (Required for Class)
- [ ] Campus Bookstore (individual Study)
- [ ] Toll Free # to Publisher
- [ ] Professional Meeting
- [ ] Publisher's Representative
- [ ] Other _____

Please define your primary professional involvement:

- [ ] Academic: Professor
- [ ] Academic: Student
- [ ] Academic: Researcher
- [ ] Industry: Administrator
- [ ] Industry: Researcher
- [ ] Industry: Technician
- [ ] Government: Administrator
- [ ] Government: Researcher
- [ ] Government: Technician

Check your areas of interest.

200 ☑ **SFI**

201 ☐ Agriculture	209 ☐ Communication Sciences	217 ☐ Information Sciences
202 ☐ Anthropology	210 ☐ Dentistry	218 ☐ Mathematics
203 ☐ Artificial Intelligence	211 ☐ Economics	219 ☐ Medical Sciences
204 ☐ Astronomy	212 ☐ Education	220 ☐ Pharmaceutical Sciences
205 ☐ Atmospheric Sciences	213 ☐ Engineering	221 ☐ Physics
206 ☐ Biological Sciences	214 ☐ Geology/Geography	222 ☐ Political Sciences
207 ☐ Chemistry	215 ☐ History/Philosophy Science	223 ☐ Psychology
208 ☐ Computer Sciences	216 ☐ Industrial Science	224 ☐ Social Sciences
226 ☐ OTHER _____	(please specify)	225 ☐ Statistics

Of which professional scientific associations are you an active member?

_____ _____ _____ _____ _____ _____

_____ _____ _____ _____ _____ _____

Would you like to be sent information about the SANTA FE INSTITUTE and its workshops?

- [ ] Yes
- [ ] No

fold and staple

BUSINESS REPLY MAIL
FIRST CLASS PERMIT NO. 828   REDWOOD CITY, CA 94065

Postage will be paid by Addressee:

ADDISON-WESLEY
PUBLISHING COMPANY, INC.®

Advanced Book Program
350 Bridge Parkway
Redwood City, CA 94065-1522